Wildlife Population Health

Craig Stephen
Editor

Wildlife Population Health

Springer

Editor
Craig Stephen
School of Population and Public Health
University of British Columbia
Nanoose Bay, BC, Canada

ISBN 978-3-030-90509-5 ISBN 978-3-030-90510-1 (eBook)
https://doi.org/10.1007/978-3-030-90510-1

This Springer imprint is published by the registered company Springer Nature Switzerland AG.
The registered company address is: Gewerbestrasse 11, 6330 Cham, Switzerland

Preface

This book was written and assembled when the COVID-19 pandemic prevented the authors from venturing out to work with animals and communities to protect fish and wildlife health and when extremes of weather and fires brought the climate change crisis into greater focus than ever before. It was written amongst a growing sense that acting on new problems as they emerged was a strategy that could no longer keep up with a rapidly changing world and a growing extinction crisis. It was written to help empower people to act.

The purpose of this book is not to describe the growing list of diseases and harms that lead to the demise of wild population. There is a rich literature on the etiologies and manifestations of wildlife diseases. There is, however, a large gap in what we know causes wildlife health and what we can do to protect it. This book aims to provide tools, perspectives, and approaches that can help people act to protect wildlife health and narrow the gap between knowing and doing.

This book strives to expose, explain, and show the value of moving beyond managing wildlife health as a biomedical and technical battle against diseases, pollutants, and pathogens to a socio-ecological, systems-based approach intended to promote and protect the assets wild animal need to stay well in times of unprecedented changes to their world.

Vancouver, BC, Canada — Craig Stephen

Acknowledgments

This book benefited greatly from reviews provided by John Berezowski, Janet Carter, Colleen Duncan, Erin Fraser, Tiggy Grillo, Jennifer Lynch, Thomas O'Connell, Jane Parmley, Barnett Rattner, LeAnn White, and feedback between the editor and the authors. The editor would like to thank the authors for their thoughtful approaches and for rising to the challenges of writing chapters that help the reader move beyond the status quo.

About This Book

The book begins by outlining what wildlife population health means, especially in a twenty-first century context. Part 1 focuses both on how health can be defined as well as how population health can be practiced. Part 2 gathers some core concepts needed to think about how to understand and manage wildlife population health such as the challenges of applying epidemiological methods in free-ranging animals or in establishing cause–effect relationships and the benefits of paying attention to the human dimensions of wildlife health. Part 3 focuses on problems that fish and wildlife health managers will undoubtedly have to confront more frequently in the Anthropocene such as the effects of urbanization, climate change, and emerging diseases. The fourth part of the book adapts health promotion and harm reduction concepts developed in human population health management for application in fish and wildlife health management. Part 4 emphasizes what we can do to implement an effective program of action for wildlife health protection.

A definition of wildlife was not imposed on the authors. Each author approached wildlife health reflecting their own background and experience. In general, the ideas and approaches discussed in this book apply to birds, reptiles, fish, and mammals that range free and live in their natural environments. The extent of human influence on wildlife animals varies significantly from species to species and place to place, but the core concepts and ideas in this book are applicable across context, circumstances, and definitions of wildlife.

Contents

Editor and Contributors

About the Editor

Craig Stephen, DVM, PhD is a veterinarian and epidemiologist who has worked at the interface of human, animal, and environmental health for 30 years. As former executive director of the Canadian Wildlife Health Cooperative, he both managed a national wildlife disease surveillance program and worked with government partners to forge a new vision of wildlife health as an asset that needs protection. Dr. Stephen has over 200 peer-reviewed and technical reports. His work evolved from finding and describing emerging environmental threats around the globe, including infectious diseases, to helping build the circumstances that allow for interspecies and intergenerational health equity. He edited and co-wrote the book *Animals, Health, and Society: Health Promotion, Harm Reduction, and Health Equity in a One Health World* and co-edited *One Health: The Theory and Practice of Integrated Health Approaches*. He currently operates a One Health and EcoHealth practice while retaining Clinical Professorships at the School of Population and Public Health (University of British Columbia) and School of Veterinary Medicine (Ross University). Craig was honored in 2020 with an award for Canada's provincial, territorial, and federal directors of wildlife for his work to advance the notion of wildlife health.

Contributors

Ryan Barry Nunavut Impact Review Board, Nunavut, Canada

John Berezowski Scotland's Rural College, Aberdeen, UK

Heath Fenton Ross University School of Veterinary Medicine, Basseterre, Saint Kitts and Nevis

Christa Gallagher Ross University School of Veterinary Medicine, Basseterre, Saint Kitts and Nevis

Colin Gillin State Wildlife Veterinarian, Salem, OR, USA

Paul Grant Fisheries and Oceans Canada, Vancouver, BC, Canada

Claire Jardine University of Guelph, Guelph, ON, Canada

Catherine Machalaba EcoHealth Alliance, New York, NY, USA

Helen Masterman-Smith Charles Sturt University, Bathurst, NSW, Australia

Andrew Peters Institute for Land, Water and Society, Charles Sturt University, Bathurst, NSW, Australia

Bob Petrie Department of Natural Resources, Government of Nova Scotia, Nova Scotia, Canada

John Rafferty Charles Sturt University, Bathurst, NSW, Australia

Colin Robertson Department of Geography and Environmental Studies, Wilfrid Laurier University, Waterloo, ON, Canada

Todd Shury Wildlife Health and Management, Parks Canada Agency, Saskatoon, SK, Canada
Western College of Veterinary Medicine, Saskatoon, SK, Canada

Jonathan Sleeman U.S. Geological Survey's National Wildlife Health Center, Madison, WI, USA

Craig Stephen School of Population and Public Health, University of British Columbia, Vancouver, BC, Canada

Matilde Tomeseli Canadian High Arctic Research Station, Polar Knowledge Canada, Cambridge Bay, Nunavut, Canada

Joy Wade Fundy Aqua Services, Nanaimo, BC, Canada

Thierry Work U.S. Geological Survey's National Wildlife Health Center Honolulu Field Station, Hawaii, USA

List of Figures

List of Tables

Part I

Core Concepts of Wildlife Population Health

What Is Wildlife Health?

Craig Stephen

Abstract

Health, although universally valued, does not have a universally shared definition. How we define a management goal, in this case health, sets the boundaries of our actions to meet that goal. Without a shared understanding of how to recognize health, it can be extremely hard to inspire cooperative action toward a shared health goal. This chapter explores three prevailing ways wildlife health has and can be defined: (1) as the absence of disease; (2) as capacity derived from interacting individual, environmental, and social factors that allows the individual or population to cope with all demands of daily life; or (3) as a social construct where individuals or population meet our social and scientific expectations for how they exist and persist in an environment. Concepts of population health can overlap with ideas of robustness and resilience. The clarity we provide for our population health goals will influence how society thinks about and acts on keeping wildlife healthy in a rapidly changing world.

Keywords

Wildlife · Health · Population · Determinants · Resilience

1 Introduction

There is little doubt that human activities are harming wild animals (Pimm et al. 2014). It is easy to make a list of the threats and hazards confronting wildlife. Emerging infections, pollutants, climate change, habitat loss, and overexploitation are commonly on that list. It is more difficult to list the attributes of a healthy population. Desires for healthy individuals, healthy populations, and healthy species feature prominently in many government policies, resource management guidelines, and social goals for wildlife. But it is rare for such aspirations to prescribe how to recognize health. The Commission of Inquiry into the Decline of Fraser River Sockeye Salmon in Canada, for example, (Cohen 2012) aimed to explain why this iconic species returned to its natal river in alarmingly low numbers. The Commission's final report used the word health 400 times. Never did it describe how to recognize a healthy salmon or a healthy population or provide a definition of health. Similarly, an evaluation of policies and legislation governing wildlife health in one Canadian province in 2018 found, of the 40 documents mentioning wildlife health, 63% failed to define health

C. Stephen (✉)
School of Population and Public Health, University of British Columbia, Vancouver, BC, Canada

C. Stephen (ed.), *Wildlife Population Health*,
https://doi.org/10.1007/978-3-030-90510-1_1

and there was no common definition across the remaining documents (Sinclair 2020). It makes it exceedingly difficult to manage toward the goal of "healthy wildlife" when there is no clarity or consistency on how to recognize when the goal has been met.

Health is a word with multiple meanings, which can be assessed differently by different people in the same circumstance. Take for example, the case of a captive elephant in a zoo. This animal may show no clinical or subclinical pathophysiological abnormality, be normal in form and function, be able to reproduce, and can fully exploit the resources offered to her. Yet she is unable to fulfill her evolutionarily destiny or exhibit all behaviors typical of her wild peers. Whereas one person may conclude she is biologically healthy, another may see impediments to her ability to express normal behaviors as an affront to her welfare and therefore, unhealthy. Many pieces of biomedical, sociological, ecological, and individual knowledge can be assembled to define health, but there is no guarantee that people will assemble the components in the same way.

Wildlife is a valued public asset that is typically managed for the benefit of society, now and into the future. Without a clear vision of how to recognize when a healthy state has been achieved, how can wildlife managers be sure they are on target to sustain healthy wildlife as a legacy for this and subsequent generations? The Inquiry into the Decline of Fraser River Sockeye Salmon noted that the lack of a health standard prevented scientists and regulators from assessing risks to wild salmon and taking informed preventive actions (Cohen 2012). Similar impediments occur for those wishing to assemble the existing knowledge of a species or population's health based on available literature. A scoping literature review covering 10 years of wildlife health publications found that 37% of the papers had ambiguous or missing definitions of health, while 56% defined health only as the absence of a set of diseases (Sinclair 2020). When different researcher or manager uses different criteria to establish if a population or species is healthy, can those various sources of information be accumulated into a shared, common understanding? It seems unsatisfactory to have competing or ambiguous definitions of wildlife health lead us to the point where successful health management is in the eye of the beholder.

There are generally three types of wildlife health definitions. The first is that health is the absence of diseases or etiologic agents. The second is that health is a capacity derived from the cumulative effect of interacting individual, environmental, and social factors that allows the individual or population to cope with all demands of daily life. The third definition sees health as the social construct where individuals or populations meet our social and scientific expectations for how they exist and persist in an environment. This chapter introduces and examines the implications of the different ways we do or do not define health.

2 Health as the Absence of Disease

For nearly 100 years, wildlife health has most often been defined in literature and legislation as the absence of diseases or the absence of specific etiological agents like pathogens, parasites, and sometimes pollutants. This reflects both the origins of fish and wildlife health (which started with a focus on parasite ecology) as well as regulatory preoccupations (which have been dedicated to prevent the movement of infectious diseases across borders). Interest in the presence and effects of environmental toxins and contaminants grew as environmental pollution came under regulatory and public scrutiny in the last 50 years. Socially impactful infectious diseases like avian cholera, anthrax, and chronic wasting disease in charismatic and hunted species focused wildlife regulators on infectious disease management. The era of emerging and re-emerging infections spawned the One Health approach in the early 2000s which heightened interest in zoonotic infections in wildlife. All these trends reenforced the attention of wildlife health on diseases and etiological agents.

The disease-centric preoccupation in wildlife health research and management is pragmatically reasonable because of disease-related outcomes, like die-offs, food safety threats, and trade barriers, demand responses from wildlife managers. However, the definition of health as the absence of disease lags 80 years behind modern concepts of human health, 40 years behind herd health in domestic animals, and is inconsistent with emerging concepts of wildlife health in terms of resilience and sustainability. A definition of wildlife health as the absence of disease can be criticized on several fronts (Box 1 adapted from Stephen 2014).

Box 1 Six Criticisms of Defining Wildlife Health as the Absence of Disease

Some level of disease or infection is normal, and parasites and pathogens are ubiquitous in wildlife populations, therefore, freedom from disease or infection is not a plausible standard.
Logistical challenges and lack of validated tests often prevent the fulfillment of epidemiological criteria for establishing freedom from specific diseases in free-ranging populations.
The absence of disease as a standard does not define the threshold of dysfunction, disruption, or infection when an animal changes from being healthy to diseased along the clinical course from exposure to death or recovery.
An absence of disease standard does not recognize that a population can be deemed healthy (often based on measures of abundance, productivity, public safety, and profitability) but still have individual members that harbor disease-causing agents or are diseased.
A focus on pathogens and pathology results in a situation wherein wildlife health is defined by what is dysfunctional and unacceptable rather than on the positive attributes of the animals.
An absence of disease standard is inconsistent with modern ideas of health as a coping capacity arising from socioecological interactions and neglects the harms that most significantly threaten fish and wildlife populations.

Basing the definition of health on the presence or absence of a subset of selected disease or etiologic agents (as is often the case in animal health regulations) does not reflect health as it is experienced by living organisms. Any living being is simultaneously challenged by multiple stressors, vulnerabilities, and threats and is benefiting from multiple assets. For example, a deer in the winter in northern Canada could simultaneously be under thermal stress due to extremely cold weather, have reduced access to food due to snow cover, be infested with parasites, live in a region contaminated with the prions that cause chronic wasting disease and carry a tissue burden of cadmium. But this animal might have good fat stores, live in a valley where it can avoid extreme weather, have a genetic endowment that supports a robust immune function, be in a region where hunting pressures do not exist, and is already pregnant and able to produce a fawn in the Spring. Its health is the combination of all these variables. Health, as experienced by a living organism, is a cumulative effect.

There is a growing consensus that the unprecedented social and environmental changes that are accompanying exponential human population growth in the Anthropocene are creating pressures that seriously impact wildlife health (Acevedo-Whitehouse and Duffus 2009). The United Nations' Intergovernmental Science-Policy Platform on Biodiversity and Ecosystem Services concluded in 2018 that we were losing species 1000 times faster than the natural rate of extinction. The 2016 Living Planet Index attributed these losses to human-induced factors including habitat degradation, invasive species, climate change, pollution, unsustainable freshwater use, and species overexploitation. Accelerating environmental and human behavioral changes have been creating new ecological niches that drive microbial evolution and epidemiological shifts that have fostered the emergence of new infectious diseases. These multiple, concurrent challenges reveal that achieving wildlife health by looking at one disease at a time by separate science, policies, and actions will be impossible. Threats to fish and wildlife populations, such as habitat changes due to natural resource use, globalization of trade, disease, and land-use pressures, do not merely exist in parallel but are intertwined and cumulative. The synergisms between environmental and social changes can pose some of the biggest threats to biodiversity (Laurance and Useche 2009) and are causing multiple and overlapping syndemics that cannot be captured by focusing only on pathology, pathogens, or contaminants in isolation. A syndemic is the occurrence of two or more states that adversely interact with each other, negatively affecting the course of each state, enhancing vulnerability, and amplifying deleterious impacts (Sharma 2017). The syndemic framework highlights the complexity of diseases and conditions as they are experienced in nature and the array of factors that give rise to them. It forces us to abandon seeing health only as the absence of a prescribed set of diseases and seek a new model that better reflects how life is lived. It compels us to think of health from a broader socioecological perspective. It highlights that an interdisciplinary approach to a single problem is insufficient to protect health in a sustainable way.

In a world of concurrent problems, unique solutions for each problem are neither feasible nor effective (Fried et al. 2012). Wildlife health research and management need to evolve from an interdisciplinary approach to single problems to one that is "interprobleminary"—an approach that examines the interactions and implications of multiple problems occurring simultaneously in a place or population. While it is necessary and important to have standards to prevent diseases and the spread and impacts of pathogens, parasites, and pollutants, this approach is insufficient to address the goal of promoting and protecting wildlife health.

3 Health as Capacity to Cope

To be healthy, an individual or population needs to have a minimal set of resources, functions, and capabilities that operate within an environment that enables them to cope with challenges to meet expected endpoints (Nordenfelt 2011). This conception of health underpins public health and human health promotion as well as domestic animal herd health. The population health approach, pioneered in human health, views health as the product of interacting social factors, the physical environment, and individual attributes and behaviors. These factors are referred to as determinants of health. These determinants interact in ways that allow a healthy individual or population to satisfy needs, change or cope with the environment and meet social and scientific expectations (Stephen 2014). In this perspective, health is the capacity provided by the interacting individual, environmental, ecological, and social attributes and circumstances rather than the lack of a diseased state or a state of deficit. The determinants of health provide wildlife the resources they can draw from to tolerate, persist through, and recover from disturbances, stressors, and change. In this sense, health is akin to ecological concepts of resilience (as described by Falk et al. 2019).

Descriptions of wildlife health in terms of resilience became more prominent in the early 2000s (ex. Deem et al. 2001; Hanisch et al. 2012; Stephen 2014). Wittrock et al. (2019a) nominated determinants of fish and wildlife health to explicitly link health and resilience (Fig. 1). They identified six themes of determinants: (1) the biologic endowment of the individual and population; (2) the animal's social environment; (3) the quality and abundance of resources providing the animal's needs for daily living; (4) the abiotic environment in which the animal lives; (5) sources of direct mortality; and (6) changing human expectations. Each theme is comprised of a subset of factors, circumstances, or conditions. The specific variables that are used to describe each factor can vary by the unique needs of each species. For example, terrestrial attributes will be more immediately and significantly impactful to a zebra than a tuna. Migration fidelity may not be relevant to nonmigratory species but may be replaced by the capacity to move with their natural range. Selection of the right set of factors and variables requires knowledge of a species ecology tailored to the circumstances of the population being managed. How each variable is measured is a subject for scientific inquiries that determine the validity and reliability of indicators for each variable. How each indicator is weighed and the acceptable thresholds for healthy populations will be a combination of scientific inquiry and social expectations.

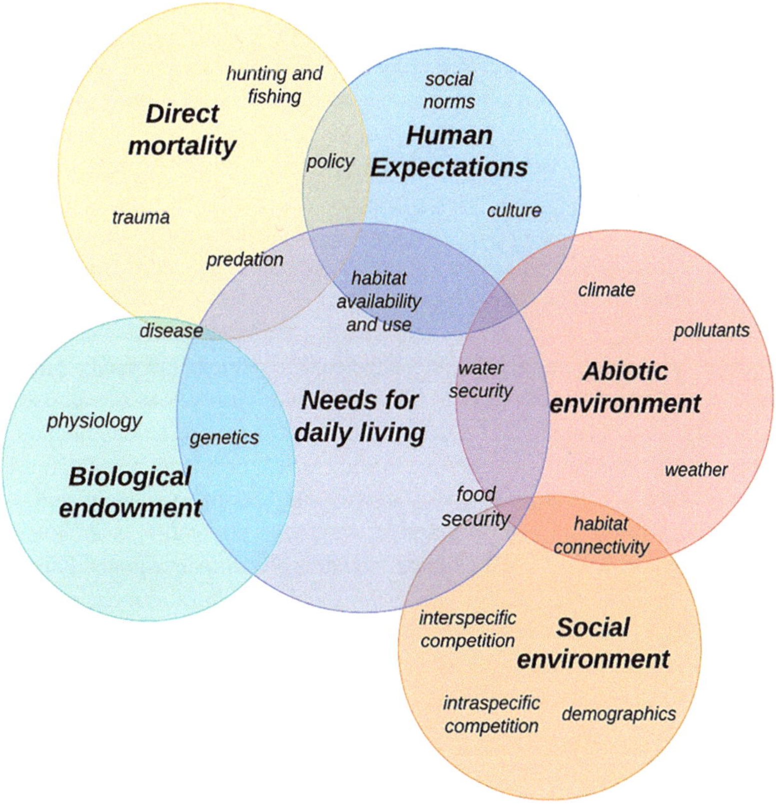

Fig. 1 A determinants of health model for fish and wildlife adapted from Wittrock et al. (2019a)

The determinants of the health model is not intended to be used as a measurement tool with which a set of variables can be summed up to declare if a population is healthy or not. Rather it provides a framework for planning, policy development, and research (Pourbohloul and Kieny 2011). It helps identify factors that affect a population's vulnerability before harms occur, without needing to rely on the occurrence of death or disease to signal the need for action. This approach can not only help conceive of the breath of factors that increase susceptibility to disease but also direct attention to the major drivers of population declines and extinctions (Wittrock et al. 2019a). The determinants of health approach help us to think of health and welfare as a sense of coherence between the capacity to identify, benefit, and use resources to deal with stress and the reality of current living conditions (Stephen and Wade 2018). This parallels the view that animal welfare is compromised when adaptations possessed by the animal make an imperfect fit to the challenges it faces in the circumstances in which it lives (Fraser et al. 1997). Understanding the ecologic, biological, and social context of wildlife health can be as, or sometimes more, important than having technically sound disease information when developing health management plans (as illustrated in Nishi et al.'s (2006) review of bison tuberculosis management).

Because the determinants of health approach requires attention to a wider suite of mechanisms to promote and protect health, it is usually beyond the scope of one organization or discipline to fulfill all requirements for a comprehensive scope of activity, especially when both public and private interests intersect. The historic focus on pathogens and pathology in wildlife health has impeded linking people in wildlife health programs with those who deal with wildlife determinants of health found in ecology, management, and sustainability. The work of Wittrock et al. (2019a, b) and others is helping bridge these worlds, allowing for more strategic, cooperative, and coordinated investment of time and resources on combating the drivers of disease and protecting the capacities for wildlife to cope with the greatest challenges to their health.

4 Health as a Social Construct

When using the definition of health as the capacity to access the needs for daily living, cope with and adapt to changes and stress, and meet expectations, one must accept that health is a social construct because we assign the expectations to the animals. We decide what we expect from a healthy wild animal, including which variables reflect health, which thresholds define acceptability, and how we weigh and value the different determinants of health.

A social construct is an idea that has been created and accepted by the people in a society. The idea that society determines if something is healthy seems at odds with the approach to health and disease that I was taught when trained as a veterinarian and infectious disease epidemiologist in the 1980—early1990s. I was taught that a disease is a measurable, physiological problem that is an identifiable biological state. Society does not determine if an elk has pneumonia, or a bat has white-nose syndrome. Rather, I was taught that the combination of an etiological agent and the body's response to that agent determined the disease state. My training made it was hard for me to understand that health is a socially defined capacity rather than a measurable biological state. But, when we will reflect on the example of the elephant in the zoo described earlier, it must be conceded that health is not an objective biological state. People having access to the exact same data can come up with different conclusions with respect to health. Let us revisit the case of the disappearing sockeye salmon. In a regulatory sense, there are two ways that salmon are considered healthy. For agencies concerned with controlling the spread of diseases, the absence of a disease model predominates. For agencies responsible for fisheries management, health is defined as the occurrence of enough animals for us to kill and harvest as part of a fishery (Wittrock et al. 2019b). There are also cultural perspectives that consider a population of sockeye salmon healthy if they can fulfill traditional consumptive and spiritual roles. In addition, there are ecological perspectives of salmon as keystone species that support both terrestrial and marine ecosystems. How people interpreted and valued different population outcomes and threats depends on if they viewed these wild animals as a commodity, as a cultural legacy, as an ecological linchpin, or as entities deserving care and respect for their own sake. Health is context-specific and influenced by the species, population, stakeholder group, and expected performance goals. Failure to incorporate ethical and cultural dimensions of wildlife health issues can lead to ineffective and unacceptable wildlife health management decisions (Stephen 2017).

We often turn to the biological and medical sciences to help find an objective answer to the question, "are these animals healthy?" But science too is affected by social norms and the norms of the discipline. For example, it has been proposed that universal validity of the resilience approach presupposes a notion of ecosystems that is based on a particular, but not universally accepted, cultural idea of individuality and society (Kirchhoff et al. 2010). Indigenous worldviews of resource management in many countries emphasize interrelationships between people, other species, land, and spiritual domains. Western scientific approaches have commonly overlooked or segregated elements of indigenous knowledge, resulting in different understandings of nature. Scientific knowledge diverges and evolves as data are collected that either refute or support accepted paradigms. Heightened understanding can build impediments to future progress as increased specialization and the parallel evolution of seemingly independent subdisciplines create silos of knowledge and fixed ways of interpreting data (Graham and Dayton 2002). Scientific paradigms are part community-defining exemplars of practice and part private, individualized ways of thinking (Wendel 2008).

The dividing line between health and diseases and the combinations of capacities, attributes, and attitudes that define a healthy state has varied over time and between cultures, disciplines, and subdisciplines (Walsh et al. 2013). Early Greeks concepts of health promoted by Hippocrates as well as some Indigenous and Asian traditions see health not as the physical well-being of an individual, but as the well-being of the whole community in partnership with their environments from a whole-life perspective. The path to separating an individual's health from the environment began in western cultures in earnest when taboos against dissection opened new opportunities to understand anatomy, physiology, and pathology. These advances shifted the focus of health care from keeping individuals healthy to reducing suffering and aiding in recovery from specific diseases. Darwinism tied the meaning of life to physical survival. This helped expand the idea of health as something that allowed animals to tolerate and resist not just biological hazards, but also environmental influences. With the advent of epidemiology and demography, data on trends and patterns of disease helped to identify circumstances and situations of living that predisposed to the disease. As we began to ask why an individual, group, or community stays well despite stressful situations and hardships, ideas about the determinants of health emerged. With the advent of health promotion and population health, we came back to Indigenous ways of knowing and the lessons of Hippocrates that saw health as a product of our interactions with the world around us. This paragraph is a snapshot of the history of western concepts of health over millennia, but each of the perspectives briefly described above can still be found today. Different cultures and disciplines emphasize different aspects of this history in their descriptions, investigations, and management of health. None of these perspectives are right or wrong, but it is important to recognize which perspective someone holds to understand what they mean when they say, "this is healthy".

Our attitudes toward wildlife health also vary across species. Social factors influence our concern for, attention to, and care for animals including (1) the extent to which we are responsible for harm to them; (2) the extent to which the animals are under our stewardship; (3) the severity of the problems that cause harms and (4) cultural and economic factors, including the popularity of the species (Kirkwood and Sainsbury 1996). It would be rare, for example, to find a person who found it acceptable to hunt deer by snagging the animal on a hook, dragging it behind a vehicle, and suffocating it before butchering the animal. But that is akin to what we do when we go fishing. Different cultural traditions have been associated with differences in how animal health and welfare are viewed. Buddhism and Hinduism, for example, do not share the Judeo-Christian view that people have dominion over animals. These fundamental differences influence attitudes toward wildlife use, abuse, and health.

The idea of well-being can help link health and society. Well-being is the unimpaired flourishing, free of obstacles, to live in a way that conforms with expectations, opportunities, and abilities. It is a state of being with others and the natural environment that arises where needs are met, where individuals and groups can act to pursue goals, and where they are satisfied with their way of life (McCrea et al. 2014). The determinants of health work together to produce a state of living that leads to health. Health, in turn, provides the raw materials for well-being. Well-being not only encompasses basic health needs such as adequate food, safety, and lack of disease but also considers how people think and feel about their life situation, or the, in the case of wildlife, how people think and feel about the situation of the animals or environments they care for. Well-being implies successful biological function, positive experiences, and freedom from adverse conditions.

5 The Population Health Approach

The term population health can be used in two ways. The first is to describe the health of a particular population of animals living in an area. The second is to describe an approach to health management. The population health approach is concerned with measuring and examining systematic differences in outcomes across populations, interactions and pathways among determinants of health, and the influence of different determinants and interventions over time and throughout the life cycle of the population (Kindig and Stoddart 2003). It deals not only with adverse outcomes like disease but also with the positive determinants of health. An underlying population health assumption is that to improve health, a population needs access to and be able to use its determinants of health. Efforts are directed at root causes of good health, along a continuum of care from helping populations recover from harms to preventing harms from occurring to ensuring populations can cope with hazards before harms occur.

Population health is firmly linked to a socioecological model which sees health as a series of interconnected, co-dependent, and interacting factors; in other words, as a system (Diez Roux 2011). This creates challenges for those wishing to study health. The fact that ecosystems are ever-changing, and human systems undergo ongoing transformations makes socioecologic systems inherently unknowable, unpredictable, and not well suited to research seeking a mechanistic truth in a reductionist fashion. Instead, the population health approach recognizes that addressing the complex interplay between the determinants of health requires cross-sectoral collaborations that use multiple strategies within multiple settings. It sees health as a cumulative effect requiring a combination of health-protecting and promoting actions. Population health management, therefore, is a collaborative enterprise that continually creates and improves physical and social environments, which provide the raw material for wildlife to be healthy and for a healthy relationship between wildlife and society.

The population health guiding philosophy is that action on root causes has greater potential for health gains even if the root causes are difficult to change. The approach allows decision-makers to consider and respond to risk using a broader perspective in a consistent and comprehensive manner. This can help prioritize risk management and effectively deploy resources toward the most important drivers of risks and harms as well as facilitate increased consultation with experts, the public, and other stakeholders, and rights holders. The population health approach helps risk managers to: (1) identify subpopulations for which a health risk is a bigger concern; (2) incorporate a wider suite of social, biological, and ecological information into risk assessments where needed and appropriate; (3) consider a wider variety of risk management options, (4) be more alert to unanticipated impacts of risk management options; (5) make better use of multifaceted risk management strategies; (6) involve a variety of partners in helping to manage risk; and (7) consider the effectiveness of risk management from a wider perspective (GoC 2020).

6 What Is Wildlife Population Health

The question, "what is wildlife health" is not merely of philosophical interest. Without understanding how to frame a health problem so that it resonates with those people who need to act to protect the health, it can be extremely hard to inspire cooperative action toward a health goal (Kellermanns et al. 2008). Health is not like physics where we can measure and weigh different attributes to find universal laws and make definitive proclamations on a state of nature. Health is characterized by uncertainty, complexity, and divergent values. There is no indisputable point at which one can declare objectively something is healthy in a way that everyone will agree. Because of the contextual nature of health, it is not possible to create a single index of health that is acceptable to all (Jayasinghe 2011).

A modern definition of wildlife health should emphasize three features: (1) health is the result of interacting biological, social, and environmental determinants that promote and maintain health as a capacity to cope with change over time; (2) health cannot be measured solely by what is absent (i.e., lack of disease or hazards) but rather by characteristics of the animals and their ecosystem that affect their vulnerability and resilience to a suite of interacting social and environmental harms; and (3) wildlife health is not a biologic state but rather a dynamic human social construct based on social expectations and scientific knowledge (Stephen 2014). Including measurements and management of hazards, along with measures and management of the determinants of population vulnerability and resilience accommodates a management approach that works proactively to maintain health, rather than just responding to adverse outcomes, such as death or disease. Stephen (2014) did not, or could not, provide the desired level of details that a health manager might need to design a health program. Health management strategies must be tailored to specific situations, developed in collaboration with stake- and right holders (Briggs 2008). Stakeholders and end users are ultimately the ones who will decide whether a population is healthy. It is, therefore, critical that stakeholders, resource users, rights holders, and decision-makers be included in the process of defining the goals and boundaries of health programs (Hancock et al. 1999; Briggs 2008).

7 Summary

The authors of this book bring with them their own experiences and expertise to their reflections of what we can do to protect wildlife health. No single definition of health was imposed on them. Instead, each was asked to consider what wildlife health managers, practitioners, and researchers will need to know to confront twenty-first-century challenges. The book was designed with population health principles in mind and strives to empower the reader with ideas and approaches that can result in disruptive changes to improve our ability to actively make changes for the betterment of wildlife health, however, one might define it.

References

Acevedo-Whitehouse K, Duffus AL (2009) Effects of environmental change on wildlife health. Philos Trans Roy Soc 364(1534):3429–3438

Briggs DJ (2008) A framework for integrated environmental health impact assessment of systemic risks. Environ Health 7(1):1–7

Cohen B (2012) The uncertain future of Fraser River sockeye volume 3: recommendations–summary–process. Public Works and Government Services Canada, Ottawa

Deem SL, Karesh WB, Weisman W (2001) Putting theory into practice: wildlife health in conservation. Conserv Biol 15(5):1224–1233

Diez Roux AV (2011) Complex systems thinking and current impasses in health disparities research. Am J Public Health 101(9):1627–1634

Falk DA, Watts AC, Thode AE (2019) Scaling ecological resilience. Front Ecol Evol 24(7):275

Fraser D, Weary DM, Pajor EA, Milligan BN (1997) A scientific conception of animal welfare that reflects ethical concerns. Anim Welf 6:187–205

Fried LP, Piot P, Frenk JJ, Flahault A, Parker R (2012) Global public health leadership for the twenty-first century: towards improved health of all populations. Glob Public Health 7(sup1):5–15

GoC (2020) Implementing the population health approach. Government of Canada. https://www.canada.ca/en/public-health/services/health-promotion/population-health/implementing-population-health-approach/implementing-population-health-approach.html#risk

Graham MH, Dayton PK (2002) On the evolution of ecological ideas: paradigms and scientific progress. Ecology 83(6):1481–1489

Hancock T, Labonte R, Edwards R (1999 Nov) Indicators that count! Measuring population health at the community level. Can J Public Health 90(1):S22–S26

Hanisch SL, Riley SJ, Nelson MP (2012) Promoting wildlife health or fighting wildlife disease: insights from history, philosophy, and science. Wildl Soc Bull 36(3): 477–482

Jayasinghe S (2011) Conceptualising population health: from mechanistic thinking to complexity science. Emerg Themes Epidemiol 8(1):1–7

Kellermanns FW, Floyd SW, Pearson AW, Spencer B (2008) The contingent effect of constructive confrontation on the relationship between shared mental models and decision quality. J Organiz Behav 29(1): 119–137

Kindig D, Stoddart G (2003) What is population health? Am J Public Health 93(3):380–383

Kirchhoff T, Brand FS, Hoheisel D, Grimm V (2010) The one-sidedness and cultural bias of the resilience approach. Gaia-Ecol Perspect Sci Soc 19(1):25–32

Kirkwood JK, Sainsbury AW (1996) Ethics of interventions for the welfare of free-living wild animals. Animal Welfare-Potters Bar 5:235–244

Laurance WF, Useche DC (2009) Environmental synergisms and extinctions of tropical species. Conserv Biol 23(6):1427–1437

McCrea R, Walton A, Leonard R (2014) A conceptual framework for investigating community wellbeing and resilience. Rural Soc 23(3):270–282

Nishi JS, Shury T, Elkin BT (2006) Wildlife reservoirs for bovine tuberculosis (Mycobacterium bovis) in Canada: strategies for management and research. Vet Microbiol 112(2–4):325–338

Nordenfelt L (2011) Health and welfare in animals and humans. Acta Biotheor 59(2):139–152

Pimm SL, Jenkins CN, Abell R, Brooks TM, Gittleman JL, Joppa LN, Raven PH, Roberts CM, Sexton JO (2014) The biodiversity of species and their rates of extinction, distribution, and protection. Science 344:6187

Pourbohloul B, Kieny MP (2011) Complex systems analysis: towards holistic approaches to health systems planning and policy. World Health Organ 89:242–242

Sharma A (2017) Syndemics: health in context. Lancet 389(10072):881

Sinclair D (2020) Exploring opportunities to modernize ontario's approach to wildlife health through understanding literature, Stakeholders, Networks, Legislation, and Policy. Doctoral dissertation

Stephen C (2014) Toward a modernized definition of wildlife health. J Wildl Dis 50(3):427–430

Stephen C (2017) Wildlife health 2.0: bridging the knowledge-to-action gap. J Wildl Dis 53(1):1–4

Stephen C, Wade J (2018) Wildlife population welfare as coherence between adapted capacities and environmental realities: a case study of threatened lamprey on Vancouver Island. Front Veter Sci 24(5):227

Walsh FJ, Dobson PV, Douglas JC (2013) Anpernirrentye: a framework for enhanced application of indigenous ecological knowledge in natural resource management. Ecol Soc 18:3

Wendel PJ (2008) Models and paradigms in Kuhn and Halloun. Sci Educ 17(1):131–141

Wittrock J, Anholt M, Lee M, Stephen C (2019b) Is Fisheries and Oceans Canada policy receptive to a new Pacific salmon health perspective? Facets 4(1): 615–625

Wittrock J, Duncan C, Stephen C (2019a) A determinants of health conceptual model for fish and wildlife health. J Wildl Dis 55(2):285–297

Wildlife Population Health Strategies

Craig Stephen

Abstract

The practice of wildlife population health strives to understand why some populations are healthier than others. It relies on four key strategies to foster conditions conducive to health by addressing underlying social, individual, and environmental determinants that constrain capacities, resources, and abilities to achieve and sustain health gains. These four strategies are (1) build management and health policies that protect the determinants of wildlife health; (2) focus on creating healthy environments that support sustainable access to the determinants of health; (3) strengthen collaborative action; and (4) reorient wildlife health from only disease management to a providing a continuum of care. This continuum of care involves maintaining health capacity, reducing vulnerability, reducing persistent harms, and promoting recovery from realized harms. Wildlife population health practice and research encourage intersectoral activities that consider contextual factors and community influences that enable or obstruct the necessary action to achieve and sustain healthy wild populations.

Keywords

Population health · Wildlife · Management · Working upstream · Continuum of care · Vulnerability · Harm reduction

1 Introduction

The term "population health" can refer to the descriptive notion of producing a measurement or declaration on the average level and distribution of health within a population. It also refers to a complex and expanding set of perspectives, methods, and practices vital to securing health benefits for populations. In this book, it provides a conceptual framework for thinking about why some populations are healthier than others. A population health perspective acknowledges that veterinary medicine and biomedical sciences explain a relatively small part of a wildlife population's health. It reflects a public health tradition that began in the West in the nineteenth century that highlights the relationships between health, society, and environments. Chapter 1 described population health as being concerned with measuring and examining systematic differences in health outcomes across populations, interactions and pathways among determinants of health, and the influence of different determinants and interventions over time and throughout the life cycle of the population (Kindig and Stoddart 2003). Population health

C. Stephen (✉)
School of Population and Public Health, University of British Columbia, Vancouver, BC, Canada

C. Stephen (ed.), *Wildlife Population Health*,
https://doi.org/10.1007/978-3-030-90510-1_2

approaches have been highly influential in shaping human health over the last two centuries but have only recently begun to be applied to wildlife health. This chapter provides more detail on the concept of population health within a wildlife health context.

Serious disruptions, be they environmental, ideological, or political, spill over into human deprivation, disease, and death (Szreter 1999). Similar disruptions, too often at the hands of human activity, deprive wildlife of access to their determinants of health, affect patterns of disease, and lead to excess suffering and premature death (Acevedo-Whitehouse and Duffus 2009; Paquet and Darimont 2010). Population health management fosters conditions conducive to health by addressing underlying social, individual, and environmental determinants that constrain animals' abilities to achieve and sustain health gains. Traditional wildlife health approaches have focused on risks to individuals and clinical or pathological factors related to diseases. A population health perspective sees that the answer to, "why is this population healthier than another" is about more than differences in disease status.

Population health strategies are designed to affect whole populations of animals to promote health as the capacity to access the needs for daily living, cope with and adapt to changes and stress, and meet human expectations. Information and understandings derived from population health research provide a rational basis for allocating resources that balances health protection and promotion with disease prevention and treatment. Population health practitioners view health and disease outcomes from a broad perspective and tend to include additional considerations, such as policy, social agendas, sustainable development, and resource allocation. The goal is to have a broad understanding of what ultimately causes the problems arising from a disease (as opposed to focusing only on the disease) and to find levers that can be managed to prevent the next problem, speed recovery from the current one, or mitigate harms most effectively, efficiently, and sustainably. To achieve these aims, population health requires collaboration between the core population sciences like epidemiology and ecology with biomedical sciences like pathology, physiology, and microbiology and with social sciences concerned with human phenomena that influence wildlife health and the collaborations needed to manage it.

2 Risk, Determinants, and Working Upstream

We are just beginning to understand how to forecast an animals' susceptibility to harm ahead of time, but much uncertainty remains. We have typically dealt with this uncertainty by calculating the statistical probability of an individual with a certain set of characteristics (called risk factors) experiencing an adverse health outcome. Some risk factors act directly (such as exposure to the rabies virus leading rabies) while other work indirectly (like climate change which affects waterfowl ecology that alters avian influenza transmission pathways, which leads to spillover of the virus to poultry which then experience a disease outbreak). These statistical associations might reflect a true causal relationship or might be a marker for some underlying or unknown factor (see Chap. 6 for more on causation). For example, age can be a statistically significant risk factor for certain diseases, but age itself is not the risk; instead, age reflects underlying physiological, behavioral, or social changes associated with age that may be leading to the disease.

Risk factors themselves have causes. These can be found in the determinants of health. Health determinants refer to underlying characteristics and circumstances that ultimately shape the health of individuals and populations (see Chap. 1). They can be thought of as the causes of causes. The word determinants in this usage do not mean that a determinant ordains an outcome. Rather the word determinant is used in the sense of the Latin *de termine* meaning the end. This gives rise to the metaphor of population health practitioners working upstream. To illustrate this metaphor, consider a town where you learn that waterfowl are being routinely found on a beach with clinically significant amounts of oil covering their body.

One response to this discovery is to recruit wildlife rehabilitators who can find affected birds on the beach, clean the birds' plumage, and care for them until they are fit for release. A second response is to monitor the water and shores for signs of oil and put in barriers to keep oil from contacting the birds. A third response would be to move upstream from the pond to see where the oil is entering the water system, preventing it from reaching the pond. Finally, one could look further upstream to the sources of the oil and try to find incentives or methods to prevent oil from escaping at the source or to find alternatives to the use of oil at the source. Population health practitioners find opportunities upstream and downstream to protect and promote health.

The causal model of Hendra virus in flying foxes (*Pteopus* spp) in Australia proposed by Plowright et al. (2011) illustrates the role of upstream drivers in wildlife diseases. In this model, the authors document how landscape use and climate change policies lead to a cascade of changes that influenced the viral ecology in a way that led to an emerging human disease. Flying foxes evolved to exploit rainforest pollen, nectar, and fruits. As rainforests were converted for forestry and urbanization, the bats moved to peri-urban and urban settings where their food sources were more concentrated and consistent in gardens. This new food source reduced the need for bats to undertake long-distance foraging forays or migrate. This in turn affected their aggregation behaviors and population densities, thus changing the setting for virus transmission. The growing overlaps between the bats, domestic animals, and people in urban and peri-urban environments created new ways for the virus to spill from the unaffected bats into horses and then people, with fatal effects. Climate change further impacted food availability through the effects of drought and forest fires. A final component of this model was the compilation of seasonal stressors (ex. pregnancy, births, food stress) that altered herd immunity in the bats. While it is not outside the realm of possibility to attack this problem by innovative vaccine technology and delivery systems, this model shows additional upstream avenues for intervention from urban planning, to home garden design, to land-use practices, to climate change policies. A population health approach would promote a suite of interventions along this continuum.

3 The Continuum of Care

Wildlife population health can be thought of as having four main goals, although each blur into and influences the other (Box 1). These goals are the basis of a continuum of actions people can take to manage wildlife population health.

Box 1 The Continuum of Goals and Targets for Action for Wildlife Population Health

	Continuum of wildlife population health management			
Goal	Maintain health	Reduce vulnerability	Reduce harms	Promote recovery
Target	Upstream determinants of health that prevent risks from arising and ensure access to the circumstances and resources needed to allow the population to meet our expectations	Individual, social and environmental attributes that affect exposure and sensitivity to risks that have arisen	Early warning of individual, ecolgical or social harms and motivations and capacities for early interventions to minimize and contain the distribution and impact of harms	Upstream and downstream determinants of health that build capacity to cope with, adapt to or bounce back from harms

3.1 Maintaining Health

A population health manager maintains healthy populations by making it easier for people to make good choices that lead to conditions that avoid the emergence of new risks and allow animals to access their determinants of health. The primary strategy of this goal is to equip people in the position to influence wildlife determinants of health with the knowledge, skills, attitudes, and circumstances to make healthy choices easier. In the human health sector, this strategy falls under the auspices of health promotion programs. Health outcomes like longevity or lack of disease are not the end goals in this part of the continuum of care. This phase of population health management helps us understand and influence circumstances that enable people to make decisions that keep wildlife healthy and keep risks at a tolerable level rather than target biomedical or technical interventions to take animals out of a diseased state.

Aldo Leopold, considered by many to be the father of wildlife ecology, said, "The most important characteristic of an organism is that capacity for internal self-renewal known as health" (Leopold 1949). Leopold argued that an overly medical approach is less important in promoting health than are environmental and population factors (as cited in Hanisch et al. 2012). Health exists at multiple scales, from the individual to group, population, community, and ecosystem level, and from local to regional to global scales. Health is affected by and affects relationships within and across these scales. Population ecology, community ecology, natural resource management, and allied fields have historically been viewed outside of the realm of wildlife health, yet it is these disciplines that most often have direct influence and understanding of how changes in determinants of health across multiple scales influence the size, productivity, distribution, and sustainability of wild populations. Promoting and maintaining wildlife health, therefore, primarily takes place outside of the traditional wildlife health sectors because the animal, social and environmental determinants of health are generally found outside of the scope of practice of most wildlife health programs and policies. A cross-sectoral collaborative approach is needed to link those with the authority to manage wildlife health with those having the knowledge and tools to manage population vulnerability and resilience (Stephen 2014).

Wildlife's access to its determinants of health is highly influenced by environmental and resource management decisions, which, in turn, are affected by our values, economics, and politics. Virtually everything we try to influence to maintain wildlife health requires someone to make a decision or to act in a certain way. Social forces and social relationships are critical elements of how people determine the value of wildlife and what they are willing to do to maintain wild populations' health. Those being asked to make a choice or implement an action to maintain healthy wildlife populations need to actively participate in the health promotion process. But, the knowledge, culture, resources, skills, and processes people bring with them to a wildlife health situation are varied, making it challenging to develop agreement on health goals and targets. Wildlife population health managers need to be pragmatic and adapt to varying expectations and knowledge.

There are often huge gaps between the available knowledge about what keeps wildlife healthy and actions toward achieving and maintaining it. Simply telling people about evidence and urging them to change what they do is ineffective (Levin 2008). Understanding the gaps between knowing and doing requires an understanding of what affects the choices people have, what enables or impedes their willingness or ability to act, and how well they can access and understand the information available to them (see Chap. 23 for more on implementation). Moving knowledge into action requires a relentless dedication to understanding the user's needs and strategies and tools to engage, inform, and motivate them under the circumstances they find themselves.

3.2 Reducing Vulnerability

Being at risk to a threat is not the same as being vulnerable to that threat. Vulnerability is a complex concept. Most simply put, it is a function of exposure to a hazard (or risk factor), sensitivity to harms from that hazard (or risk factor), and capacity to cope with and adapt to those harms. Exposure is affected by landscape and management factors that influence contact between an animal and a hazard/risk factor and affect the type, magnitude, and rate of contact. Exposure can be reduced by affecting its frequency, intensity, and/or duration. Sensitivity can be influenced at the individual, population, or socioecological system level. Sensitivity might be affected by the characteristics of the population (such as the presence of compromised sub-populations), individual susceptibility, the presence of concurrent stressors, and/or how human activities affect the aforementioned. Adaptive capacity will be influenced by biological characteristics, management decisions, social support to undertake actions that reduce the burden of a specific adverse health outcome, and the underlying health and resilience of the population (Stephen et al. 2018). Interacting environmental and social factors influence exposures and sensitivities, and various biological, social, political, and economic forces shape adaptive capacity. Vulnerability and its determinants are, therefore, dynamic, vary by place and problem, and are system-specific (Smit and Wandel 2006).

Vulnerability can be reduced by targeting liabilities that make wildlife more prone to hazard exposure and sensitive to harms as well as building capacities to deal successfully with harms, hazards, and risks that are encountered (McEntire 2011). Strategies to promote adaptive capacity overlap with strategies to maintain health because population health is the foundation for the ability to recover from or adjust easily to stressors and risks. Exposure is most often managed by building barriers. These might be physical barriers (such as in the case of wildlife fences in Africa to prevent the spread of pathogens to domestic animals; e.g., Ferguson and Hanks 2012); they might be administrative barriers such as prohibitions on the trade or importation of certain species (such as bans on salamander importation to prevent the translocation of *Batrachochytrium salamandrivorans*; e.g., Gray et al. 2015), or they may be land use protocols (such as the need to separate domestic sheep from wild sheep to prevent the sharing of respiratory pathogens in North America; e.g., USDA 2012). Often, immunization is the first tool imagined when thinking of ways to reduce wildlife sensitivity to diseases. But there is an extremely limited number of vaccines available for wildlife and many challenges in effective vaccine delivery. Moreover, many of the threats to wildlife health do not involve infectious diseases. Sensitivity can also be modified by managing concurrent stressors that increase the likelihood that exposure will result in harm. For example, the high levels of immunotoxic contaminants in marine mammals have been implicated as a contributor to viral disease outbreaks in marine mammals (Ross 2002). A small, threatened population in a genetic bottleneck that is not benefiting from public investment in its conservation and is combating other diseases would, for example, be more sensitive to the impacts of a new disease than a large, genetically diverse population receiving significant public funding to protect its determinants of health; even if both populations have equal susceptibility to the disease.

An all-hazards approach to managing vulnerability recognizes and integrates exposure, sensitivity, or adaptive capacity variables that are common across multiple hazard types, and then supplements these common elements with hazard-specific subcomponents to fill gaps as required. A rapid threat assessment of chronic wasting disease (CWD) in woodland caribou in northern Canada, for example, concluded that the ecology and behavior of woodland caribou may reduce exposure and sensitivity to the causative prions, but the multiple hazards they are already facing may compromise population capacity to cope with the disease if introduced to their range (Zimmer et al. 2019). The addition of one more stressor to the many others these caribou face (including habitat alteration from human land-use activities and forest fire, predation, and climate change (Environment Canada 2012)) may prevent the caribou from coping with the additional burden of a new disease. A similar rapid assessment of the threat of snake fungal disease in southern Canada concluded that cofactors influencing the spread and effects of this disease overlapped with the major challenges facing snakes in Canada (e.g., climate, habitat loss, habitat degradation) (Stephen et al. 2017). In both cases, avenues to reduce vulnerability to disease could be found in strategies to address other stressors threatening the population, with some modifications to account for the unique characteristics of an infectious disease threat.

3.3 Reducing Harms

Standard animal disease control methods very often cannot be acceptably, effectively, or efficiently applied in free-ranging wildlife. Vaccination and mass treatment are impeded because of the lack of available and proven effective medications or vaccines plus problems in their delivery. Changing social values are increasingly excluding depopulation and selective slaughter as options. Quarantine and isolation are not feasible for most free-ranging animals. Wildlife health managers are often only left with modifying human uses and interactions with wildlife, managing wildlife-domestic animal interactions, or environmental modifications to reduce wildlife disease harms. However, most wildlife health management efforts have historically focussed on biophysical harms from disease rather than seeing harm as a socioecological phenomenon because most interest has been on the disease rather than *illness* arising from the effects of the disease.

Illness, in a human health care setting, refers to how the patients and those around them, perceive the origin and significance of a health event; how it effects the patient's behavior or relationships with others; the meaning they give to that experience, and the steps taken to remedy this situation (Helman 1981). Diseases are something the patient's organs and body systems have. Most cases of wildlife disease are accompanied by social reactions (e.g., desires to eliminate wildlife-associated disease threats to agricultural profits or public health, regulatory requirements to preserve a population or animal welfare concerns about how a disease is managed). These social reactions, more than the disease itself, cause problems that need to be managed. Wildlife disease could, therefore, be seen as a social illness.

Social values and preferences affect how we allocate resources to the protection of one species over another (Kirkwood and Sainsbury 1996). Wildlife health managers are often confronted with the social illness caused by a wildlife disease but only relatively recently the human dimensions of wildlife disease management have been subject to systematic inquiry and implementation (see Chap. 19). Most attention remains on understanding, tracking, and interfering with the pathological processes of the disease rather than managing the problems linked to the illnesses (arising from social perceptions and impacts) from the disease.

Many wildlife health illnesses involve multiple concurrent harms. Chronic wasting disease (CWD) of cervids illustrates this point. This invariably fatal prion disease of deer, elk, and moose causes wasting, neurological dysfunction, and eventual death of infected animals. It has been deemed as a significant threat to the future vitality of free-ranging deer and elk in North America (WAFWA 2017). Because of its similarities to the zoonotic bovine spongiform encephalopathy, public health agencies advised precaution in consuming cervids affected with or hunted from CWD-positive areas (despite no direct evidence of human CWD to date). These advisories discouraged hunting and in doing so harmed rural incomes from hunting and lowered confidence in consuming cervids for subsistence needs. These food safety concerns have implications for rural food security and Indigenous rights to access safe and sustainable wildlife. International trade in cervid products is impeded by this disease. CWD therefore causes individual animal harms, conservations harms, harms to agricultural trade, increased social conflict,

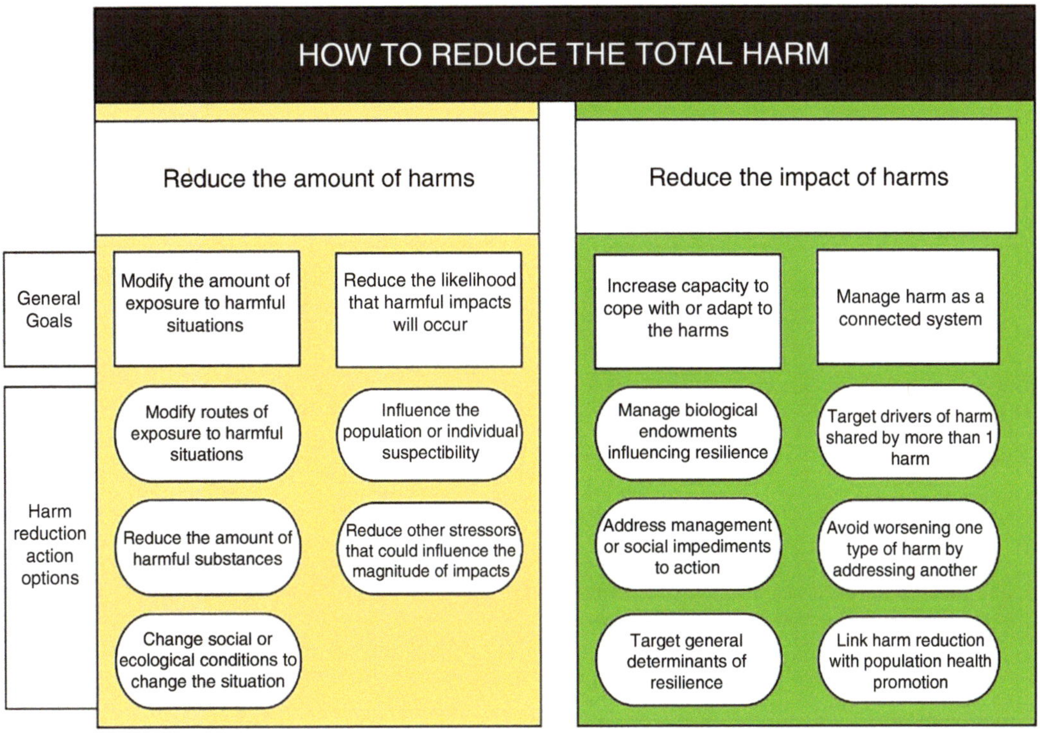

Fig. 1 Generic overview of options for harm reduction action

psychological harm through increased fear of natural resources, altered ecosystem functions, cultural harms by discouraging traditional hunting, and political conflicts linked to debates on how to control this disease. Despite this breath of harm, wildlife health research and many policies have been unevenly interested with controlling or eliminating the prion which serves as the etiologic cause of CWD. A harm reduction approach would not focus only on eradicating the CWD prion from populations or environments, (although recognizing the importance of trying to do so) but rather would also use existing knowledge, resources, and values to promote collaborative approaches that made incremental gains to reduce ecological, population, or social harms related to the illness.

Harm reduction is both a goal and a process. As a goal, it wants to reduce harm to individuals, populations, communities, species, or environments, whether through clinical care, preventive actions, or health promotion. Total harm faced by a population can, in general, be reduced by reducing the total amount of harm (which is achieved by reducing exposure or sensitivity), or by reducing the total impact of harm (achieved by increasing capacity to cope or reducing cumulative effects) (Fig. 1). Harm reduction, therefore, targets the same factors as vulnerability management but does so after the occurrence of the harm.

Harm reduction, as a process, focuses on developing local relationships and collaboration for collective actions leading to incremental improvements in health. Many of the underlying causes of harm cannot be quickly eliminated because of the often-slow pace of scientific, social, and political change. The necessary scientific advances, technology, or regulations needed to eliminate harm can take considerable time to be achieved. Different expectations for how much and what types of harm should be attacked can discourage actions especially when the science remains inconclusive. Action, too often, is delayed due to scientific uncertainties or social conflicts. Harm reduction processes promote relationships, structures, and circumstances to make incremental gains toward safer situations by reducing health, social, and/or ecological harms to individuals, communities, and ecosystems, without relying on the elimination of the hazard. It relies on five principles (Stephen 2020):

1. *It is collaborative.* Individuals, agencies, companies, and communities affected by or affecting harm need to be involved in cocreating harm reduction strategies tailored to a specific situation. Multidisciplinary pathways are sought to overcome barriers to implementing recommendations, foster collaboration on shared goals, and reduce conflict to enable actions and opportunities to prevent, mitigate, or cope with harms. Collaboration helps people see different aspects of the problem and, by exploring these differences, find solutions that go beyond their own perspectives. Successful collaborators see that their role and responsibilities for helping to reduce harm may extend beyond their interests.
2. *It creates a trusting environment for collective actions.* Trust is essential for collaboration and can be built by being honest in negotiations, communicating purposefully and regularly, behaving in accordance with agreements, and not taking advantage of others or events when the opportunity is available. There needs to be a shared goal(s) of where the group wants to go and a hierarchy of achievable steps that, taken one at a time, can lead to a safer and healthier situation. Some degree of negotiation will be needed to create a shared vision that will help collaborators see how working toward collective interests will meet the interests of themselves or their organization. Trust, commitment, and a deeper understanding of the value of collective action are gained by focussing on a series of incremental smalls wins toward the long-term goal.
3. *It is oriented to finding pragmatic solutions.* While not ruling out the long-term prohibitions or elimination of hazardous situations, gains that are feasible within the current circumstances and state of knowledge are sought rather than relying on the creation of a preferred future before acting. The focus is on finding strengths, possibilities, and opportunities to reduce harms rather than emphasizing the discovery of the cause of harm or attributing blame to others. It is about working with what we have and who we have today to make incremental improvements.
4. *It is local.* Many forms of risk management emphasize top-down actions. Harm reduction emphasizes bottom-up, locally developed planning. This reflects its focus on working in the context in which the harms are occurring. Harm reduction emphasizes action plans that adapt generic recommendations to local circumstances to produce gains that can be built on over time to lessen present harms while preparing for tomorrow's risks. Incorporating the context in which environmental, organizational, and personal factors interact increases the likelihood of finding shared priorities for action.
5. *It is integrative.* Social, ecological, and individual harms are interrelated. Seeking consensus on biological harms without accounting for social harms can increase conflict and delay actions. The usual approach of examining one type of harm in isolation from another reduces the chances of finding common pathways or opportunities to reduce or eliminate risks and harms. Examining them together helps to build consensus on goals and find actions that may have benefits across domains. Harm reduction shifts the focus for change from technical and biological matters alone to include social innovations and opportunities.

3.4 Promoting Recovery

Recovery interventions focus on populations that have already been harmed. Recovery, therefore, must be attentive to what has happened as well as what we hope to happen. The goal is to provide the circumstances and capacities needed to minimize suffering, limit the impact of ongoing harm, reduce any lasting effects, and restore ecological and social functions. Recovery efforts should prevent further deterioration and recurrence.

A critical first step in recovery is to develop a shared vision of what constitutes success. Different interests will have different views on what successful recovery looks like. It may be defined by some as a specific abundance or distribution of a population. Others may define it as the presence or absence of certain diseases or risk factors. Still, others might view recovery when a population fulfills its utilitarian role in local economies or cultures. A shared vision provides a foundation to connect different goals and objectives to a larger purpose. Recovery planning provides an opportunity to improve a population's health and resiliency. It has the potential to inspire goals beyond the restoration of the status quo while attending to the harms caused by a particular situation or disease. Relationships built in the harm reduction process can facilitate a shared understanding to develop a common vision for what successful recovery will look like. Population health assessments and hazard vulnerability assessments provide data that show the gaps between the population's current status and its desired state to help develop and assess goals, priorities, and strategies.

There is no single blueprint for local recovery action. Short-term recovery will focus on ensuring the hazardous or risky circumstances are under control and that populations can cope with the immediate impacts. Longer-term recovery mirrors the first phase of maintaining health in that it focuses on rebuilding the capacity for populations to cope with the stressors of their new reality and to bounce back. Given the importance and the broad range of the determinants of health, recovery requires the efforts of and coordination among a wide array of sectors and community stakeholders. Each place and problem will dictate what approach for collaboration will work best for that context. A population cannot be returned to health through the efforts of any single sector or stakeholder group without considering how these efforts affect and are affected by those of others. Accordingly, population health recovery requires buy-in and coordination among sectors and stakeholders, and focused, collective planning.

In tandem with the need for a shared vision of successful recovery is the need for agreement on the indicators used to assess progress toward that vision. Recovery assessments and surveillance should provide a systematic examination of the indicators that identify key problems and assets that might affect recovery. Not only should these indicators track features of the populations (e.g., abundance, vulnerability, seroprevalence, etc.) but also should monitor the social, economic, and structural forces that enable or impede recovery action. Indicators should help to prioritize subpopulations or locations that are uniquely vulnerable or are not receiving adequate care or investment to ensure recovery. Recovery indicators should track functional status, comorbidities, and other contextual factors that allow one to understand impacts on populations and find avenues for effective interventions.

4 Summary

For many wildlife health problems, actions across the population health continuum of care are needed to achieve a meaningful degree of prevention and health protection. The phrase "population health" in this book is used to convey a way of conceiving health that includes the whole range of health determinants, many of which fall outside of the usual realm of wildlife health practice. Successful population health action requires people and institutions willing and able to learn across disciplines and be adaptive to the unique circumstances and context of the population(s) of concern.

Wildlife population health requires cooperation across multiple sectors and combines diverse but complementary approaches. It relies on four key strategies (adapted from WHO 1986): (1) build wildlife management and health policies that protect the determinants of health; (2) focus on creating healthy environments that support sustainable access to the determinants of health;

(3) strengthen collaborative action; (4) and reorient wildlife health from only disease management to a providing a continuum of care.

Although no single program can address the wide range of influences on population health, the population health approach helps to orient programs away from more isolated and categorical approaches to more integrated ones. It illustrates the need for intersectoral activities and makes obvious the limitations put onto programs that do not consider contextual factors and community influences that enable or obstruct the necessary action to achieve and sustain healthy wild populations.

References

Acevedo-Whitehouse K, Duffus AL (2009) Effects of environmental change on wildlife health. Philos Trans Roy Soc 364(1534):3429–3438

Canada E (2012) Recovery strategy for the woodland Caribou *(Rangifer tarandus caribou)*, Boreal population, in Canada. In: Species at risk act recovery strategy series. Environment Canada, Ottawa, p 138

Ferguson K, Hanks J (2012) The effects of protected area and veterinary fencing on wildlife conservation in southern Africa. Parks 18(1):49–60

Gray MJ, Lewis JP, Nanjappa P, Klocke B, Pasmans F, Martel A, Stephen C, Parra Olea G, Smith SA, Sacerdote-Velat A, Christman MR (2015) Batrachochytrium salamandrivorans: the north American response and a call for action. PLoS Pathog 11(12): e1005251

Hanisch SL, Riley SJ, Nelson MP (2012) Promoting wildlife health or fighting wildlife disease: insights from history, philosophy, and science. Wildl Soc Bull 36(3): 477–482

Helman CG (1981) Disease versus illness in general practice. J R Coll Gen Pract 31(230):548–552

Kindig D, Stoddart G (2003) What is population health? Am J Public Health 93(3):380–383

Kirkwood JK, Sainsbury AW (1996) Ethics of interventions for the welfare of free-living wild animals. Animal Welfare-Potters Bar 5:235–244

Leopold A (1949) A Sand County almanac, and sketches here and there. Oxford University Press, New York

Levin B (2008) Thinking about knowledge mobilization. A discussion paper prepared at the request of the Canadian Council on Learning and the social sciences and Humanities Research Council. Available from http://en.copian.ca/library/research/ccl/knowledge_mobilization/knowledge_mobilization.pdf. Accessed May 2020

McEntire D (2011) Understanding and reducing vulnerability: from the approach of liabilities and capabilities. Disaster Prevent Manage 20(3):294–313

Paquet PC, Darimont CT (2010) Wildlife conservation and animal welfare: two sides of the same coin. Anim Welf 19(2):177–190

Plowright RK, Foley P, Field HE, Dobson AP, Foley JE, Eby P, Daszak P (2011) Urban habituation, ecological connectivity and epidemic dampening: the emergence of Hendra virus from flying foxes (Pteropus spp.). Proc R Soc B Biol Sci 278(1725):3703–3712

Ross PS (2002) The role of immunotoxic environmental contaminants in facilitating the emergence of infectious diseases in marine mammals. Hum Ecol Risk Assess Int J 8(2):277–292

Smit B, Wandel J (2006) Adaptation, adaptive capacity and vulnerability. Glob Environ Chang 16(3):282–292

Stephen C (2014) Toward a modernized definition of wildlife health. J Wildl Dis 50(3):427–430

Stephen C (2020) Harm reduction for reciprocal care. In: Stephen C (ed) Animals, health and society: health promotion, harm reduction and health equity in a one health world. CRC Press, Boca Raton

Stephen C, Shirose L, Snyman H (2017) Snake fungal disease in Canada rapid threat assessment. A report from the Canadian wildlife health cooperative. http://www.cwhc-rcsf.ca/technical_reports.php

Stephen C, Wittrock J, Wade J (2018) Using a harm reduction approach in an environmental case study of fish and wildlife health. Ecohealth 15(1):4–7

Szreter S (1999) Rapid economic growth and 'the four ds' of disruption, deprivation, disease and death: public health lessons from nineteenth-century Britain for twenty-first-century China? Trop Med Int Health 4(2):146–152

USDA (2012) Recommendations for domestic sheep and goat management in wild sheep habitat. Prepared by the Wild Sheep Working Group Western Association of Fish and Wildlife Agencies. https://www.fs.usda.gov/Internet/FSE_DOCUMENTS/stelprdb5385708.pdf

WAFWA (Western Association of Fish and Wildlife Agencies) (2017) Recommendations for adaptive Management of Chronic Wasting Disease in the west. WAFWA wildlife health committee and mule deer working group. WAFWA, Edmonton, Alberta, Canada and Fort Collins, Colorado. https://wafwa.org/wpdm-package/recommendations-for-adaptive-management-of-chronic-wasting-disease-in-the-west/

World Health Organization (1986) WHO the Ottawa charter for health promotion. https://www.who.int/healthpromotion/conferences/previous/ottawa/en/

Zimmer M, Zimmer P, Stephen C (2019) Rapid threat assessment CWD in woodland Caribou in western and northern Canada. A report from the Canadian wildlife health cooperative. http://www.cwhc-rcsf.ca/technical_reports.php

Population Health Core Competencies and Scope of Practice

Craig Stephen

Abstract

Core competencies describe the knowledge, skills, and attitudes necessary for the practice of a discipline or an approach. Wildlife population health practitioners need technical knowledge but that alone is insufficient to improve health. Skills and perspectives that promote collaborative solutions by developing shared visions for action in an evidence-based and ethical manner are essential, as is the ability to see health problems in their social and ecological context. Wildlife population health can only be delivered by highly collaborative interprofessional teams focussed on fulfilling society's interest in assuring conditions in which wildlife can be healthy. This chapter proposes seven core wildlife population health competencies: (1) being a specialized generalist, (2) being an agent of change, (3) being an advocate for wildlife health, (4) embracing partnerships and collaboration, (5) using evidence-based practice, (6) being able to find entry points into complex problems, and (7) leaning toward a systems approach.

Keywords

Population health · Wildlife · Core competencies · Change · Evidence-based practice

1 Introduction

There is a growing trend for professions and disciplines to create lists of core competencies required to effectively operate within a discipline's paradigm or scope of practice. Core competencies are the essential knowledge, skills, and attitudes necessary for the practice of the discipline or to succeed in a job or task. They are used to guide education and training to build workforce capacity and to develop and implement policies, practices, and research to meet the goals and responsibilities of the discipline. This chapter proposes wildlife population health core competencies. In doing so, this chapter outlines key aspects of the scope of practice of wildlife population health.

Wildlife population health works at the edges, intersections, and overlaps of different types of knowledge because population health is the cumulative effect of interacting determinants of health and because those interactions are interpreted through different disciplinary and personal perspectives (see Chaps. 1, 6, and 19). Wildlife population health practice asks us to shift from only describing and analyzing disease

C. Stephen (✉)
School of Population and Public Health, University of British Columbia, Vancouver, BC, Canada

C. Stephen (ed.), *Wildlife Population Health*,
https://doi.org/10.1007/978-3-030-90510-1_3

to collaboratively developing actions to protect health. The skills and knowledge to fulfill that task surpass those needed to detect, diagnose, and assess diseases or to understand how diseases or determinants result in individual or population impacts. There is a lack of consensus or evidence upon which to build internationally acceptable wildlife health core competencies (Stephen et al. 2018). Although publications on core competencies are available for wildlife management, public health, One Health, natural resource management, veterinary medicine, and other allied fields, none have been systematically created for wildlife health.

Table 1 summarizes core competencies from five disciplines that can inform wildlife population health competencies. There are shared themes of competency that operate across these disciplines. These disciplines were highlighted because they focus on solutions-driven research and management and, like wildlife population health, those solutions are influenced by diverse types of information and stakeholders. Table 1 shows that technical knowledge is insufficient to effect change to improve health. Skills and perspectives that promote collaborative solutions by developing shared visions for action in an evidence-based and ethical manner are essential. The ideas of advocating for, leading, and facilitating change permeate Table 1 as do the need to be able to see the whole and to see problems within their social and ecological context. This table of competencies drives to the conclusion that wildlife population health can only be delivered by highly collaborative interprofessional teams focused on fulfilling society's interest in assuring conditions in which wildlife can be healthy. Foundational knowledge of the drivers and determinants of wildlife health along with the ability to measure and assess health outcomes and program performances must be accompanied by an aptitude for integration and collaboration.

As wildlife population health encompasses several distinct jobs or positions within a variety of types of organizations, specific competencies may be more or less relevant for individuals based on their responsibilities. Individuals should initially focus on developing competencies in the areas most relevant to their jobs but are encouraged to explore the breadth of competencies to expand their ability to deliver on the responsibilities and obligations for healthy wild populations.

At the writing of this book, there were no published wildlife population health core competencies. The core competencies highlighted in this chapter are drawn from other elements of wildlife health that are presented in other chapters in this book, the competencies found in Table 1, and the author's experience as an advocate for and practitioner of population health. Core competencies are usually developed through a consultative and participatory process. This chapter hopes to set a foundation for subsequent consultation and collaboration to define and validate wildlife population health competencies.

2 Core Competencies for Wildlife Population Health

2.1 Being a Specialized Generalist

A workforce with a diverse knowledge base that can communicate and integrate knowledge from multiple disciplines is needed for the twenty-first century because the increasingly complex problems societies are facing need to be solved by synthesizing knowledge and skills from multiple areas (Zhang and Shen 2015). There has been a drive over the last century to describe expertise based on the depth of knowledge but the twenty-first-century workforce must be able to work across a breadth of knowledge. Borrowing from ecology, we can define a specialist as someone who thrives in a narrow set of conditions and a generalist as someone who thrives in a wide variety of conditions and can make use of a variety of different resources. A specialized generalist transgresses boundaries to advance cooperation and conversations about facts, practices, and values between researchers, managers, decision-makers, and society at large. Specialized generalists focus on applied knowledge and

Table 1 A synthesis of core competencies from fields relevant to the practice of wildlife population health

Domain of competence	Public health (PHAC 2019)	One Health (Frankson et al. 2016)	Global health (Barry et al. 2009)	Sustainable development (Roy et al. 2020)	Wildlife management (DeLany Jr 2004)
KNOWLEDGE	Has key knowledge and critical thinking skills related to public health science	Adept at systems thinking, can work across boundaries of knowledge and recognize impacts throughout a system		Capable of systems thinking and critical reflection	Strong foundation in wildlife management principles and techniques
ASSESSMENT	Collects, assesses, and analyses information to support evidence-based recommendations and assessments		Assesses needs and assets that lead to the identification and analysis of the determinants that affect health. Determines the reach, effectiveness, and impact of programs and policies		Accurately measures, analyses and records relevant indices and measurements
PROGRAMS	Implements and evaluates policies and programs to effectively choose options and to plan, implement and assess policies and programs	Works across boundaries, leads and manages resources and people and understands roles and responsibilities	Develops measurable goals and objectives in response to assessment of needs and assets. Identifies knowledge-based strategies. Carries out effective and efficient, and ethical strategies		Able to work within existing laws, regulations, and policies
PARTNERING	Forms partnerships, collaborations, and advocacy to influence and work with others in pursuit of a common goal	Identifies shared goals and values, establish trust, and builds strategic and collaborative teams	Works collaboratively across disciplines, sectors, and partners to enhance the impact and sustainability of programs and policies	Forms collaborative, interdisciplinary partnerships to craft useful knowledge	

(continued)

Table 1 (continued)

Domain of competence	Public health (PHAC 2019)	One Health (Frankson et al. 2016)	Global health (Barry et al. 2009)	Sustainable development (Roy et al. 2020)	Wildlife management (DeLany Jr 2004)
SOCIO-CULTURAL	Interacts with diverse knowledge, groups, and communicates	Has value driven personal attributes, honesty, and integrity	Develops and delivers programs that are culturally sensitive and ethical	Defines problems and solutions considering social context and perceptions.	Develops personal environmental ethics and philosophy and acts as an ethical manager
LEADERSHIP	Builds capacity, improves performance, and enhances the quality of the working environment	Establishes a vision and strategy for change. Diplomatically negotiates and resolves conflicts to promote collaborative action	Provides strategic direction and opportunities for participation in developing policy, mobilizing and managing resources for health promotion, and building capacity.		
ADVOCACY	Advocates for policies and services that promote and protect health of individuals and communities	Advocates for change	Advocates with and on behalf of individuals and communities to improve health and build capacity for actions		
COMMUNICATION	Can exchange ideas, opinions, and information	Communicates lessons learned		Effectively communicates ideas to diverse audiences	Writes simple technical reports
CHANGE		Supports and leads teams to make changes that solve problems	Enables change and empowers individuals and communities to improve health	Leverages multiple forms of knowledge and systems to promote change that solves problems	

innovation by combining special expertise with collaborative capacities (Parkes 2020).

Population health science is not its own discipline. It integrates knowledge, theory, and tools from multiple disciplines to develop a broad understanding of the multifactorial pathways that produce health so that more effective solutions can be found (Bachrach et al. 2015). Transdisciplinary research and practice are appropriate when searching for science-based solutions to problems in a real-world context where factual uncertainties, values, and societal stakes affect decisions (Wiesmann et al. 2008). Transdisciplinarity, therefore, is relevant to much of population health management. Specialized generalists are specialists who can look at problems from different disciplinary perspectives. Figure 1 adapts a transdisciplinary model for evidence-based practice (Satterfield et al. 2009) to introduce some guiding themes for the specialized generalist. Figure 1 shows how wildlife population health decisions must consider contextual factors, societal needs, evidence, and expertise. Specialized generalists help to integrate these factors and support decisions that will be feasible, effective, and acceptable.

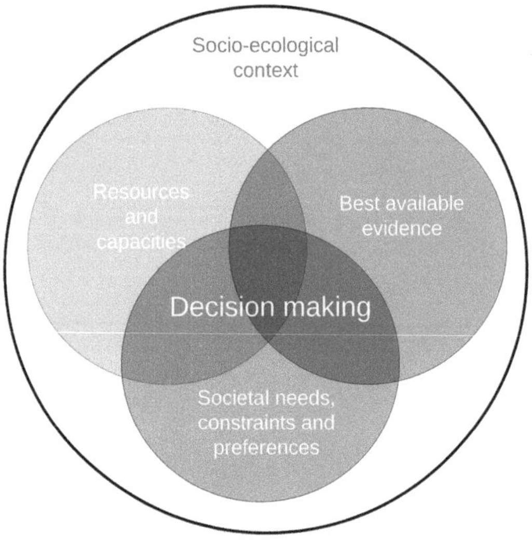

Fig. 1 A conceptual model that frames how the specialized generalist assembles information and perspectives to inform evidence-based decisions to promote and protect wildlife population health

Being a specialized generalized does not mean you need to be a master of everything, but rather that you have a unique suite of knowledge, attitudes, skills, tools, and processes needed to put general principles or shared knowledge into practice in different contexts, settings, and circumstances (Parkes 2020). You have a discipline-based speciality that allows you to engage in some aspect of the wildlife health problem but also have additional skills that enable you to explore insights from many different perspectives and recognize patterns that point to shared needs or coordinated actions.

2.2 Being an Agent of Change

Table 2 summarizes the attributes and assets of effective agents of change. A change agent needs the skill and capacity to stimulate, facilitate, and coordinate change. Only telling people about evidence and urging them to change is ineffective (Levin 2008). The gaps between knowing what to do to promote wildlife health and inspiring the actions to do it can be wide and difficult to breach. Never have we been able to produce and share so much information, making the knowing versus not knowing gap less pressing than the gap between knowing but not doing. Despite the

Table 2 Change management competencies (adapted from Higgs and Rowland (2000))

Can create the case for change and secure partnerships and support
Can scope the breadth, depth, and sustainability of a change strategy
Can help others develop capacity and confidence to achieve the change goals
Can develop a vision and message that advocates and inspires action for change
Can formulate and guide the implementation of a credible change plan with appropriate goals, resources, metrics, and evaluation processes
Can use multiple forms of knowledge to develop individual, group, and organizational capabilities for change
Has a personal, yet objective, commitment to the change
Can identify an incremental path to the change and can link that path to the desired solutions and goals

considerable time, efforts, and resources that have been dedicated to understand how and why certain problems occur, comparatively little effort has been dedicated to understand how to implement knowledge in policy or practice to protect or improve wildlife health (see Chap. 23 on implementation). Closing the knowing-to-doing gap requires processes, conditions, and relationships that enable trusted access, exchange, use, and evaluation of knowledge to support decisions and actions.

To create change, one needs a theory of change. A theory of change describes how and why the desired change is expected to happen in each context. It bridges what we know, what we want to achieve, and the activities it will take to get there. A theory of change helps identify the approach that should be taken to effectively address the causes of problems that hinder progress. Its construction requires the involvement of knowledge creators, planners, beneficiaries, and stakeholders to develop consensus on the shared goals and help people to see how sharing their knowledge contributes to long-term positive impacts. Figure 2 is a simplified theory of change used to advocate federal and provincial governments for a Pan-Canadian wildlife health program.

A well-developed theory of change shows the connections between short-term action and long-term goals, making explicit what we know, what we assume, and what we can feasibly do. A theory of change also considers impediments to change. These can include lack of access to enough high-quality evidence, lack of interest among potential knowledge users, low trust in the evidence, lack of capacity or ability to find and interpret evidence, lack of support for knowledge mobilization, strong forces that resist change, and pressures of various kinds pushing against the available evidence. The growing volume of research evidence, lack of time to read and thoughtfully review the evidence, financial disincentives, organizational barriers (e.g., lack of facilities or equipment), and peer group barriers (e.g., social norms that are not in line with the desired action) can further impede change (Stephen 2020).

This competency requires pragmatic skills in program planning and implementation because change does not happen by theory alone. Planning and evaluation include everything from assessing needs, setting goals and objectives, planning activities, securing resources for implementation, and measuring outcomes. Planning benefits from a careful situational assessment that examines the population of interest and issues that may affect implementation, including stakeholder wants, needs, and assets (Public Health Ontario 2015). A situational assessment helps to identify strategies and activities that are feasible with available resources and create meaningful goals and outcome objectives.

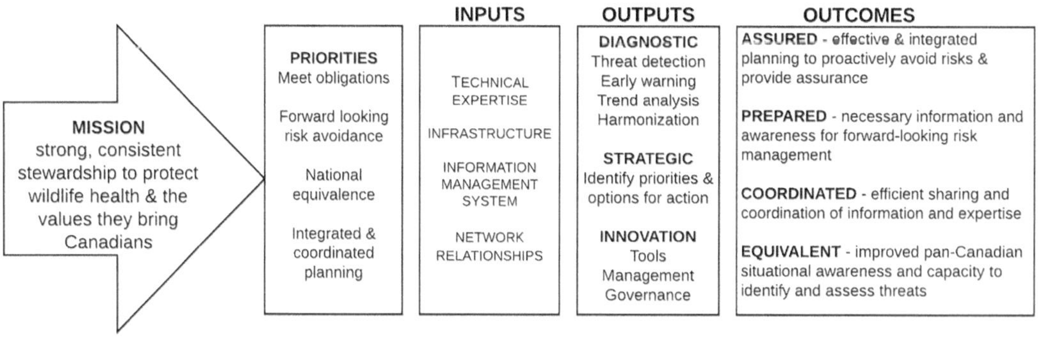

Fig. 2 An abbreviated theory of change diagram used to advocate federal and provincial governments for a Pan-Canadian wildlife health program

2.3 Being an Advocate

Wildlife health objectives are often contested by some group(s) such as governments, industry, community groups, as well as from within the wildlife health field itself. Protracted and highly organized opposition can impede action. An advocate offers independent support to those who cannot be heard (in this case wildlife) and ensures people can make an informed decision by laying bare what is at risk, implications of decisions, the values in play, and the evidence at hand.

The issue of whether wildlife health researchers and practitioners should be advocates is fraught with debate. On the one hand, some people worry that if the evidence is not moved into action by those who generate the evidence, the evidence could be misinterpreted, misused, or not used. On the other hand, there are fears that if one takes an advocacy position, one's evidence might be taken as biased and skewed toward a personal a priori perspective. The role of population health advocates is not to push forward their personal agenda but instead to strategically plan how to mobilize knowledge to action that addresses shared wildlife health goals. An advocate is tasked with structuring an argument in favor of wildlife health but within the reality of the social context in which actions will be taken and from a solid basis in evidence. There are eight questions to keep in mind when developing an advocacy strategy (Table 3).

An important skill for effective advocacy is being able to tailor advocacy strategies and messages that meet people at their stage of willingness to change. People go through five steps before adopting a new idea or innovation: (1) they need to become aware of the new idea, (2) they need to become motivated and able to find out more, (3) they need to see how the change applies to their own needs and circumstances, (4) they decide to try (or reject) the new idea, and (5) they need to confirm that their decision helped to meet their goals to continue its application (Kaminski 2011). An action or change is more likely to be adopted and spread if its advantages can be demonstrated to those who are being asked to

Table 3 Guiding questions when developing a wildlife population health advocacy strategy (adapted from Chapman 2004)

Question	Rationale
What is the shared objective of the action you are advocating?	Advocacy strategies need to be clearly relevant to the objectives you are trying to achieve
What are your advocacy objectives?	Advocacy objectives support ways to introduce compelling facts and perspectives or introduce different voices to enhance the authenticity or power of an argument for the shared health objective
How can a win-win situation be argued for?	People may not want to change on behalf of wildlife without seeing how that change fits their values or assists them in meeting their goals. Cooperation works better than coercion when aiming to make sustainable change
To whom are key decision-makers accountable?	Advocates need to understand what key decision-makers worry about, who endorses their policies and how to access and influence them
How will you frame the issue?	Framing involves making some aspects more salient to promote a particular problem definition, causal interpretation, moral evaluation, and/or recommendation. The goal is to establish the dominant perception of the problem circulating among decision-makers
What is your "quick pitch"	Wildlife health issues can be technical and tedious to decision-makers. Develop the short, concise story you can tell that conveys your key message (s) in an accessible and understandable manner
What is the best way to spread your message?	Make sure your messages get to your target decision-makers using methods and processes that they are familiar with and respect
Who is in your coalition for change?	Using credible and accessible voices that can be called on and are willing to speak up at strategically important times can amplify your message

adopt it, is consistent with social norms, and can be feasibly applied. Advocates are more likely to prompt a change if they are connected to and understand those they are trying to help change. Understanding and empathizing with the change targets help to communicate the need for and the value of change. The advocate and their targets need to be able to hear from, respond to, and influence each other (Lunenburg 2010). Wildlife population health advocates, therefore, need to understand the people and groups they wish to influence as much, or even more, than the wildlife populations they serve.

Advocates can be informed by theories of behavior change, such as the theory of planned behavior, the transtheoretical model, and the health belief model to uncover insights into how to help people move from one set of actions to another. The Knowledge-to-Action framework is another helpful resource (Graham et al. 2006). It was developed to help, create, and sustain evidence-based actions. It is an explicit process to determine what knowledge needs to be translated, how it is translated, by whom, when, and why. This framework highlights that a critical first step in turning knowledge into action is to cultivate trust and relationships between knowledge creators and knowledge users to establish a common understanding of needs and processes.

2.4 Embracing Partnerships and Collaboration

Wildlife population health requires partnerships with a diversity of collaborators and stakeholders because the practice crosses disciplinary, species, and subject boundaries. Wildlife population health practitioners are tasked with the challenge of engaging teams of diverse academic, community, and government specialists and facilitating a transdisciplinary approach among people most often comfortable working within their own speciality or worldview. Partnerships and collaboration offer powerful opportunities for joint actions that provide sustainable benefits (Stemshorn and Zussman 2012). Wildlife health practice must go beyond academic interdisciplinarity to engage directly with the production and use of knowledge outside the academic environment to support action-oriented work. The scope of partnerships will depend not just on the issues at hand but on the willingness, freedom, and capacity of key players to share roles, responsibilities, and resources.

Mascia et al. (2003) (as quoted by Schultz 2011) said that "conservation is a human endeavor: initiated by humans, designed by humans, and intended to modify human behavior." Given this very human aspect of protecting wildlife, wildlife health science would be well served to work with social scientists, politicians, and communities. The ability to manage constructive multi-stakeholder negotiations to resolve conflicting interests is a necessary partnership competency. Perhaps most critical to effective partnering is true collaborative goal setting, shared leadership, transparent and explicit expectations of partners that match their capacity, and clear guidelines on responsibilities, reporting, and information sharing (Stephen and Stemshorn 2016).

2.5 Use Evidence-Based Practice

Wildlife population health practitioners are obliged to use the best available evidence to inform policies and practices. Evidence-based practice means that people use all available evidence in an objective and balanced way rather than only relying on evidence that supports their view or distorts evidence to suit their needs. Considering the definitions of wildlife population health espoused in Chaps. 1 and 2, being able to gather, assess, integrate, and apply diverse data, opinions, and values to promote coordinated action are core competencies for evidence-based practice. There are, however, pressures working against an evidence-based approach; many species are understudied, ecological and epidemiological variations can limit generalizations of lessons learned elsewhere, different people interpret evidence differently based on their disciplines and values, organized interest groups can favor some types of information and discredit others, and political pressures could focus on short-term action before evidence is assembled (Brownson et al. 2009). Evidence-based wildlife population health practice requires us to wed information on the species, the harms, and the hazards, with the social situations in which the problem is embedded and the realities of a specific real-world environment which includes uncertainties, conflicting values, and lack of experience managing many wildlife health problems.

To meet the challenges of evidence-based practice, wildlife health practitioners must be able to access and critically appraise evidence. In some cases, the evidence in question may be within the practitioner's field, but in other situations, partnerships or consultations with others may be needed to assess the reliability and validity of the evidence at hand. In many fields, there are accepted criteria for assessing evidence, but in others, a participatory process may be needed to establish the criteria for acceptable evidence. Due to prevailing unknowns, especially for understudied species or problems, an accepted standard for biological, ecological, and social evidence to be considered when making decisions and how to accommodate varying levels of reliability, validity, uncertainty, and representativeness of the evidence may need to be established de novo for new wildlife health situations. This may include guiding how to deal with uncertain or conflicting evidence and establishing causal relationships. The lack of a shared vision among proponents and opponents of the evidence can doom groups to get stuck in a "more evidence is needed" rut that creates new facts that are scientifically interesting and important but fail to satisfy the evidence expectations of decision-makers and stakeholders. Groups can get stuck in seeking evidence to support their perspectives rather than working together to resolve critical uncertainties that prevent action.

Evidence-based practitioners need to be able to interpret and communicate the acquired evidence in formats that are accessible and understandable to target audiences (Albarqouni et al. 2018). Before accessing and evaluating evidence, the practitioner or researcher needs to be clear on the questions being asked. As there are multiple needs and perspectives surrounding wildlife health issues, the problems bothering one group may not be the same as another, hence, evidence may not be tailored to all critical uncertainties impeding action. Being able to recognize the importance of and strategies for identifying and prioritizing uncertainties or knowledge gaps and finding the means to develop consensus on investigative priorities in practice are key skills. Both qualitative and quantitative approaches will be needed to resolve uncertainties.

2.6 Finding "Entry-Points"

Disciplinary knowledge and experience are necessary for interdisciplinary problem-solving (Zhang and Shen 2015). Disciplinary perspectives act as "conceptual gateways" that provide an entry point to begin to characterize a complex problem (Yukawa 2015). Having confidence in one's own discipline can build comfort in working with other disciplines.

It is difficult to prescribe the necessary discipline(s) one must master to be proficient in multi- to transdisciplinary fields of practice. The historic focus on the absence of disease as the defining feature of wildlife health gave prominence to disease-oriented fields such as pathology, microbiology, toxicology, and disease ecology as essential disciplinary foundations for wildlife health practitioners or researchers. The expanded notion of wildlife health articulated in chapter one broadens the scope of foundational disciplines for wildlife population health. Epidemiology, ecology, critical assessment of the knowledge, and program evaluation all can serve as good disciplinary entry points for wildlife population health practitioners. Wildlife population health practitioners need to be able to knowledgeably speak about the health status of populations, the quality and distribution of the determinants of health and risk factors, strategies for health promotion and disease control, as well as the factors that influence the deployment and impact of wildlife health services and resources. They need not be an expert in all these factors but should have sufficient knowledge to critically assess information and evidence and apply knowledge to practice.

2.7 Leaning Toward the Systems Thinking

Many of the most pressing wildlife health issues are not purely technical problems. Monodisciplinary, linear, causal chains of reasoning have led to great understandings about wildlife health but have exposed only part of the information needed to understand complex problems that are entwined in diverse social, economic, political, cultural, and value systems (Kreuter et al. 2004). Population health practitioners need to be attuned to changes in science, changes in stakeholder goals and values, and the impact of actions intended to protect wildlife. Effective health protection and planning need to integrate the ecological, social, and individual context to promote sustainable action.

Systems thinking is a common core competency for solutions-oriented fields operating in the 'real world' such as conservation and health management. Systems thinking is used in the human health sector to accelerate a more realistic understanding of what works, for whom, and under what circumstances (de Savigny and Adam 2009). Systems thinking emphasizes interrelationships rather than things that underlie complex situations. There are three core systems thinking concepts. First, a system is made up of parts, relationships, functions, and products. Understanding their relationships is as much or more important than knowing the parts. Second, because there are feedbacks, synergies, and antagonisms in a system, systems relationships can be complex and dynamic and do not act in a linear predictable fashion. Surprises are to be expected. Third, the goal is not to elucidate the intricacies of all parts, products, functions, and relationships in a system but rather to know them well enough to find entry points that allow you to find places where a small intervention can have a meaningful impact. Systems approaches should help disentangle the dynamics that generate wildlife health problems, incorporate and reconcile diverse perspectives, prompt discussion on trade-offs, and foster the integration of evidence and knowledge into decision-making (Mahajan et al. 2019). Seeing the whole of an issue rather than some of its parts also helps with other core competencies such as finding entry points and building partnerships. A more in-depth discussion of systems and complexity in wildlife health is found in Chap. 17.

3 Summary

This chapter proposes skills, knowledge, and attitudes needed to apply population health thinking to wildlife health. The proposed core competencies correspond with those for other fields striving to find solutions under complex, changing, and messy real-world settings. The candidate competencies reflect the need for wildlife population health practitioners to be transdisciplinary and systems-oriented, effective in helping disparate groups find common ground, and able to critically assess and mobilize information for action.

A primary limitation of this chapter is that the competencies were developed based on analogy and argument rather than empirical work supporting their validity. Some readers will undoubtedly disagree with some of the nominated competencies, others will see gaps or deficiencies or recommend amendments and additions. This is because each of us will approach wildlife population health from our own disciplinary backgrounds and experiences. The necessary competencies must be further developed and validated through systematic evaluation and participatory processes to achieve consensus.

References

Albarqouni L, Hoffmann T, Straus S, Olsen NR, Young T, Ilic D, Shaneyfelt T, Haynes RB, Guyatt G, Glasziou P (2018) Core competencies in evidence-based practice for health professionals: consensus statement based on a systematic review and Delphi survey. JAMA Netw Open 1(2):e180281

Bachrach C, Robert S, Thomas Y (2015) Training in interdisciplinary health science: current successes and future needs. Commissioned by the Institute of Medicine Roundtable on Population Health Improvement

Barry MM, Allegrante JP, Lamarre MC, Auld ME, Taub A (2009) The Galway consensus conference: international collaboration on the development of core competencies for health promotion and health education. Glob Health Promot 16(2):05–11

Brownson RC, Fielding JE, Maylahn CM (2009) Evidence-based public health: a fundamental concept for public health practice. Annu Rev Public Health 30: 175–201

Chapman S (2004) Advocacy for public health: a primer. J Epidemiol Community Health 58(5):361

de Savigny D, Adam T (2009) Systems thinking for health systems strengthening. In: Alliance for health policy and systems research. WHO, Geneva. https://apps.who.int/iris/bitstream/handle/10665/44204/9789241563895_eng.pdf;jsessionid=A2AB40D37368CB72A5A32A0A19093947?sequence=1. Accessed 23 Aug 2021

DeLany Jr BW (2004) Entry-level job skills needed by wildlife management professionals. LSU doctoral dissertations. 1515. https://digitalcommons.lsu.edu/gradschool_dissertations/1515. Accessed 23 Aug 2021

Frankson R, Hueston W, Christian K, Olson D, Lee M, Valeri L, Hyatt R, Annelli J, Rubin C (2016) One health core competency domains. Front Public Health 4:192

Graham ID, Logan J, Harrison MB, Straus SE, Tetroe J, Caswell W, Robinson N (2006) Lost in knowledge translation: time for a map? J Contin Educ Health Prof 26(1):13–24

Higgs M, Rowland D (2000) Building change leadership capability: 'the quest for change competence'. J Chang Manage 1(2):116–130

Kaminski J (2011) Diffusion of innovation theory. Canadian J Nursing Inform 6(2):1–6

Kreuter MW, De Rosa C, Howze EH, Baldwin GT (2004) Understanding wicked problems: a key to advancing environmental health promotion. Health Educ Behav 31(4):441–454

Levin B (2008) Thinking about knowledge mobilization. A discussion paper prepared at the request of the Canadian Council on Learning and the social sciences and humanities research council. Prepared for the Canadian council on Learning, p 32

Lunenburg FC (2010) Managing change: the role of the change agent. Int J Manage Business Admin 13(1):1–6

Mahajan SL, Glew L, Rieder E, Ahmadia G, Darling E, Fox HE, Mascia MB, McKinnon M (2019) Systems thinking for planning and evaluating conservation interventions. Conserv Sci Pract 1(7):e44

Mascia M, Brosius JP, Dobson TA, Forbes B, Horowitz L, McKean M, Turner N (2003) Conservation and the social sciences. Conserv Biol 17:649–650

Parkes M (2020) Working together for WHOLE systems: approaching Well-being and health, while oriented to living-systems and equity. In: Stephen C (ed) Animals, health and society: health promotion, harm reduction and health equity in a one health world. CRC Press, London

Public Health Agency of Canada (2019) Core competencies for public health in Canada. https://www.canada.ca/en/public-health/services/public-health-practice/skills-online/core-competencies-public-health-canada.html. Accessed 23 Aug 2021

Public Health Ontario (2015) At a glance: the six steps for planning a health promotion program. Ontario Agency for Health Protection and Promotion, Toronto

Roy SG, de Souza SP, McGreavy B, Druschke CG, Hart DD, Gardner K (2020) Evaluating core competencies and learning outcomes for training the next generation of sustainability researchers. Sustain Sci 15(2):619–631

Satterfield JM, Spring B, Brownson RC, Mullen EJ, Newhouse RP, Walker BB, Whitlock EP (2009) Toward a transdisciplinary model of evidence-based practice. Milbank Q 87(2):368–390

Schultz PW (2011) Conservation means behavior. Conserv Biol 25(6):1080–1083

Stemshorn B, Zussman D (2012) Financing for public veterinary services to ensure that they meet international standards. Rev Sci Tech Off Int Epiz 31(2):681–688

Stephen C (2020) Bridging the knowing to doing gap to support one health action. In: Stephen C (ed) Animals, health and society: health promotion, harm reduction and health equity in a one health world. CRC Press, London

Stephen C, Sleeman J, Nguyen N, Zimmer P, Duff JP, Gavier-Widén D, Grillo T, Lee H, Rijks JM, Ryser-Degiorgis MP, Tana T (2018) Proposed attributes of national wildlife health programmes. Revue Scientifique et Technique-Office Int des Epizoot 37:3

Stephen C, Stemshorn B (2016) Leadership, governance and partnerships are essential one health competencies. One Health 5:161–163

Wiesmann U, Biber-Klemm S, Grossenbacher-Mansuy W, Hadorn GH, Hoffmann-Riem H, Joye D, Pohl C, Zemp E (2008) Enhancing transdisciplinary research: a synthesis in fifteen propositions. In: Handbook of transdisciplinary research. Springer, Dordrecht, pp 433–441

Yukawa J (2015) Preparing for complexity and wicked problems through transformational learning approaches. J Educ Libr Inf Sci 56(2):158–168

Zhang D, Shen J (2015) Disciplinary foundations for solving interdisciplinary scientific problems. Int J Sci Educ 37(15):2555–2576

Part II

Foundations of Wildlife Population Health Management

Epidemiological Study Design in Free-Ranging Wildlife: Theoretical and Practical Considerations

Todd Shury and Claire Jardine

Abstract

Epidemiologic studies have long provided evidence for understanding and developing management actions related to diseases and health, including zoonoses, for both human and domestic animal populations. There is an urgent need to understand interactions between human, wildlife, and domestic animal populations using epidemiological approaches in order to manage emerging diseases, climate change, and biodiversity loss. Challenges associated with applying epidemiological techniques to wild animal populations, including issues associated with sample collection and diagnostic tests, and lack of knowledge about wildlife populations are reviewed and practical solutions to these problems are summarized.

Keywords

One Health · Wildlife epidemiology · Wildlife management · Wildlife disease · Confounding · Biases · Validity

1 Introduction

Epidemiology is the study of disease and other health-related events in populations and the factors that determine their occurrence (Thrusfield et al. 2017). Epidemiologic studies have long provided key evidence for understanding and developing management actions related to noninfectious and infectious diseases, including zoonoses, for both human and domestic animal populations. There is a clear need to understand interactions between human, wildlife, and domestic animal populations within a One Health context due to the increasing frequency of disease emergence over the past two decades (Cunningham et al. 2017; Stephen 2021). However, there are several challenges, associated with applying epidemiological techniques to wild animal populations, including issues associated with sample collection and diagnostic tests, and lack of knowledge about wildlife populations. Many of the assumptions and population characteristics needed by epidemiological methods are much more challenging to objectively measure and apply in wild populations compared to domestic animals or people. Although such studies are

T. Shury (✉)
Wildlife Health and Management, Parks Canada Agency, Saskatoon, SK, Canada

Western College of Veterinary Medicine, Saskatoon, SK, Canada
e-mail: todd.shury@pc.gc.ca

C. Jardine
Ontario Veterinary College, University of Guelph, Guelph, ON, Canada

C. Stephen (ed.), *Wildlife Population Health*,
https://doi.org/10.1007/978-3-030-90510-1_4

challenging and no single study can generate all the required information about a wildlife population, it is important to remember that it is the cumulative effort that counts, with progress being made in incremental successive studies over time (Stallknecht 2007). Wildlife epidemiological studies and methods must embrace multiple lines of evidence to triangulate toward a better understanding and be interdisciplinary in all its forms, including indigenous ways of knowing. In this chapter our objectives are to: (1) describe some of the major issues and problems experienced by researchers and health professionals when attempting to apply epidemiological principles and methods to free-ranging wildlife populations, and (2) help researchers and practitioners to understand how these methods and tools can be applied effectively in free-ranging populations.

2 Goals and Objectives for Epidemiological Studies in Wildlife Populations

Epidemiological approaches can be used to describe, understand, and ultimately inform the health management of free-ranging wildlife populations. Descriptive approaches are often used initially to determine the "what" and the "who": which etiologic agents are involved in causing morbidity and mortality and which demographic groups are primarily affected? Excellent references exist for understanding patterns of disease occurrence and how to collect such data effectively (Delahay et al. 2009; Thrusfield et al. 2017; Wobeser 2006). Once patterns of disease and health in a population are understood, one is often interested in then generating or testing hypotheses about the 'why' and 'how': why are certain parameters more, or less, involved in causing disease or health outcomes, and how can these factors be managed? This often involves more analytical epidemiological approaches that use observational, retrospective, or prospective study designs (cross-sectional, cohort, and case-control designs).

To manage and understand health implications and move from knowledge to action (Stephen 2021), researchers often want to implement control measures in an adaptive management framework. These approaches often involve before-after-control-impact (BACI) designs (Conner et al. 2007; Shaffer and Buhl 2016; Rytwinski et al. 2015), or experimental designs where different treatments are implemented in similar populations (Cassidy 2015; Delahay et al. 2009). Regardless of the approach taken, it is important to understand the value of evidence triangulation and the importance of putting knowledge gained into action (Stephen 2021). Evidence from multiple different pathways of study is often extremely valuable for understanding complex ecological phenomena. For example, using two-eyed seeing approaches (Kutz and Tomaselli 2019), quantitative and qualitative approaches can provide complementary and corroborating information that allow us to more fully understand both human values and wildlife health determinants in a holistic framework. A two-eyed seeing approach is a collaborative, iterative, adaptive process that bridges multiple knowledge systems including western science and indigenous ways of knowing to co-generate knowledge within a rigorous, transparent, and appropriate process of knowledge acquisition and verification (Kutz and Tomaselli 2019).

3 Issues with Epidemiologic Studies in Wildlife Populations

When attempting to understand determinants of health in free-ranging fish and wildlife populations, there are many barriers that create challenges with the interpretation of surveillance outputs and epidemiologic data. These include lack of validated diagnostic tests, issues associated with sample collection, and a host of population-level issues related to determining causation. Despite these challenges, it is still feasible and quite reasonable to apply epidemiologic methods to the study of wildlife health. One just needs to understand the limitations beforehand and know how these limitations can be addressed using proper study design. Some of the major challenges and limitations when studying free-ranging wildlife populations are described below (Table 1), followed by epidemiological methods and approaches that work best for studying free-ranging wild populations.

Collecting biological samples for wildlife health studies is challenging from a cost and welfare perspective. Live-capture studies depend on being able to capture and collect samples that do not compromise the health of the captured animal (Stallknecht 2007). These studies can be very expensive and time-consuming, so many studies rely on sampling free-ranging populations using convenience sampling frames; those samples that are easily acquired through existing means such as hunting, fishing, or citizen science efforts. This means these samples are often biased and not representative of the populations being sampled (Wobeser 2006). Sample sizes for studies involving wild populations are often small, due to the high cost of capturing and sampling remote populations, or when working with small, endangered populations with few individuals. Even with large populations of relatively common species, detecting pathogens at low prevalence requires enormous effort over long periods of time during which, many of these population characteristics change rapidly.

Random sampling is a core assumption for many epidemiological study designs but it is often very difficult or impossible to achieve in free-ranging populations. To achieve a random sample, one must know a population's underlying spatial and temporal variability to design sampling strategies that ensure all animals in the population have an equal (non-zero) probability of being sampled. Despite evidence that a species' social structure can have enormous implications on disease transmission and reservoir status, the social structures of many wildlife populations are unknown, and understanding their social network is a very expensive undertaking, so it is often restricted to small, localized populations. Population characteristics such as age structure, sex ratio, recruitment, and home range and connectivity are often undetermined in wild populations, and one often has to infer these characteristics from convenience samples.

Table 1 Common issues arising from sampling and collecting from free-ranging wildlife populations

Capture/sampling issues
• Representativeness and random sampling are extremely difficult
• Capture, sample acquisition is very costly leading to a small sample size
• Method of capture often produces another set of biases (selection bias) and influences animal behavior post-capture
• Spatial location/home range data are costly and difficult to obtain
• Wildlife capture is not considered culturally/ethically acceptable by some groups
Diagnostic test issues
• Lack of validated tests for a particular species
• Sensitivity and specificity of diagnostic tests are often unvalidated/unknown. Predictive values are not calculable due to lack of data (e.g., prevalence)
• Strains/species of pathogens are often novel or never before detected, and therefore lack a diagnostic test
• Cross reactions with similar pathogens
• Reduced or nonexistent laboratory capacity in underdeveloped countries
• Lack of "normal" reference ranges for biochemical, hematologic tests, contaminants
Population issues
• Underlying population structure (age, sex, location, herd/flock status, social structure) is often unknown
• Interspecies interactions and cumulative effects difficult to quantify
• Reservoir status is very difficult to determine (spillover host, reservoir, non-susceptible hosts)
• Comorbidities are often not measured due to cost and challenges with diagnostic testing
• Sublethal effects or cumulative effects are very difficult to discern in wild populations
• Ecological processes work differently at different geographical scales, making generalization difficult

Issues with diagnostic tests also impose challenges. Validated diagnostic tests are often not available for wild species and tests developed for domestic species must often be deployed (Thomas et al. 2021). Unfortunately, this means that data on sensitivity and specificity of these tests often do not exist, so test performance (their positive or negative predictive values) is often unknown before taking it into the field. Latent class analyses and other techniques have been used to estimate test performance when these parameters are unknown when sampling from multiple populations (Richomme et al. 2019; Shury et al. 2015). Many tests developed for domestic species can cross react with similar pathogens found in wild populations, but determining which pathogen a positive test indicates requires additional cost and laboratory expertise to tease apart confounding results. One example involves *Borrelia turicatae*, which causes tick-borne relapsing fever and causes cross-reactions on a test meant to determine exposure to *Borrelia burgdorferi*, the causative agent of Lyme disease (Gettings et al. 2019).

Many countries lack the veterinary infrastructure to allow diagnostic testing in wild populations. Many of these exist in biodiversity hotspots where zoonotic spillover of pathogens to human populations may be higher. International efforts which help to establish veterinary infrastructure such as the PREDICT program (https://p2.predict.global/) have made major improvements in diagnostic capacities in recent years.

Understanding the health of populations can also be challenged by a lack of established reference ranges for parameters commonly used to assess health in domestic animals and humans, such as biochemical tests and hematology. Zoological institutions have been great sources of information in this regard in the past 30 years and there are many more established reference ranges available now. Many studies often build in a validation sample set from healthy animals to determine proper reference ranges as part of the study design. There can, however, be substantial differences between health parameters for species kept in captivity versus their free-ranging counterparts (see McAdie 2018).

4 Study Designs and Approaches for Free-Ranging Wildlife Populations

There are two broad approaches that can be applied to the study of wildlife health and disease: experimental and observational studies. Experimental studies are used relatively infrequently in free-ranging wildlife because they require the investigator to control and manipulate the exposure variable of interest (discussed below). Most wildlife health studies rely on observational studies, including descriptive and analytic studies. For a more in-depth review of current standards for reporting observational research refer to the STROBE (Strengthening the Reporting of Observational Studies in Epidemiology) guidelines (http://www.strobe-statement.org) (Sargeant et al. 2016).

4.1 Descriptive Studies

Descriptive studies include case studies and case series. If one is describing a new case of a disease in a new species or location and reports this finding, this is typically described in a case report. If multiple animals having similar criteria and causes are found, this is a case series. If one goes out to look for a particular disease or presence of an etiological agent or risk factor in a population and just intends to summarize the data, these are typically also included as descriptive studies and are very common in the wildlife field (Kaur et al. 2008; Parra et al. 2006; Salb et al. 2014). This type of information is very valuable and can help to guide the efficient management of outbreaks. For example, Salb et al. (2014) summarized anthrax outbreak data from wild wood bison in northern Canada, utilizing outbreak information collected over a 46-year period, demonstrating that outbreaks had declined over time, that outbreaks peaked in early July, and primarily involved bulls. These types of studies allow researchers to identify hypotheses that can be investigated using analytic or experimental studies.

4.2 Analytic Studies

Analytic studies formally compare results between groups that differ with respect to their exposures or outcomes. Analytic or explanatory studies are used to better understand how different variables such as age, sex, or geographic area affect and interact with other variables such as disease status. They typically seek to understand causation or how an outcome variable (disease status) is related to other variables of interest.

4.2.1 Cross-Sectional Studies

Cross-sectional studies look across populations at one point in time to record information about their subjects without manipulating the study environment. They can be classified as descriptive or analytical, depending on whether the outcome variable is only being documented and measured or assessed for potential associations with exposures or risk factors. They may be concerned with single or multiple variables of interest. Cross-sectional studies are one of the most common epidemiological study designs used in wildlife as no prior knowledge (e.g., health/disease status, age, and sex) is needed about subject animals before capture and sampling. Repeated cross-sectional studies are studies repeated over time using different individuals in the population. Longitudinal studies are cross-sectional studies that involve the same individuals repeatedly captured and sampled over time. These types of studies are relatively easier to implement relative to other types of epidemiological study designs, such as cohort studies, but researchers have less control over how animals are categorized or grouped and it is not possible to show proof of causal association with these studies, as one is measuring both the outcome and a set of explanatory variables at the same time. As a result of exposure and outcome being measured at the same time, causality cannot be reliably inferred from cross-sectional studies (see Chap. 6), unlike with cohort and case-control designs. The outcome being measured in cross-sectional studies is typically prevalence of a disease or health parameter being studied as one often does not know the actual population structure or exactly how individuals comprise a population of wild animals (i.e., the denominator), making calculation of rates (incidence, rate ratio) difficult or impossible.

4.2.2 Case-Control Study Designs

Case-control studies are a very powerful method of studying health determinants in human and domestic animal populations but are rarely used in wildlife populations as some population characteristics need to be known beforehand. Typically, with these studies, a subset of the population with a particular outcome (the cases) is compared to another subset of the population that has not experienced the outcome (controls). No intervention is attempted and no attempt is made to alter the course of the disease. Most often, case-control studies are retrospective as they look back in time to compare how frequently the exposure to a risk factor is present in each group to determine the relationship between the risk factor and the disease. Prospective case-control studies are less common and rarely possible for free-ranging wildlife as they involve following the sample group over time while monitoring their health and exposures. Cases emerge when animals develop the disease or condition under investigation as the study progresses. Challenges in identifying cases, finding retrospective exposure information, or prospectively following animals over the time limit the application of these methods.

Examples, where case-control studies have been successful, include closely monitored populations such as lowland gorillas (Haggblade et al. 2019), African buffalo (le Roex et al. 2013), and sea otters (Shockling Dent et al. 2019). This was possible in the case of lowland gorillas because there was a preexisting dataset of 132 "clinical interventions" available over a 20-year period, several, but not all of which involved snaring and subsequent treatment. This type of design was only possible in this case because of very close monitoring of this population with controls being clinical interventions other than snaring. In the paper published by Shockling Dent et al. (2019), a similar retrospective case-control design worked effectively as there was a large necropsy database from a closely monitored population where findings from nasal mite-infested sea otters (cases) could be compared with a large number of controls who did not have the nasal mites at necropsy, allowing researchers to determine that older sea otters were 9.4 times more likely to be infested with mites than younger otters.

4.3 Cohort Study Design

A cohort study selects animals based on exposure and then studies the development of disease in the exposed and unexposed groups of animals. Cohort studies can be logistically difficult because of the need to identify animals initially free of the outcome and then follow them over time to determine the development of the outcome (Caswell et al. 2018). This can be particularly challenging for wildlife studies. Nonetheless, cohort studies have been used to study wildlife disease. For example, Miller et al. (2008) used a cohort study design to compare the annual survival of prion infected and apparently uninfected adult mule deer. They found that prion infection dramatically lowered the survival of free-ranging adult mule deer (*Odocoileus hemionus*). Cohort studies are powerful because exposure is identified before the outcome, which confirms that the proposed cause preceded the development of the outcome.

4.4 Experimental Designs

Experimental studies, where the investigator manipulates the exposure variable of interest, are of primary importance in understanding pathogenesis, validating diagnostics, and providing a necessary perspective for interpreting data (Stallknecht 2007). Experimental studies are often conducted in laboratory settings which have limited applicability in real-world situations, but they can also be carried out in wild settings. There are numerous examples that experimentally manipulate various factors using rodents (Behnke et al. 2001; Dantzer et al. 2020; Sweeny et al. 2020), but fewer with birds, large mammals, and other taxa. For example, several recent studies have explored the relationship between gastrointestinal nematodes and bovine tuberculosis in wild African buffalo (Beechler et al. 2017; Ezenwa et al. 2010; Jolles and Ezenwa 2015; Seguel et al. 2019), but these types of studies in large mammals are relatively rare.

Despite challenges, experimental approaches have also been successfully applied in the field. An increasing number of studies have used antiparasitic treatment experiments of wildlife hosts to assess the impacts of parasites on health and fitness (reviewed by Pedersen and Fenton 2015). For example, Newey and Thirgood (2004) experimentally reduced parasite burdens in mountain hares (*Lepus timidus*) to test the hypothesis that parasites reduce hare fecundity. They found that treatment with ivermectin significantly reduced the abundance of *Trichostrongylus retortaeformis* and increased the fecundity of the hares. Experimental studies allow us to assess cause and effect in a way that is not possible using observational studies alone (Pedersen and Fenton 2015).

The combination of both lab and field experimental studies can lead to the discovery of additional relationships that are relevant to epidemiology (Stallknecht 2007). A great example of this can be found in Ezenwa et al. (2010) where experimental data was combined with longitudinal field studies to make inferences about co-infections and population health for African Buffalo.

4.5 Other Approaches

Although it is beyond the scope of this chapter to delve into the world of mathematical modeling, we wanted to highlight the role those mathematical models can play in identifying knowledge gaps, assessing possible management strategies, and understanding the spatial and temporal factors of disease emergence (Alexander et al. 2012; Ryser-Degiorgis 2013). Despite limitations, including limited data for wildlife populations, a broad range of modeling approaches has been applied to support decision-making in wildlife management problems (McCallum 2016). For example, modeling different vaccination strategies for preventing brucellosis in bison in the Greater Yellowstone Ecosystem (GYE) suggested that vaccinating all female bison captured during boundary operations for bison leaving the park combined with remote darting of female bison in the park would be the most effective alternative (Treanor et al. 2010). Moreover, although it was clear from the modeling that brucellosis could not be eradicated in GYE bison populations using vaccination, it could be useful to reduce prevalence over a 30+ year time period.

Molecular epidemiology has been a rapidly expanding field in the last two decades due to improvements in computer analysis power and rapidly decreasing costs of whole-genome sequencing methodologies. As a result, there have been many advances in understanding the global distribution and epidemiology of infectious pathogens in recent years involving wildlife populations. A thorough discussion of this field is beyond this chapter, but interested readers are referred to the following references as examples of this for more information (Carlson et al. 2019; Thompson and Ash 2016; Wong et al. 2019).

4.6 Approaches to Spatial Data

Spatial health analysis focuses on mapping diseases, risk factors, and other health outcomes and analyzing them in comparison to two or more variables. Due to the importance of how wildlife assemble and move in shared places, the interactions of people, domestic animals, and wildlife in spaces and the effects of landscape features on health outcomes, spatial analysis is growing in importance. It is beyond the scope of this chapter to delve into this topic in detail, but more is available in Chap. 14, as well as the following references (White et al. 2018; Pfeiffer and Hugh-Jones 2002; Cunningham et al. 2021; Moustakas 2017; Baratchi et al. 2013).

5 Validity, Bias, and Confounding

To make inferences about a study population, one generally must take a sample of the population of interest, measure a set of variables, and then analyze those variables to infer what the health of the population may be or understand how the health parameters influence an outcome of interest (typically disease prevalence or rate). The ways in which we undertake these steps influence how far our studies systematically (as opposed to randomly) deviate our measurements or observations from the truth. In free-ranging wildlife populations, many biases result from our data collection methods because logistic and financial considerations drive us to compromise our studies away from epidemiological study ideals and assumptions (Table 2). Reducing the amount of, and understanding the nature and direction of biases, during and after data collection, allow us to measure these inferences and understand the validity of our data more precisely.

Bias affects the validity of a study. Validity generally refers to how well our sample population (generally referred to as the target population) reflects the true nature of the overall (source) population about which we are making inferences. Internal validity refers to how unbiased our inferences about the association between an exposure (e.g., geographic location or home range) and a health outcome (e.g., disease status or serological exposure) truly are for the study population. External validity refers to how generalizable our findings about these associations are to other populations, situations, or species. Poorly designed studies often lead to erroneous associations, resulting in indefensible conclusions and poorly targeted management interventions. Having valid measures are very important to ensure management approaches are targeted appropriately and can be used to ensure the effectiveness of management interventions over time.

Selection bias is likely the most common and important source of bias in most wildlife studies and results from the target population not accurately reflecting the attributes of the source

Table 2 Examples of types of biases common to studies of health and disease in free-ranging wildlife populations

Categories of bias	Definition	Examples
Measurement	Inaccurate measurements or misclassification of study subjects which does not represent the true situation	Bias may result if tests or tools used to measure disease or health metrics are inaccurate. For example, animals may be systematically misclassified as diseased or not diseased if the sensitivity and/or specificity of the test is not 100% (see examples in Lachish and Murray 2018)
Selection	Subjects selected for study do not represent the larger population	Bias may result if animals from your study population are more or less likely to have the disease of interest than the population as a whole. For example, selection biases should be considered when using samples from opportunistically collected animals (e.g., road-killed, hunter killed, or rehabilitated animals) to estimate the prevalence of disease in the larger population (see Conner et al. 2000, e.g., with Chronic Wasting Disease)
Confounding	The outcome variable is impacted by the effects of another variable associated with both the exposure of interest and the outcome variable	Confounding may occur if the impacts of extraneous variables on the outcome of interest are not considered or measured. In a recent study commenting on the impacts of plastic pollution on wildlife health, Roman et al. (2020), highlight the need to explore factors other than plastic, such as nutrition, that might influence the response

population. Many datasets acquired from wildlife are convenience samples (acquired when animals are reported by the public or staff) or are acquired through hunting, fishing, or other consumptive means that do not offer all animals in the population an equal likelihood of being sampled.

Confounding is a source of bias in wildlife studies which must be recognized and controlled for either during sampling or post hoc during data analysis. It results from situations where one or more variables are associated with both the exposure of interest and the outcome variable. When these confounding variables are not measured, we can get a biased relationship between other variables and the outcome. For example, suppose you wanted to understand the relationship between tuberculosis status and survival. You logically would want to know about the outcome variable of survival, the animals' tuberculosis status, and a determined number of other health-related variables that could influence survival. Increasing age is often associated with increasing exposure to infectious diseases, so it is likely associated with TB status. But age is also likely associated with the probability of survival as the older an animal gets the less likely it is to survive the coming year. Age would, therefore, act as a confounder and muddle the relationships between survival and TB status in this example. Fortunately, there are many sampling methodologies that can account for confounding during sampling, as well as many analytical methods to control these relationships.

6 Overcoming Biases and Other Challenges

Despite the problems identified above, there is still value in using epidemiological approaches to understand health in wildlife populations. There are many strategies that can be employed, but sources of variability or error must first be identified so they can be dealt with. Previously published information on the species or population of interest from other geographic areas can often give a sense of these sources. Pilot studies using potential diagnostic tests on smaller subsets of individuals can help to refine parameter estimates a priori. Matching on known confounders such as age, sex, and species can also be used to control for confounding and should be considered during study design and analysis. Matching can be done through either frequency or individual matching (thorough explanations can be found in Dohoo et al. (2009)).

Once potential sources of bias or error have been identified, appropriate sampling strategies can be devised to reduce sources of error and bias. Identification of important covariates is an important first step to ensure parameters that can be measured are measured during capture or field sampling. The preparation of causal diagrams is an excellent way to visualize and identify potential covariates (Greenland et al. 1999; Dohoo et al. 2009). Sampling strategies to reduce bias involve hierarchical or probability sampling. For cohort studies, ensuring that exposed and unexposed groups are essentially as similar as possible in all important parameters of interest is important to be able to reduce bias to the extent possible.

Stallknecht (2007) also provides several excellent suggestions for both conducting and analyzing wildlife data that include: (1) developing integrated plans (use a variety of approaches involving both field and lab studies), (2) archive samples (important for future studies and molecular epidemiology approaches), (3) maintain quality control (integrate different diagnostic tools, understand data limitations), (4) interpret data carefully (need to question results because wildlife systems are complex), (5) don't restrict yourself to traditional approaches (unique challenges require unique approaches; innovate), and (6) don't be intimidated (individual studies contribute to our overall understanding). Similarly, Lachish and Murray (2018) suggest several considerations to reduce uncertainty and bias in disease ecology studies including: (1) rigorous identification of sources of uncertainty (use of pilot studies, a priori information), (2) employing rigorous sampling strategies (probability sampling, hierarchical levels, control for measured covariates), (3) statistical adjustment of parameter estimates for observation error where possible (mark recapture, occupancy models, simulations, sensitivity analyses), and (4) acknowledging remaining uncertainty (temper inferences and conduct sensitivity analyses).

6.1 Mixed Methods and Participatory Epidemiological Approaches in Wild Populations

Participatory epidemiology involves the participation of communities or human populations in studying the wildlife health parameters of interest. This is discussed in more detail in Chap. 5. Combining different epidemiological approaches, both quantitative and qualitative, to better understand complex wildlife population systems can be very powerful and lead to strong outcomes when communities are directly involved in understanding wildlife health. One example is the management and control of bovine tuberculosis in the area around Riding Mountain National Park in Manitoba, Canada. Understanding farmer and rancher attitudes toward elk and deer management (Brook 2015; Brook et al. 2013; Brook 2008, 2010; Brook and McLachlan 2009; Brook and McLachlan 2006) on their lands along with field epidemiological investigations (Nishi et al. 2006; Shury 2015; Shury et al. 2014; Shury and Bergeson 2011) allowed a holistic view of the bovine tuberculosis problem in the region and narrowed down potential solutions that would be ultimately successful in eradicating the disease amongst sympatric cattle and wildlife populations over time. Similar research in northern Canada involving local indigenous communities to understand caribou and muskoxen diseases have been very successful in building bridges between scientists and community members and providing key solutions to understand emerging pathogens and parasites and how these are associated with climate change (Forde et al. 2016; Hoberg et al. 2008; Keatts et al. 2021; Tomaselli et al. 2019). The common denominator amongst these examples is the importance of transdisciplinary research leading to concrete impacts on the health of both humans and wildlife.

7 Conclusion

Hopefully, we have managed to demonstrate how epidemiological approaches can provide necessary tools to describe, monitor, and understand the role of health outcomes in wild animal populations. Although there are numerous challenges that need to be overcome to understand and manage health in wildlife populations, one should not be intimidated to undertake such research, as it is critically important for society as we face major biodiversity and climate crises. Epidemiological approaches have and will continue to provide important and much-needed data for managing future pandemics and to prevent and manage risks to both human and wildlife populations in coming decades.

References

Alexander KA, Lewis BL, Marathe M, Eubank S, Blackburn JK (2012) Modeling of wildlife-associated zoonoses: applications and caveats. Vector-Borne Zoonotic Dis 12(12):1005–1018

Baratchi M, Meratnia N, Havinga PJ, Skidmore AK, Toxopeus BA (2013) Sensing solutions for collecting spatio-temporal data for wildlife monitoring applications: a review. Sensors (Basel) 13(5): 6054–6088. https://doi.org/10.3390/s130506054

Beechler BR, Jolles AE, Budischak SA, Corstjens PL, Ezenwa VO, Smith M, Spaan RS, van Dam GJ, Steinauer ML (2017) Host immunity, nutrition and coinfection alter longitudinal infection patterns of schistosomes in a free ranging African buffalo population. PLoS Negl Trop Dis 11(12):e0006122

Behnke JM, Bajer A, Sinski E, Wakelin D (2001) Interactions involving intestinal nematodes of rodents: experimental and field studies. Parasitology 122(S1): S39–S49

Brook R, McLachlan S (2006) Factors influencing farmers' concerns regarding bovine tuberculosis in wildlife and livestock around Riding Mountain National Park. J Environ Manage 80(2):156–166

Brook RK (2008) Elk-agriculture conflicts in the greater Riding Mountain ecosystem: building bridges between the natural and social sciences to promote sustainability. https://mspace.lib.umanitoba.ca/handle/1993/8037

Brook RK (2010) Incorporating farmer observations in efforts to manage bovine tuberculosis using barrier fencing at the wildlife–livestock interface. Prev Vet Med 94(3–4):301–305

Brook RK (2015) Farmer perceptions of hay yard barrier fence effectiveness in mitigating bovine tuberculosis transmission risk among elk (Cervus canadensis), white-tailed deer (Odocoileus virginianus) and cattle (Bos taurus). Canadian Wildlife Biol Manage 4(1): 18–30

Brook RK, McLachlan SM (2009) Transdisciplinary habitat models for elk and cattle as a proxy for bovine tuberculosis transmission risk. Prev Vet Med 91(2–4): 197–208

Brook RK, Vander Wal E, van Beest FM, McLachlan SM (2013) Evaluating use of cattle winter feeding areas by elk and white-tailed deer: implications for managing bovine tuberculosis transmission risk from the ground up. Prev Vet Med 108(2–3):137–147

Carlson CJ, Kracalik IT, Ross N, Alexander KA, Hugh-Jones ME, Fegan M, Elkin BT, Epp T, Shury TK, Zhang W, Bagirova M (2019) The global distribution of bacillus anthracis and associated anthrax risk to humans, livestock and wildlife. Nat Microbiol 4(8): 1337–1343

Cassidy A (2015) "Big science" in the field: experimenting with badgers and bovine TB, 1995-2015. Hist Philos Life Sci 37(3):305–325

Caswell J, Bassel L, Rothenburger J, Gröne A, Sargeant J, Beck A et al (2018) Observational study Design in Veterinary Pathology, part 1: study design. Vet Pathol 55(5):607–621

Conner M, Miller M, Ebinger M, Burnham K (2007) A meta-BACI approach for evaluating management intervention on chronic wasting disease in mule deer. Ecol Appl 17(1):140–153

Conner MM, McCarty CW, Miller MW (2000) Detection of bias in harvest-based estimates of chronic wasting disease prevalence in mule deer. J Wildl Dis 36(4): 691–699

Cunningham AA, Daszak P, Wood JL (2017) One health, emerging infectious diseases and wildlife: two decades of progress? Philos Trans Roy Soc 372(1725):20160167

Cunningham CX, Comte S, McCallum H, Hamilton DG, Hamede R, Storfer A, Hollings T, Ruiz-Aravena M, Kerlin DH, Brook BW, Hocking G, Jones ME (2021) Quantifying 25 years of disease-caused declines in Tasmanian devil populations: host density drives spatial pathogen spread. Ecol Lett 24:958–969

Dantzer B, van Kesteren F, Westrick SE, Boutin S, McAdam AG, Lane JE et al (2020) Maternal glucocorticoids promote offspring growth without inducing oxidative stress or shortening telomeres in wild red squirrels. J Experiment Biol 223:1

Delahay RJ, Smith GC, Hutchings MR (2009) Management of disease in wild mammals. In: Management of disease in wild mammals. Springer, Tokyo

Dohoo I, Martin W, Stryhn H (2009) Veterinary epidemiologic research. VER Inc, Charlottetown

Ezenwa V, Etienne R, Luikart G, Beja-Pereira A, Jolles A (2010) Hidden consequences of living in a wormy world: nematode-induced immune suppression

facilitates tuberculosis invasion in African buffalo. Am Nat 176(5):613–624
Forde TL, Orsel K, Zadoks RN, Biek R, Adams LG, Checkley SL et al (2016) Bacterial genomics reveal the complex epidemiology of an emerging pathogen in Arctic and boreal ungulates. Front Microbiol 7:1759
Gettings J, Lopez J, Krishnavahjala A, Armstrong B, Thompson A, Yabsley M (2019) Antibodies to Borrelia turicatae in experimentally infected dogs cross-react with Borrelia burgdorferi serologic assays. J Clin Microbiol 57:9
Greenland S, Pearl J, Robins JM (1999) Causal diagrams for epidemiologic research. Epidemiol 10(1):37–48
Haggblade MK, Smith WA, Noheri JB, Usanase C, Mudakikwa A, Cranfield MR, Gilardi KV (2019) Outcomes of snare-related injuries to endangered mountain gorillas (gorilla beringei beringei) in Rwanda. J Wildl Dis 55(2):298–303
Hoberg EP, Polley L, Jenkins EJ, Kutz SJ, Veitch AM, Elkin BT (2008) Integrated approaches and empirical models for investigation of parasitic diseases in northern wildlife. Emerg Infect Dis 14(1):10
Jolles A, Ezenwa V (2015) Ungulates as model systems for the study of disease processes in natural populations. J Mammal 96(1):4–15
Kaur T, Singh J, Tong S, Humphrey C, Clevenger D, Tan W et al (2008) Descriptive epidemiology of fatal respiratory outbreaks and detection of a human-related metapneumovirus in wild chimpanzees (*pan troglodytes*) at Mahale Mountains National Park. Western Tanzania Am J Primatol 70(8):755–765
Keatts LO, Robards M, Olson SH, Hueffer K, Insley SJ, Joly DO et al (2021) Implications of zoonoses from hunting and use of wildlife in north American arctic and boreal biomes: pandemic potential, monitoring, and mitigation. Front Public Health 9:451
Kutz S, Tomaselli M (2019) "Two-eyed seeing" supports wildlife health. Science 364(6446):1135–1137
Lachish S, Murray K (2018) The certainty of uncertainty: potential sources of bias and imprecision in disease ecology studies. Front Veter Sci 5:90
le Roex N, Koets AP, van Helden PD, Hoal EG (2013) Gene polymorphisms in African buffalo associated with susceptibility to bovine tuberculosis infection. PLoS One 8(5):64494
McAdie M (2018) Indicators of individual and population health in the Vancouver Island marmot (Marmota vancouverensis). M.Sc. thesis, Thompson Rivers University
McCallum H (2016) Models for managing wildlife disease. Parasitology 143(7):805–820
Miller MW, Swanson HM, Wolfe LL, Quartarone FG, Huwer SL, Southwick CH, Lukacs PM (2008) Lions and prions and deer demise. PLoS One 3(12):e4019
Moustakas A (2017) Spatio-temporal data mining in ecological and veterinary epidemiology. Stoch Env Res Risk A 31:829–834
Newey S, Thirgood S (2004) Parasite–mediated reduction in fecundity of mountain hares. Proc R Soc Lond Ser B Biol Sci 271:413–415
Nishi J, Shury T, Elkin B (2006) Wildlife reservoirs for bovine tuberculosis (*Mycobacterium bovis*) in Canada: strategies for management and research. Vet Microbiol 112(2–4):325–338
Parra A, García A, Inglis NF, Tato A, Alonso JM, Hermoso de Mendoza M et al (2006) An epidemiological evaluation of *Mycobacterium bovis* infections in wild game animals of the Spanish Mediterranean ecosystem. Res Vet Sci 80(2):140–146
Pedersen A, Fenton A (2015) The role of antiparasite treatment experiments in assessing the impact of parasites on wildlife. Trends Parasitol 31(5):200–211
Pfeiffer DU, Hugh-Jones M (2002) Infectious diseases of wildlife: detection, diagnosis and management (part one)-diagnostics-geographical information systems as a tool in epidemiological assessment and wildlife disease management. Revue Scientifique et Tech-Office Int des Epizoot 21(1):91–102
Richomme C, Courcoul A, Moyen J, Reveillaud E, Maestrini O, de Cruz K et al (2019) Tuberculosis in the wild boar: frequentist and Bayesian estimations of diagnostic test parameters when *Mycobacterium bovis* is present in wild boars but at low prevalence. PLoS One 14(9):e0222661
Roman L, Gilardi K, Lowenstine L, Hardesty BD, Wilcox C (2020) The need for attention to confirmation bias and confounding in the field of plastic pollution and wildlife impacts: comment on "clinical pathology of plastic ingestion in marine birds and relationships with blood chemistry". Environ Sci Technol 55(1):801–804
Ryser-Degiorgis MP (2013) Wildlife health investigations: needs, challenges and recommendations. BMC Vet Res 9(1):1–7
Rytwinski T, van der Ree R, Cunnington G, Fahrig L, Findlay C, Houlahan J et al (2015) Experimental study designs to improve the evaluation of road mitigation measures for wildlife. J Environ Manage 154: 48–64
Salb A, Stephen C, Ribble C, Elkin B (2014) Descriptive epidemiology of detected anthrax outbreaks in wild wood bison (Bison bison athabascae) in northern Canada, 1962–2008. J Wildl Dis 50(3):459–468
Sargeant J, O'Connor A, Dohoo I, Erb H, Cevallos M, Egger M et al (2016) Methods and processes of developing the strengthening the reporting of observational studies in epidemiology - veterinary (STROBE-vet) statement. J Vet Intern Med 30(6):1887–1895
Seguel M, Beechler B, Coon C, Snyder P, Spaan J, Jolles A et al (2019) Immune stability predicts tuberculosis infection risk in a wild mammal. Proc Biol Sci 286:1912
Shaffer JA, Buhl DA (2016) Effects of wind-energy facilities on breeding grassland bird distributions. Conserv Biol 30(1):59–71
Shockling Dent C, Miller M, Batac F, Dodd E, Smith W, Pesapane R et al (2019) Pathology and epidemiology of nasopulmonary acariasis (*Halarachne* sp.) in

southern sea otters (*Enhydra lutris nereis*). Int J Parasitol Parasites Wildlife 9:60–67

Shury TK (2015) The epidemiology of bovine tuberculosis (Mycobacterium bovis) in the greater Riding Mountain ecosystem. https://ecommons.usask.ca/handle/10388/ETD-2015-01-1966

Shury TK, Bergeson D (2011) Lesion distribution and epidemiology of Mycobacterium bovis in elk and white-tailed deer in South-Western Manitoba. Canada Veter Med Int 2011:5

Shury TK, Bergeson D, Surujballi O, Lyashchenko KP, Greenwald R (2014) Field evaluation of three blood-based assays for elk (Cervus canadensis) naturally infected with *Mycobacterium bovis*. Prev Vet Med 115:3–4

Shury TK, Nishi JS, Elkin BT, Wobeser GA (2015) Tuberculosis and brucellosis in wood bison (*Bison bison athabascae*) in northern Canada: a renewed need to develop options for future management. J Wildl Dis 51:3

Stallknecht D (2007) Impediments to wildlife disease surveillance, research, and diagnostics. Curr Top Microbiol Immunol 315:445–461

Stephen C (2021) Animals, health and society : health promotion, harm reduction and health equity in a one health world. CRC Press, Boca Raton

Sweeny AR, Thomason CA, Carbajal EA, Hansen CB, Graham AL, Pedersen AB (2020) Experimental parasite community perturbation reveals associations between sin Nombre virus and gastrointestinal nematodes in a rodent reservoir host. Biol Lett 16(12):20200604

Treanor JJ, Johnson JS, Wallen RL, Cilles S, Crowley PH, Cox JJ, Maehr DS, White PJ, Plumb GE (2010) Vaccination strategies for managing brucellosis in Yellowstone bison. Vaccine 28:F64–F72. https://doi.org/10.1016/j.vaccine.2010.03.055

Thomas J, Balseiro A, Gortázar C, Risalde MA (2021) Diagnosis of tuberculosis in wildlife: a systematic review. Vet Res 52(1):1–23

Thompson RC, Ash A (2016) Molecular epidemiology of giardia and cryptosporidium infections. Infect Genet Evol 40:315–323

Thrusfield M, Christley R, Brown H, Diggle PJ, French N, Howe K et al (2017) Veterinary epidemiology. Wiley, Hoboken

Tomaselli M, Elkin B, Kutz S, Harms NJ, Nymo HI, Davison T et al (2019) A transdisciplinary approach to Brucella in muskoxen of the western Canadian Arctic 1989–2016. Ecohealth 16(3):488

White LA, Forester JD, Craft ME (2018) Dynamic, spatial models of parasite transmission in wildlife: their structure, applications and remaining challenges. J Anim Ecol 87(3):559–580

Wobeser GA (2006) Essentials of disease in wild animals. Blackwell, New York, p 243

Wong AC, Li X, Lau SK, Woo PC (2019) Global epidemiology of bat coronaviruses. Viruses 11(2):174

Participatory Epidemiology and Surveillance for Wildlife Health

Matilde Tomaselli

Abstract

The veterinary application of qualitative participatory approaches, also known as participatory epidemiology and participatory surveillance, is well established for livestock diseases, having contributed to significant advances in disease control. These approaches hold considerable promise for improving wildlife health assessment; however, their application is still in its infancy. To unlock their full potential for wildlife health, participatory approaches developed for livestock diseases need to be adapted to meet the unique challenges of working with free-ranging wildlife and documenting reliable data. This chapter offers a valuable contribution to develop the emerging field of participatory wildlife health surveillance by introducing concepts and methods of participatory approaches for livestock diseases, illustrating their value for wildlife, and drawing lessons for their effective application to wildlife health from the examples of the integration of local ecological knowledge in wildlife co-management. The chapter includes a practical framework for the implementation of participatory wildlife health surveillance to both harvested and non-harvested wildlife populations and discusses potential benefits and limitations of this approach, including differences and synergies with citizen science initiatives. Readers will learn how engaging with local knowledge holders allows for a shared understanding of wildlife health, the timely identification of problems, and the development of shared solutions that improve decision-making around wildlife health.

Keywords

Participatory epidemiology · Participatory surveillance · Interviews · Triangulation · Existing veterinary knowledge · Local knowledge · Scientific knowledge · Local experts · Ethnoveterinary knowledge · Ecological knowledge

1 Participatory Epidemiology and Participatory Surveillance: Origins, Concepts, Methods, and Applications

Participatory epidemiology (PE) and participatory surveillance (PS) are qualitative participatory approaches to veterinary epidemiological research and disease surveillance that have respectively contributed to significant improvements in the understanding and control of livestock diseases in low-income countries (Catley and Mariner 2002; Catley 2003;

M. Tomaselli (✉)
Canadian High Arctic Research Station, Polar Knowledge Canada, Cambridge Bay, Nunavut, Canada
e-mail: matilde.tomaselli@polar-polaire.gc.ca

C. Stephen (ed.), *Wildlife Population Health*,
https://doi.org/10.1007/978-3-030-90510-1_5

Jost et al. 2007; Mariner et al. 2011; Catley et al. 2012; Goutard et al. 2015; Allepuz et al. 2017; Alders et al. 2020). These approaches originated in the 1990s as the veterinary application of participatory rural appraisal (PRA) (Catley et al. 2012), which had evolved a decade earlier from rapid rural appraisal (RRA) as a bottom-up multidisciplinary strategy to improve rural development through community empowerment (Chambers 1983, 1994a, b, c, Pretty 1995).

Participatory epidemiology and PS (formerly referred as PDS, participatory disease searching, or surveillance) access community knowledge systems, particularly the knowledge of livestock owners about the diseases affecting their animals, including clinical presentation, gross pathology, epidemiological features of the disease, and associated risk factors, as well as treatment (knowledge collectively referred to as ethnoveterinary knowledge or existing veterinary knowledge) (Mariner and Paskin 2000). This knowledge is gathered in the field through a combination of multiple participatory appraisal techniques and tools, ranging from semi-structured interviews (individual and group) to interactive scoring and visual exercises (e.g., proportional piling, ranking and scoring, seasonal calendars, Venn diagrams) and direct observations, which allow for the generation of both qualitative and quantitative "epidemiological intelligence" (Mariner and Paskin 2000; Alders et al. 2020). Techniques are applied in a flexible way through an appraisal supported by iterative analyses, making the assessment progressively more relevant to the local context or situation (Jost et al. 2007). Like PRA, PE, and PS rely on the process of "triangulation," cross-checking information using multiple methods and sources, to improve data quality and reliability (Pretty 1995; Mariner and Paskin 2000; Jost et al. 2007; Catley et al. 2012). Triangulation is applied both "within-method" using a number of qualitative research methods (e.g., interviewing multiple participants in individual and group settings, direct observations, examination of secondary documentation, etc.) and "across-method" coupling qualitative research methods with conventional veterinary assessments and diagnostics (i.e., clinical and pathological examinations, field and laboratory testing) (Mariner and Paskin 2000; Catley et al. 2012; OIE 2014; Alders et al. 2020). In general, in PE and PS systems the overall data gathering process is guided by the PRA key principle of optimizing trade-offs between the cost of learning and usefulness of information, also referred to as "optimal ignorance" and "appropriate imprecision" (Chambers 1994a).

Central to the application of PE and PS systems is the concept of active participation of local resource users or experts (Box 1). When PE and PS are fully applied, the knowledge of local stakeholders is used not only as a source of epidemiological data on animal health but also to define local priorities that help improve or shape veterinary programs and intervention measures (Mariner and Paskin 2000; Jost et al. 2007; OIE 2014; Alders et al. 2020). Although this last form of stakeholders' contribution can be overlooked, it is important to recognize that in PE and PS local stakeholders should participate beyond providing missing epidemiological data. Failing to do so can hinder the effectiveness of participatory programs, and some argue that PE and PS should not be qualified as "participatory" without the full meaningful participation of local stakeholders (Catley et al. 2012; Alders et al. 2020).

Much of the early development and application of PE and PS were associated with efforts to eradicate rinderpest, a severe viral disease of even-toed ungulates (i.e., artiodactyls) causing up to 100% mortality in immunologically naive cattle and water buffalo and considered the animal disease with the greatest impact on human livelihoods (Mariner et al. 2012). Participatory surveillance was essential in locating the last foci of rinderpest in remote pastoralist areas of East Africa where conventional surveillance had failed to disclose disease and to guide targeted control for eradication (Mariner and Roeder 2003). Subsequently, PS was used as a tool to confirm the absence of clinical disease in several countries of Africa and Asia (Roeder 2011; Mariner et al. 2012). Building from that experience, PE and PS have since been used in both rural and urban settings in Africa and Asia to improve the surveillance and control of other

Box 1 Typologies of participation

Pretty (1995) noted that in rural development programs the term "participation" should not be used without appropriate clarification and identified seven typologies of participation that range from manipulative and passive participation at the lower end of the spectrum to self-mobilization at the opposite end. As participatory research become more and more "fashionable" this aspect continues to be central. In PE and PS programs, Catley et al. (2012) emphasized that critical consideration on the level of participation achieved is often overlooked. However, this aspect is crucial because the level of participation has a direct influence on the outcomes of programs, and opens dilemmas about the use, and misuse, of this term (Catley et al. 2012). As PE and PS become more popular and utilized, the discussion centered on the meaning of "participation" has become urgent as emphasized during the Second International PENAPH conference held in Khon Kaen (Thailand) in January, 2018.

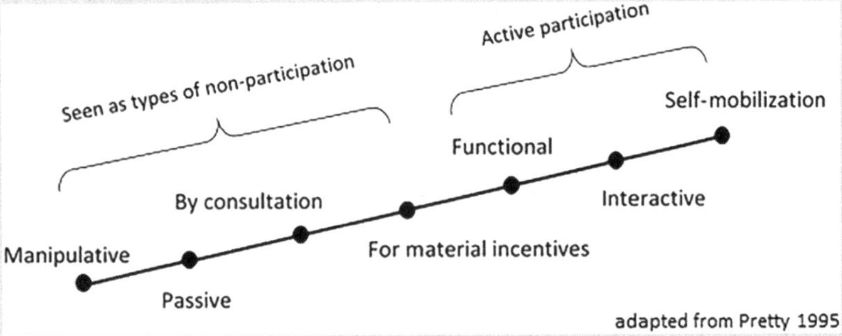

livestock diseases that have an impact on people's well-being; examples include peste des petits ruminants (PPR) and foot and mouth disease (FMD) in Pakistan (Hussein et al. 2008; Anjum et al. 2006), Rift Valley fever (RVF) in Kenya and Tanzania (Jost et al. 2010), and highly pathogenic avian influenza (HPAI) in Indonesia (Azhar et al. 2010). The use of PS proved to be cost-effective for targeted studies compared to conventional surveillance, as well as sensitive and timely for detection of different types of disease situations, ranging from rare or emerging diseases to prevalent but underreported diseases, the latter of which PE can contribute to up to a tenfold increase in case detection (Jost et al. 2007; Mariner et al. 2011). Furthermore, the participatory process has enabled local stakeholders to have a greater role in shaping disease control programs that align with local priorities (Jost et al. 2007; Mariner et al. 2011; Catley et al. 2012).

Since its initial application, PE, either in the form of surveillance or epidemiological research, has contributed to address important knowledge gaps about livestock diseases in marginalized areas; prioritizing disease and guiding better control strategies; unraveling the etiology of complex syndromes (Catley et al. 2001, 2004); and, informing models for disease transmission (Mariner et al. 2005, 2006a, b), including zoonoses (Grant et al. 2016). In the past two decades, the use of PE has increased and its applications have broadened to include, among others, participatory risk analysis, impact assessment, veterinary public health, evaluation tools for surveillance (including hunter-based surveillance—Schulz et al. 2016), and training (Catley et al. 2012; Allepuz et al. 2017). Despite this, however, PE activities continue to be implemented mainly in Africa and Asia and to be largely centered on livestock systems (Allepuz et al. 2017).

While PS is presented as a surveillance tool limited to livestock diseases in the World Organization for Animal Health's (OIE) most recent guide for animal health surveillance (OIE 2014), participatory approaches applied to wildlife health are increasingly gaining attention within the veterinary and wildlife conservation community. For example, PE has recently been proposed as a tool that can contribute to achieve PPR eradication goals and protecting biodiversity by helping to fill knowledge gaps on the epidemiology of PPR at the wildlife-livestock interface and assess the impact of PPR directly on wildlife populations, while improving community engagement for shared management decisions (Fine et al. 2020).

To date, however, the direct application of PE and PS to wildlife health assessment has been largely underrepresented, with disease transmission at the wildlife-livestock and wildlife-human interfaces being more common. For example, while studying bovine trypanosomiasis in Kenya, Catley et al. (2002) found that Orma pastoralists considered contact with wild buffalo a risk factor for FMD in their cattle, with contact between livestock and wildlife more likely to occur from January to mid-March in the Orma season "bona hageiya." Coffin et al. (2015) explored interactions between people and wildlife at the edge of a national park within a study focused on anthrax (*Bacillus anthracis*) management in Western Uganda. PE has been implemented to unravel complex socioecological interactions in the project "Lawa model," an EcoHealth approach that has been ongoing for over 10 years for the control of the foodborne carcinogenic parasitic disease opisthorchiasis (*Opisthorchis viverrini*) endemic in the Lawa Lake region of Thailand (Sirpa et al. 2017). The interaction explored included the human–wildlife interface, given the disease is transmitted through the consumption of undercooked wild fish (Cyprinid species). Only a few studies in the published literature appear to specifically apply PE methods to wildlife, in assessments that are limited to the investigation of specific morbidity or mortality events (e.g., Chen et al. 2012; Orotzco et al. 2020).

Reasons why the use of PE and PS for wildlife health assessment are still in their infancy maybe owing to the need to modify PE methods developed and applied for livestock diseases to meet the unique challenges of working with and documenting reliable data on free-ranging wildlife, and to move from a disease-centered approach to a health-centered approach. By adapting veterinary participatory methods to the wildlife context, capitalizing on lessons learned from other fields of study (i.e., use of ecological knowledge in wildlife comanagement systems), and applying robust qualitative research methods (see sect. 5.2), a recent pilot PS project developed in the Canadian Arctic for muskox health assessment attempted to address these issues, ultimately highlighting the full potential of PE and PS to strengthen wildlife health surveillance and improve decision-making (Tomaselli 2018) (see Box 2). This chapter summarizes the core lessons emerging from that project with the objective of enabling wildlife practitioners across the globe to implement PE and PS projects for wildlife health.

Box 2 Participatory Muskox Health Surveillance in the Canadian Arctic

This pilot project was developed to explore the potential of the combined application of local and scientific knowledge within a participatory framework for improving the understanding of wildlife health and its continuous assessment (Tomaselli 2018). While this program was applied in the Arctic to muskoxen, and also in part to caribou, lessons learned are valuable beyond both the targeted species and locality of implementation. Interviews of several local experts (Inuit hunters—men, women, and elders—, community residents and bush pilots) were the foundation of the system, making possible an understanding of the local context, gathering missing epidemiological data on wildlife health, informing the design of conventional surveillance components, and interpreting scientific data within context (Fig. 1). Participatory

(continued)

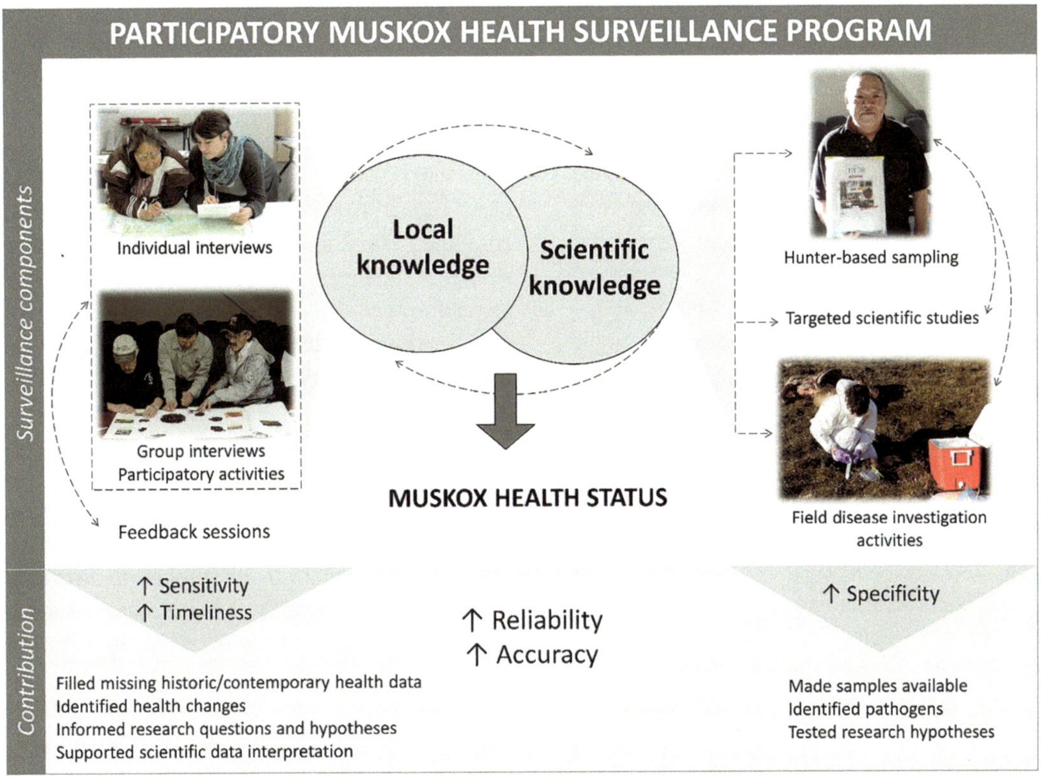

Fig. 1 Schematic representation of the participatory muskox health surveillance program initiated in the Canadian Arctic. The surveillance components developed access different knowledge systems—local and scientific knowledge—each contributing to strengthening specific aspects of the surveillance performance (sensitivity, timeliness, and specificity). Working in synergy, they achieved a more reliable and accurate muskox health status assessment than single surveillance components and knowledge systems could have attained in isolation. Dashed lines represent how the surveillance components developed relate and influence each other within the system. Figure modified from Tomaselli and Curry (2019)

epidemiology activities, adapted for wildlife health, were critical to obtain missing historic and contemporary data on population demographics, body condition status, the prevalence of diseases, causes of mortality, and disease outbreaks. For example, PE identified and characterized major declines of muskoxen and caribou before conventional surveys. It also suggested possible causes of these declines, including emerging and re-emerging diseases and mortality outbreaks that had gone largely undetected through conventional surveillance methods (Tomaselli et al. 2018a). Engaging with local hunters allowed for the establishment of successful sample collections that led to the identification of zoonotic pathogens—Orf virus and *Brucella suis* biovar 4—which were suspected based on PE (Tomaselli et al. 2016, 2018a). Samples obtained through various means (hunters' collections, archives, field investigations) were then used in targeted scientific studies to further characterize spatial and temporal trends of disease in the muskox population. The use of PE to interpret sample data was key to improve confidence in results and depth of knowledge as

(continued)

it provided historical context to better interpret scientific data and complementary syndromic information to offset diagnostics limitations (Tomaselli et al. 2019). Including local experts in the surveillance facilitated access to epidemiological data on wildlife health that would have been otherwise difficult to obtain and enriched overall data interpretation. Participatory surveillance allowed for the effective mobilization of existing local knowledge and use in synergy with scientific knowledge, improving reliability and accuracy of the surveillance output, and enabling early detection of changes (Fig. 1). Finally, working together with local knowledge holders had benefits beyond the acquisition and interpretation of wildlife health data; it promoted dialogue between parties to improve interventions for both wildlife comanagement and public health (Tomaselli et al. 2018a, b). Learning from this experience, participatory epidemiology has also now been successfully applied to explore narwhal and polar bear health, highlighting the broad potential of the application of participatory approaches for wildlife health assessment.

2 Documenting Local Knowledge on Wildlife Health: Integrating Lessons from Other Fields of Study

Since the 1990s, livestock veterinarians and veterinary epidemiologists working with ecologists and wildlife managers have been combining local knowledge[1] with data obtained through "conventional" scientific methods to address specific questions on applied issues in their respective fields. These two applications of local knowledge—ethnoveterinary knowledge of livestock owners applied to livestock disease surveillance and ecological knowledge of local resource users applied to wildlife comanagement—can be combined to capitalize on their respective strengths and create a novel and holistic approach to wildlife health surveillance that ensures direct participation of those who are most affected by changes in wildlife health and by decision-making on wildlife.

A detailed discussion on the use of ecological knowledge in wildlife comanagement systems is beyond the scope of this book chapter, however, it is important to note that this field of study developed in parallel to PE applied to livestock diseases (for additional references, see Tomaselli 2018). Although local ecological knowledge has not been mobilized as effectively as ethnoveterinary knowledge, the range of observations about wildlife documented in ecological knowledge studies includes movements and abundance of animals, behavior, and body condition, morbidity and mortality, interspecies interactions, and biotic and abiotic features of the animals' environments. These types of information are all relevant for the holistic understanding of wildlife health and its continuous assessment (see Chap. 1). Experiences from this tradition, including the barriers that remain for the effective combined use of local knowledge with scientific knowledge in comanagement systems, are relevant for guiding the application of PE and PS for wildlife health.

While the PE tradition provides the pragmatic framework for documenting local wildlife health knowledge (i.e., using PE tools and techniques) and its combination with scientific knowledge (i.e., applying "across-method" triangulation), robust qualitative methods are needed to ensure

[1] There are a multitude of names that refer to experiential-based knowledge driven by local resource use and practices, including general names such as local and traditional knowledge, indigenous knowledge, technical knowledge, folk knowledge and wisdom, and more specific names that connote specific groups of people, for example *Inuit Qaujimajatuqangit* or Inuit knowledge, or that refer to specific types of knowledge, for example ecological knowledge or veterinary knowledge, and combination of all the above. This local and experiential-based body of knowledge is herein conjointly referred as local knowledge.

data quality and reliability to avoid the information generated being dismissed as anecdotal and failing to be integrated into decision-making (Kutz and Tomaselli 2019). In the literature on local knowledge applied to wildlife comanagement, there are examples in which local knowledge improperly collected or interpreted has led to its dismissal as untrustworthy and requiring validation by scientific knowledge (Gilchrist et al. 2005; Brook and McLachlan 2005). This can lead to conflicts. Conflicts might also arise if the PRA principles of "optimal ignorance" and "appropriate imprecision" were applied to wildlife surveillance (Chambers 1994a). These principles from the PE tradition can be appropriate in the context of rural development and action-oriented research but might not be appropriate for producing rigorous accounts for understanding and assessing wildlife health that needs to be trusted by a variety of stakeholders to act upon.

Although flexibility is also a strength of participatory approaches, often it has been mistakenly interpreted as utilizing qualitative methods and techniques in a loose way and with a lack of transparency in reporting methods used (Brook and McLachlan 2008; Tomaselli et al. 2018a). Qualitative methods must be applied using a rigorous process of data collection and analyses to produce accurate accounts (Kutz and Tomaselli 2019). This is critical for promoting the transdisciplinary application of qualitative and quantitative disciplines for wildlife health and avoiding further separation between them. The use of rigorous methods will allow for comparability and a combination of results across localities, which will be important for health surveillance of wildlife populations with large home ranges that intersect multiple communities and groups of key informants.

3 Participatory Wildlife Health Surveillance in Practice

This section contains practical considerations that help guide the development of PE and PS programs on wildlife health for both harvested and non-harvested populations. Strengths and limitations of this approach are also discussed, including considerations on how a participatory framework for data collection and interpretation can improve the performance of wildlife surveillance.

3.1 Participatory Framework for Data Gathering and Interpretation

Figure 2 presents a pragmatic working framework to guide the implementation of participatory wildlife health surveillance programs—particularly the data acquisition and interpretation process—in different settings and for a variety of harvested and non-harvested wildlife populations. Here follows a detailed description on the process of documenting relevant local knowledge from local experts and its combination with scientific knowledge derived from sample analyses. Different options for accessing wildlife samples are also discussed.

Multiple stakeholders—PE practitioners, local communities/organizations, local key informants, and wildlife managers—are typically involved in participatory programs which must be designed as team efforts requiring partnerships and mutual understanding between parties (see Chap. 21). Participatory epidemiology practitioners must have a sound understanding of qualitative and PE methods and techniques (see Participatory Epidemiology Network for Animal and Public Health PENAPH website, www.penaph.net), including the ethical principles that apply when gathering knowledge from people, and core knowledge of wildlife health and diseases.

The prioritization of the local knowledge of key informants is recommended using interviews that aim to understand the local context and gather epidemiological information in the form of PE data on the wildlife population(s) under study. Understanding the local context allows practitioners to implement programs that are locally effective by capitalizing on the opportunities that are locality- and species-specific (e.g., identification of all possible key

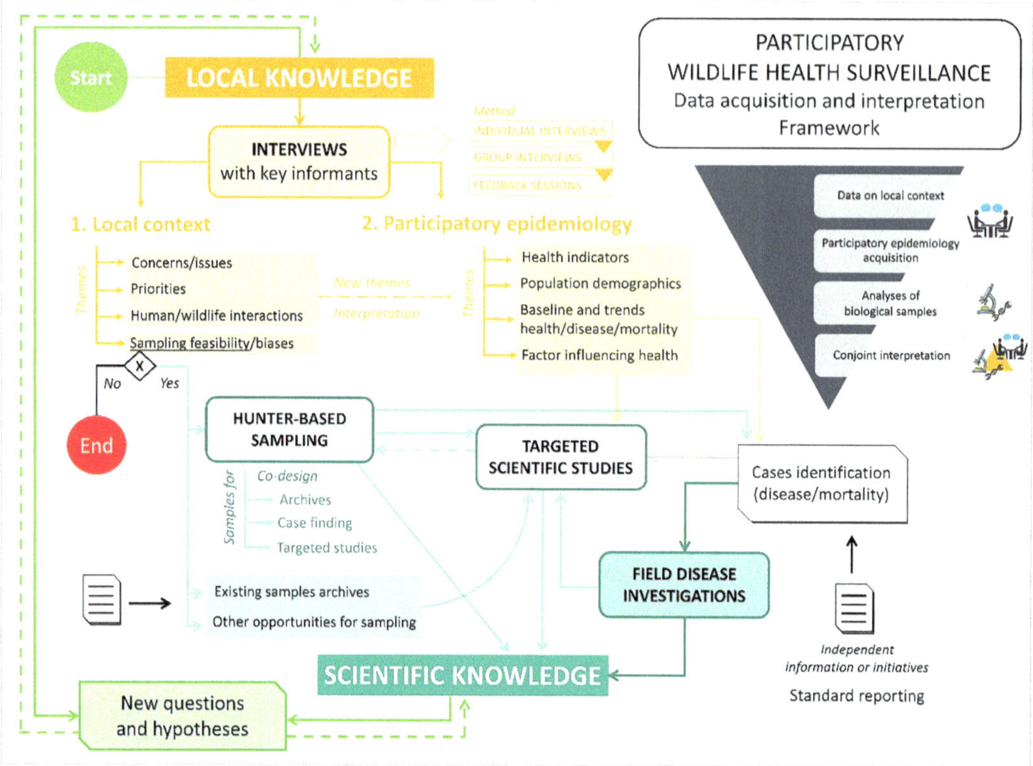

Fig. 2 Process map describing the general participatory framework for data gathering and interpretation on wildlife health. The main surveillance components or activities, which refer to either the local knowledge system (orange) or the scientific knowledge system (blue), are indicated inside rounded rectangles; solid lines represent the main connections among activities or explain features within a single activity; dashed lines refer to feedback connections within the process; green solid lines represent the flow lines generated by the surveillance output leading to a new starting point for the data gathering/interpretation process; finally, green dashed feedback lines represent the following iterations of the data gathering/interpretation process. "Sampling feasibility" is the discriminating node that determines whether hunter-based sampling may be feasible or not. Light blue rounded rectangles include surveillance activities that are subjected to the availability of biological samples and, therefore, may not be implementable (i.e., non-harvested wildlife). The process flow is also schematically summarized in the top right corner. Adapted from Tomaselli (2018)

informants, data sources, sampling opportunities, and appropriate sampling methods to use) and evaluating relevant biases associated with PE data interpretation and sampling design. Additional priorities to explore and concerns to address can also emerge while engaging local experts. Knowledge about the local context can help to shape intervention strategies for both public health protection and wildlife management. Once the local context is understood—for example, through the use of open-ended semi-structured individual interviews of key informants coupled with direct observations—then the PE data collection process can be initiated. This is typically done using participatory appraisal techniques applied in group interviews of carefully selected key informants (e.g., Tomaselli et al. 2018a).

Collaboration with local organizations is essential for the development of this initial phase including for the identification of relevant key informants. The use of robust qualitative methods is necessary to produce reliable interview-based data, including defining sample size until thematic saturation is reached, applying "within-method" triangulation (e.g., individual and group interviews), analyzing information through thematic analysis, and co-interpreting results with participants through feedback sessions (Tomaselli et al. 2018a, b; Kutz and Tomaselli 2019).

Hunter-based sampling is an effective tool to collect biological samples from harvested wildlife for immediate testing and archiving (OIE 2014). Codesigning these programs with local hunters and carefully selecting the type of samples and information needed to meet surveillance objectives are the keys to success. Generally, blood samples can be easily collected and archived even in challenging settings (e.g., filter papers used for collection and storage; see Curry et al. 2014) as well as information on and samples from abnormal lesions found in harvested animals. If additional samples can be collected and archived, referring to the PE data can help identify which tissues can be valuable in the future. When resources become available, archived samples can then be accessed for implementing targeted scientific studies. Hunter-based sampling programs are flexible surveillance tools that can be modified over time to fulfill evolving surveillance objectives. Any modifications should be discussed with the hunters participating in the program to enable its continued codesign, which ultimately influences its sustainability.

Interviews with key informants and hunter-based sampling are likely to increase the reporting of events consistent with overt disease or mortality, and the submission of abnormal tissues requiring immediate analyses. Being prepared to quickly implement field investigations and laboratory testing is essential to increase the specificity of the surveillance system and to provide timely responses. Linking local surveillance to existing broader (e.g., provincial, territorial, or national) wildlife surveillance systems can help achieve this goal. For example, in Canada, all provinces and territories can easily and rapidly access veterinary diagnostic expertise on wildlife diseases through the Canadian Wildlife Health Cooperative (see CWHC website, www.cwhc-rcsf.ca).

Knowledge derived from interviews, field disease investigations, and hunter-based sampling can together help to identify and prioritize targeted scientific studies (e.g., Tomaselli et al. 2019). This feedback among surveillance components can help to use limited available resources more effectively. Both the local knowledge and the scientific knowledge are likely to generate further questions and priorities which can be explored using the same approach that combines the knowledge of key informants (i.e., exploring new themes) with targeted sampling and diagnostics. The surveillance system, therefore, keeps evolving and has the potential to quickly adapt to local needs.

While hunter-based sampling is not feasible for non-harvested wildlife, PE and PS can still be implemented. It can be more challenging to identify individuals who have detailed knowledge about non-harvested species, making understanding the local context even more important. For example, in some areas, fishermen can provide detailed information on sea birds that often follow and congregate around fishing boats, but, at some times of the year, crop farmers can also provide observations on seagulls that swarm their fields to feed on worms as farmers till the soil.

Knowledge on the life history of the wildlife populations under study can help to identify areas where it is more likely to gather important information for population health assessments. For example, the Arctic is one of the ideal locations to document data on the health of shorebirds as this is where these seasonally migratory species breed. Shorebirds may not be harvested but people who spend lots of the spring and summertime on the land (e.g., geese harvesters, people involved in egg or berry picking, or even fishing) are likely to observe shorebirds and can serve as the key informants of the surveillance. Although it is unlikely that detailed PE could be compiled for diseases of non-harvested wildlife (i.e., lesions localized in internal organs), critical data could still be gathered on population trends and productivity, overt mortality/disease, and holistic understanding of factors associated with population health. Active sampling and laboratory analyses can be promptly implemented for non-harvested wildlife following reports of overt mortality/disease. Thus, a PS program on non-harvested wildlife can provide a data-rich output to enable evidence-based interventions.

Finally, in some cases, it may be possible to combine elements of participatory surveillance for harvested and non-harvested species. For example, wildlife populations with vast home ranges that span across multiple jurisdictions may be harvested in some geographic areas but protected in others (e.g., national parks, marine protected areas). While logistically challenging to implement, participatory surveillance efforts coordinated across localities can provide unique insights for better understanding wildlife population health.

3.2 Strengths of Participatory Approaches for Wildlife Health Assessments

The major strengths of assessing wildlife health through participatory approaches lie in the potential of producing more reliable and accurate results and enabling more timely identification of changes than is possible through the use of conventional methods alone. This in turn allows for implementing more effective and timely interventions. Capitalizing on the knowledge of local experts as a source of epidemiological intelligence develops a shared understanding of wildlife health and identifies potential problems even prior to their overt manifestation. For example, observations of decreased body condition of the animals or lower number of juveniles in the population (or any other direct or indirect indicators of reproductive success, including habitat degradation, as appropriate to the species under study) can predict impending population declines long before they could be detected through scientific studies alone. The inclusion of PE in surveillance systems can improve not only timeliness but also the sensitivity of events' detection (Tomaselli et al. 2018a). This can promote proactive management interventions rather than reactive responses. In addition, PE can enable tracking population health indicators in real time, including population demographics. For harvested wildlife that is actively managed this can allow for timely adaptation of harvest rates in response to changing trends, avoiding the risk of unsustainable harvests which would be more challenging to achieve by solely relying on scientific population estimates/censuses typically repeated at multiyear intervals.

It is important to note that increased sensitivity and timeliness apply to those events that are recognizable and are likely to be detected, many of which are context dependent. That is why understanding the local context, including how people interact with wildlife, is an integrative part of the PE data gathering and interpretation process (Tomaselli et al. 2018a, b) (Fig. 2). For example, if hunters only minimally inspect specific internal organs of harvested wildlife (maybe because they do not consume them) we cannot expect an increased sensitivity or timeliness of detection of lesions localized in those organs. Similarly, the apparent prevalence of lesions in organs that are not consumed/inspected can be significantly underestimated by hunters (Tomaselli et al. 2018a, b). The ability of the interviewer(s) to identify different disease presentations directly also influences sensitivity and timeliness (OIE 2014). The interviewer's own knowledge on wildlife health and diseases is therefore key to explore and interpret the ethnoveterinary knowledge of participants in greater depth (Tomaselli et al. 2018a).

Another major benefit of participatory systems for wildlife health assessment is that knowledge of local experts can guide the application of conventional methods of assessment and provides insights for results' interpretation. While scientific data are important to increase the specificity of surveillance systems, they have limitations of their own and it can be difficult to interpret them in context. For example, when a pathogen is identified through laboratory testing for the first time in an area it does not necessarily mean it is a new or emerging occurrence. PE data can provide the historical context in which scientific information can be better interpreted (e.g., Tomaselli et al. 2016, 2018a, 2019). In particular, the process of "across-method" triangulation enables the synergistic use of local and scientific knowledge, reducing overall uncertainty and increasing confidence and depth of knowledge (Tomaselli et al. 2019). This process can also facilitate more effective use than possible through conventional assessments alone of existing fragmented wildlife health data that are often the outcome of uncoordinated initiatives.

The process of engaging with local knowledge holders has the potential to significantly improve management (see Chap. 21). As evident from the wildlife comanagement tradition, the effort of combining local knowledge with scientific knowledge can produce collaborative partnerships between stakeholders that allow for an improved understanding of each other's perspectives, building trust, resolving conflicts, identifying shared solutions, and ultimately enhancing overall management that benefits both people and wildlife (Huntington et al. 2004; Kendrick and Manseau 2008; Berkes 2009; Huntington 2011).

Finally, from a practical perspective, using PE and PS for wildlife health assessment is likely to be more cost-effective compared to epidemiological studies and surveillance that require extensive field efforts to collect biological samples for testing. Local ecological studies have already demonstrated that capitalizing on activities of resource users that are already occurring reduces the need for undertaking expensive fieldwork (Anadón et al. 2009).

3.3 Limitations of Participatory Approaches for Wildlife Health Assessments

The considerable time commitment to build and maintain relationships and collaborations with a variety of stakeholders and the need for project personnel with transdisciplinary expertise are two major limitations to the development and implementation of participatory projects on wildlife health.

Participatory programs typically require significant time to build successful collaborations and the willingness of people to participate and share their knowledge. A project leader and team with the necessary training including qualitative research methods, PE techniques, wildlife health and diseases, wildlife sampling, and field disease investigation are critical. Cultural competence, ability to work in a team, and flexibility are other important qualities of participatory project practitioners.

Although these are not limitations per se, it is critical to consider the time commitment required to develop and maintain local collaborations, build trust among stakeholders, and train local PE practitioners and program coordinators. Considerable efforts are required to make the system continuously relevant and effective locally; in particular, interviews of key informants should be performed regularly to allow the collection of real-time PE data on the targeted wildlife, enabling the implementation of adaptive management.

Although one can see these limitations also as opportunities to create effective locally adapted programs that can timely detect and respond to changes in wildlife health, it is important to keep these aspects in mind to make sure programs are developed on a solid foundation and with an eye toward sustainability.

4 Beyond Local Experts for Wildlife Health Assessments: Participatory Epidemiology and Citizen Science Initiatives

Collaborations between wildlife professionals and the public have also provided opportunities to improve wildlife surveillance through programs that have capitalized on existing citizen science networks (Lawson et al. 2015). For example, in Britain, the Garden Wildlife Health platform (www.gardenwildlifehealth.org) can help the general public report incidents of sick or dead garden wildlife, thus providing early warning signals for emerging threats. In North America, birdwatchers engaged in the FeederWatch project (www.feederwatch.org) helped to identify the emergence of house finch conjunctivitis

(*Mycoplasma gallisepticum*) in 1994 and have since contributed to its monitoring. While a full discussion on citizen science (CS) is beyond the scope of this chapter, it is important to mention their contribution to wildlife health and clarify the differences that exist with PE.

Both PE and CS initiatives improve wildlife health assessments through the engagement of local actors, however, they are different in nature and, thus, lead to different outcomes both in terms of depth of knowledge generated and management implications. Participatory epidemiology strives for a holistic understanding of wildlife health by including local knowledge holders in programs that are designed with a bottom-up approach; while CS capitalizes on the willingness of the laypeople (often volunteers, not necessarily local experts) to be involved in science activities (which are generally designed with a top-down approach) making scientific data more accessible for analyses.

These approaches are profoundly different in the types of knowledge systems they access and in the power balances between actors. Such differences are especially evident when engaging with Indigenous peoples. This is because the value contexts and worldviews of Indigenous ways of knowing typically differ from those of the scientific knowledge system, but also because Indigenous peoples must be in control of their knowledge (Tengö et al. 2021). However, synergies between approaches are possible and even advocated for (Tengö et al. 2021). While keeping their important differences in mind, PE and CS can be used in combination to further support the application of multiple evidence-based approaches for wildlife health.

5 Summary

Participatory wildlife health surveillance is a promising new field of application of participatory epidemiology that enables the systematic use of local knowledge for historic, contemporary, and real-time assessments of wildlife population health. Common impediments to gathering and interpreting field data on wildlife health significantly limit the ability to carry out effective wildlife surveillance (see Chap. 9). Participatory approaches provide specific means to overcome some of these challenges by ensuring direct participation of local resource users and knowledge holders while promoting the development of programs and interventions that are relevant locally and which can improve comanagement systems (Tomaselli et al. 2018a).

The health of wildlife populations intersects with the health of humans, domestic animals, and the environment (Aguirre et al. 2002). Participatory approaches for wildlife health assessments are effective tools to explore health within interrelated interfaces, enabling the application of One Health and Eco Health principles which are advocated for within the veterinary and wildlife professions, yet are difficult to apply effectively (Rostal et al. 2012; Gibbs 2014).

The recent COVID-19 pandemic (SARS-CoV-2) has underscored the need for wildlife surveillance to enable timely detection and response to zoonotic emerging infectious diseases (Dobson et al. 2020; Watsa and Wildlife Disease Surveillance Focus Group 2020). Participatory approaches hold great potential to strengthen wildlife surveillance capacity, improving early detection of problems, and enabling timely interventions. They must be considered in the array of tools available for future pandemic prevention. The COVID-19 crisis has also emphasized the need to address global ecological challenges that can greatly increase the risk of future zoonotic pandemics from unsustainable wildlife practices to habitat loss and fragmentation (Dobson et al. 2020). The systematic inclusion of local people and knowledge for wildlife health promotes a platform for knowledge exchange among stakeholders, facilitating the development of effective solutions for more sustainable patterns of local resource use which can contribute to mitigating global challenges.

Like other fields of study in which participatory research is applied, it will be important in participatory wildlife health surveillance to focus on the process of participation rather than the mere application of participatory tools. Local users participate in the surveillance not only by providing relevant PE data on wildlife health but also by codesigning the system and co-interpreting its output together with wildlife professionals. The derived outcome is a surveillance program customized to local needs that can help to understand local issues under a holistic lens and foster dialogue among stakeholders for developing effective interventions. Continuing to learn from other participatory traditions, including advances realized in PE and PS applied to livestock diseases, and other relevant fields of study will be crucial for the future development of participatory wildlife health surveillance and the transdisciplinary application of knowledge systems—local knowledge and scientific knowledge—for wildlife health. For example, borrowing from strategies applied for the biological assessment of wild Pacific salmon, authors have recently proposed a novel application of PE for wildlife health within a modeling framework for conservation status assessment using a traffic light approach (Peacock et al. 2020). Incorporating new methods for mobilizing knowledge (e.g., using social platforms, mobile applications) and creating synergies with complementary initiatives, such as citizen science (e.g., Tengö et al. 2021), will also contribute to further advancements in this field.

Finally, participatory wildlife health surveillance provides a renewed opportunity for veterinarians and other wildlife health professionals to strengthen their role in wildlife health, leading collaborations across multiple disciplines and stakeholders. Recognizing that participatory epidemiology and surveillance are rarely integrated into veterinary and wildlife health education at present, it will be essential to equip future generations of wildlife health practitioners with these skills to enable them to more effectively service wildlife health moving forward.

References

Aguirre AA, Ostfeld RS, Tabor GM, House C, Pearl MC (2002) Conservation medicine: ecological health in practice. University Press, Oxford

Alders RG, Ali SN, Ameri AA, Bagnol B, Cooper TL, Gozali A et al (2020) Participatory epidemiology: principles, practice, utility, and lessons learnt. Front Vet Sci 7:532763. https://doi.org/10.3389/fvets.2020.532763

Allepuz A, de Balogh K, Aguanno R, Heilmann M, Beltran-Alcrudo D (2017) Review of participatory epidemiology practices in animal health (1980-2015) and future practice directions. PLoS One 12:e0169198. https://doi.org/10.1371/journal.pone.0169198

Anadón JD, Giménez A, Ballestar R, Pérez I (2009) Evaluation of local ecological knowledge as a method for collecting extensive data on animal abundance. Conserv Biol 23:617–625. https://doi.org/10.1111/j.1523-1739.2008.01145.x

Anjum R, Manzoor H, Zahoor AB, Irshad H, Farooq U (2006) Epidemiological analyses of foot and mouth disease in Pakistan. Int J Agric Biol 8:648–651

Azhar M, Lubis AS, Siregar ES, Alders RG, Brum E, McGrane J et al (2010) Participatory disease surveillance and response in Indonesia: strengthening veterinary services and empowering communities to prevent and control highly pathogenic avian influenza. Avian Dis 54:749–753. https://doi.org/10.1637/8713-031809-Reg.1

Berkes F (2009) Evolution of co-management: role of knowledge generation, bridging organizations and social learning. J Environ Manage 90:1692–1702. https://doi.org/10.1016/j.jenvman.2008.12.001

Brook RK, McLachlan SM (2005) On using expert-based science to "test" local ecological knowledge. Ecol Soc 10:r3

Brook RK, McLachlan SM (2008) Trends and prospects for local knowledge in ecological and conservation research and monitoring. Biodivers Conserv 17: 3501–3512. https://doi.org/10.1007/s10531-008-9445-x

Catley A (2003) Validation of participatory appraisal for use in animal health information systems in Africa. Doctoral thesis. University of Edinburgh, Edinburgh

Catley A, Alders RG, Wood JL (2012) Participatory epidemiology: approaches, methods, experiences. Vet J 191:151–160. https://doi.org/10.1016/j.tvjl.2011.03.010

Catley A, Chibunda RT, Ranga E, Makungu S, Magayane FT, Magoma G et al (2004) Participatory diagnosis of a heat-intolerance syndrome in cattle in Tanzania and association with foot-and-mouth disease. Prev Vet Med 65:17–30. https://doi.org/10.1016/j.prevetmed.2004.06.007

Catley A, Irungu P, Simiyu K, Dadye J, Mwakio W, Kiragu J et al (2002) Participatory investigations of bovine trypanosomiasis in Tana River district. Kenya Med Vet Entomol 16:1–12. https://doi.org/10.1046/j.0269-283x.2002.00346.x

Catley A, Mariner JC (2002) Where there is no data: Participatory approaches to veterinary epidemiology in pastoral areas of the Horn of Africa. International Initiative for Environment and Development, London

Catley A, Okoth S, Osman J, Fison T, Njiru Z, Mwangi J et al (2001) Participatory diagnosis of a chronic wasting disease in cattle in southern Sudan. Prev Vet Med 51:161–181. https://doi.org/10.1016/S0167-5877(01)00240-9
Chambers R (1983) Rural development: putting the last first. Longman, London
Chambers R (1994a) Participatory rural appraisal (PRA): challenges, potentials and paradigm. World Dev 22: 1437–1454
Chambers R (1994b) Participatory rural appraisal (PRA): analysis of experience. World Dev 22:1253–1268
Chambers R (1994c) The origins and practice of participatory rural appraisal. World Dev 22:953–969
Chen CC, Pei KJC, Lai YC, Mortenson JA (2012) Participatory epidemiology to assess sarcoptic mange in serow of Taiwan. J Wildl Dis 48:869–875. https://doi.org/10.7589/2011-07-189
Coffin JL, Monje F, Asiimwe-Karimu G, Amuguni HJ, Odoch T (2015) A one health, participatory epidemiology assessment of anthrax (bacillus anthracis) management in Western Uganda. Soc Sci Med 129:44–50. https://doi.org/10.1016/j.socscimed.2014.07.037
Curry PS, Ribble C, Sears WC, Orsel K, Hutchins W, Godson D et al (2014) Blood collected on filter paper for wildlife serology: evaluating storage and temperature challenges of field collections. J Wildl Dis 50: 308–321. https://doi.org/10.7589/2012-06-150
Dobson AP, Pimm SL, Hannah L, Kaufman L, Ahumada JA, Ando AW et al (2020) Ecology and economics for pandemic prevention. Science 369:379–381. https://doi.org/10.1126/science.abc3189
Fine AE, Pruvot M, Benfield CT, Caron A, Cattoli G, Chardonnet P et al (2020) Eradication of Peste des Petits ruminants virus and the wildlife-livestock interface. Front Vet Sci 7:50. https://doi.org/10.3389/fvets.2020.00050
Gibbs EPJ (2014) The evolution of one health: a decade of progress and challenges for the future. Vet Rec 174: 85–91. https://doi.org/10.1136/vr.g143
Gilchrist G, Mallory M, Merkel F (2005) Can local ecological knowledge contribute to wildlife management? Case studies of migratory birds. Ecol Soc 10:20–32
Goutard FL, Binot A, Duboz R, Rasamoelina-Andriamanivo H, Pedrono M, Holl D et al (2015) How to reach the poor? Surveillance in low-income countries, lessons from experiences in Cambodia and Madagascar. Prev Vet Med 120:12–26. https://doi.org/10.1016/j.prevetmed.2015.02.014
Grant C, Iacono GL, Dzingirai V et al (2016) Moving interdisciplinary science forward: integrating participatory modelling with mathematical modelling of zoonotic disease in Africa. Infect Dis Poverty 5:17–29. https://doi.org/10.1186/s40249-016-0110-4
Huntington H, Callaghan T, Fox S, Krupnik I (2004) Matching traditional and scientific observations to detect environmental change: a discussion on Arctic terrestrial ecosystems. Ambio 33:18–23. https://doi.org/10.1007/0044-7447-33.sp13.18
Huntington HP (2011) Arctic science: the local perspective. Nature 478:182–183. https://doi.org/10.1038/478182a
Hussain M, Afzal M, Ali Q, Taylor W, Mariner J, Roeder P (2008) The epidemiology of peste des petits ruminants in Pakistan and possible control policies. Rev Sci Tech Off Int Epiz 27:869–876
Jost CC, Mariner JC, Roeder PL, Sawitri E, Macgregor-Skinner GJ (2007) Participatory epidemiology in disease surveillance and research. Rev Sci Tech Off Int Epiz 26:537–547
Jost CC, Nzietchueng S, Kihu S, Bett B, Njogu G, Swai ES, Mariner JC (2010) Epidemiological assessment of the Rift Valley fever outbreak in Kenya and Tanzania in 2006 and 2007. Am J Trop Med Hyg 83:65–72. https://doi.org/10.4269/ajtmh.2010.09-0290
Kendrick A, Manseau M (2008) Representing traditional knowledge: resource management and Inuit knowledge of barren-ground caribou. Soc Nat Resour 21: 404–418. https://doi.org/10.1080/08941920801898341
Kutz S, Tomaselli M (2019) "Two-eyed seeing" supports wildlife health. Science 364:1135–1137. https://doi.org/10.1126/science.aau6170
Lawson B, Petrovan SO, Cunningham AA (2015) Citizen science and wildlife disease surveillance. Ecohealth 12: 693–702. https://doi.org/10.1007/s10393-015-1054-z
Mariner JC, Hendrickx S, Pfeiffer DU, Costard S, Knopf L, Okuthe S et al (2011) Integration of participatory approaches into surveillance systems. Rev Sci Tech Off Int Epiz 30:653–659
Mariner JC, House JA, Mebus CA, Sollod AE, Chibeu D, Jones BA et al (2012) Rinderpest eradication: appropriate technology and social innovations. Science 337: 1309–1312. https://doi.org/10.1126/science.1223805
Mariner JC, McDermott J, Heesterbeek JAP, Catley A, Roeder P (2005) A model of lineage-1 and lineage-2 rinderpest virus transmission in pastoral areas of East Africa. Prev Vet Med 69:245–263. https://doi.org/10.1016/j.prevetmed.2005.02.001
Mariner JC, McDermott J, Heesterbeek JAP, Thomson G, Martin SW (2006b) A model of contagious bovine pleuropneumonia transmission dynamics in East Africa. Prev Vet Med 73:55–74. https://doi.org/10.1016/j.prevetmed.2005.09.001
Mariner JC, McDermott J, Heesterbeek JAP, Thomson G, Roeder PL, Martin SW (2006a) A heterogeneous population model for contagious bovine pleuropneumonia transmission and control in pastoral communities of East Africa. Prev Vet Med 73:75–91. https://doi.org/10.1016/j.prevetmed.2005.09.002
Mariner JC, Paskin R (2000) Manual on participatory epidemiology: methods for the collection of action-oriented epidemiological intelligence. Food and Agriculture Organization of the United Nations (FAO), Rome
Mariner JC, Roeder PL (2003) Use of participatory epidemiology in studies of the persistence of lineage 2 rinderpest virus in East Africa. Vet Rec 152:641–647. https://doi.org/10.1136/vr.152.21.641

OIE (2014) (office international des epizooties). Guide to terrestrial animal health surveillance. World Organization for Animal Health, Paris

Orozco MM, Argibay HD, Minatel L, Guillemi EC, Berra Y, Schapira A et al (2020) A participatory surveillance of marsh deer (Blastocerus dichotomus) morbidity and mortality in Argentina: first results. BMC Vet Res 16:1–18. https://doi.org/10.1186/s12917-020-02533-x

Peacock SJ, Mavrot F, Tomaselli M, Hanke A, Fenton H, Nathoo R et al (2020) Linking co-monitoring to co-management: bringing together local, traditional, and scientific knowledge in a wildlife status assessment framework. Arct Sci 6:247–266. https://doi.org/10.1139/as-2019-0019

Pretty JN (1995) Participatory learning for sustainable agriculture. World Dev 23:1247–1263. https://doi.org/10.1016/0305-750X(95)00046-F

Roeder PL (2011) Rinderpest: the end of cattle plague. Prev Vet Med 102:98–106. https://doi.org/10.1016/j.prevetmed.2011.04.004

Rostal MK, Olival KJ, Loh EH, Karesh WB (2012) Wildlife: the need to better understand the linkages. In: Mackenzie J, Jeggo M, Daszak P, Richt J (eds) One health: the human-animal-environment interfaces in emerging infectious diseases. Current topics in microbiology and immunology, vol 365. Springer, Heidelberg, pp 101–125

Schulz K, Calba C, Peyre M et al (2016) Hunters' acceptability of the surveillance system and alternative surveillance strategies for classical swine fever in wild boar, a participatory approach. BMC Vet Res 12: 187–198. https://doi.org/10.1186/s12917-016-0822-5

Sripa B, Tangkawattana S, Sangnikul T (2017) The Lawa model: a sustainable, integrated opisthorchiasis control program using the EcoHealth approach in the Lawa Lake region of Thailand. Parasitol Int 66:346–354. https://doi.org/10.1016/j.parint.2016.11.013

Tengö M, Austin BJ, Danielsen F, Fernández-Llamazares Á (2021) Creating synergies between citizen science and indigenous and local knowledge. Bioscience 71: 503–518. https://doi.org/10.1093/biosci/biab023

Tomaselli M (2018) Improved wildlife health and disease surveillance through the combined use of local knowledge and scientific knowledge. Doctoral thesis, Faculty of Veterinary Medicine, University of Calgary, Calgary. http://hdl.handle.net/1880/107597

Tomaselli M, Curry P (2019) Wildlife health and disease surveillance. In: Cork S, Halliwell R (eds) The veterinary laboratory & field manual, 3rd edn. 5M Publishing, Sheffild, pp 420–432

Tomaselli M, Dalton C, Duignan PJ, Kutz S, van der Meer F, Kafle P et al (2016) Contagious ecthyma, rangiferine brucellosis, and lungworm infection in a muskox (Ovibos moschatus) from the Canadian Arctic, 2014. J Wildl Dis 52:719–724. https://doi.org/10.7589/2015-12-327

Tomaselli M, Elkin B, Kutz SJ, Harms NJ, Nymo HI, Davison T et al (2019) A transdisciplinary approach to Brucella in muskoxen of the Western Canadian Arctic 1989-2016. Ecohealth 16:488–501. https://doi.org/10.1007/s10393-019-01433-3

Tomaselli M, Gerlach C, Kutz S (2018b) Checkley SL and the Community of Iqaluktutiaq. Iqaluktutiaq voices: local perspectives about the importance of muskoxen, contemporary and traditional use and practices. Arctic 71:1–14. https://doi.org/10.14430/arctic4697

Tomaselli M, Kutz S, Gerlach C, Checkley S (2018a) Local knowledge to enhance wildlife population health surveillance: conserving muskoxen and caribou in the Canadian Arctic. Biol Conserv 217:337–348. https://doi.org/10.1016/j.biocon.2017.11.010

Watsa M (2020) Wildlife disease surveillance focus group. Rigor Widlife Dis Surveill Sci 369:145–147. https://doi.org/10.1126/science.abc0017

Causation in Wildlife Population Health

Craig Stephen

Abstract

Wildlife population health researchers and practitioners routinely need to confront the question of "what is causing this problem?" There are many impediments to answering this question due to the challenges in undertaking causal research in free-ranging wildlife and in the varying ways evidence of causation is evaluated by different disciplines and stakeholders. Wildlife health outcomes have a series of antecedent events, circumstances, and conditions that must be in place for outcomes to occur where and when they do. The cause of health outcomes should be thought of as a set of minimal conditions and events that inevitably produce the outcome. These conditions can be explored in controlled experiments or observational studies, neither of which on their own provides sufficient evidence to declare a causal effect exists under natural conditions. Researchers and practitioners can be guided by criteria for establishing cause–effect relationships, but ultimately, a pragmatic approach relying on some degree of experience and opinion is needed. Developing agreement on the question "what caused this problem" is, therefore, part biomedical science and part negotiation. The pragmatic goal of casual studies in wildlife health management is not necessary to fully understand all component causes leading to an outcome, but rather to develop agreement on what constitutes sufficient reliable evidence to produce willingness and confidence to act.

Keywords

Wildlife · Causation · Koch · Hill's criteria · Evans · Component cause · Uncertainty

1 Introduction

Many of the underlying questions that motivate wildlife health research or management are questions of "why is this happening" and "what can we do about it?" Answering these questions requires some understanding of cause-and-effect relationships. Population health practitioners need to know about causes before they intervene to prevent disease, reduce harm, and promote health. Establishing wildlife health causal relationships is fraught with limitations due to the challenges of observing the interactions of free-ranging wildlife with the many hazards, determinants, and risk factors in their environments. Generating definitive proof of causal linkages under natural conditions is complicated by the interactions of factors that can modify responses to threats and stressors, by the synergist, antagonist, and cumulative impacts of concurring stressors that are varying over time

C. Stephen (✉)
School of Population and Public Health, University of British Columbia, Vancouver, BC, Canada

C. Stephen (ed.), *Wildlife Population Health*,
https://doi.org/10.1007/978-3-030-90510-1_6

and space, variations in the prevalence, intensity, and frequency of exposures between populations and ecosystems, delays between exposures and outcomes, and the inherent unpredictability within complex systems. This chapter explores ways to think about causation in wildlife population health and provides some pragmatic suggestions to help work within the reality of incomplete casual information.

2 What Is a Cause?

The question of "what caused that" is a basic human question, one we must master to make our way in the world. For this book, a pragmatic definition of a cause is something that makes a change in wildlife health outcomes. There remains, however, significant philosophical and epidemiological debate about whether we can truly know the cause of an outcome or event. Health research often retreats to estimating proportional average contributions of risk factors and determinants rather than making definitive statements about causation. This is because the "cause" of the health outcome is not a thing, but rather an accumulation of things. Let us take, for example, the question of what causes rabies. Unarguably, the disease called rabies cannot occur without the involvement of the neurotropic rabies virus, of the Lyssavirus genus, within the family *Rhabdoviridae*. But that virus cannot be considered *the* cause of the disease. If I spilled a vial full of that virus on a table, rabies would not occur. The virus must find its way into a warm-blooded animal, often through a bite, for rabies disease to occur. So, one might consider the bite as the cause. But if that bite were afflicted on an effectively immunized host, the disease would not occur. Thus, a susceptible host is the cause. But, for that bite to occur, the infected and bitten hosts must be in proximity to each other, which requires an ecosystem that supports those hosts and favors their interaction. However, there is not just one type of ecosystem that allows this series of events to occur. Therefore, there must be a variety of suitable ecosystems which in turn require an assemblage of conditions on Earth, which requires a certain solar system and etcetera. For rabies to occur, there must be an accumulation of a variety of circumstances and settings at the same time and in the same geographical region. This simplified example reminds us that health outcomes have a series of antecedent events, circumstances, and conditions that must be in place for the outcome to occur where and when it did. It may be that no single, specific event, condition, or characteristic is sufficient by itself to produce the outcome but rather we should think of a cause as a set of minimal conditions and events that inevitably produce the outcome of interest (Rothman and Greenland 2005).

2.1 Sufficient Component Causes

The model of "sufficient component causes" helps us to come to grasp with a multicausal world. Rothman and Greenland (2005) described causation as a pie made up of many components (think of them as slices of the pie). For the outcome to occur, the entire pie must be present. Figure 1 illustrates two hypothetical causal "pies." Both require component cause A to be present (for our example, component cause A could be the rabies virus). But the combination of other host, exposure, and environmental components (B through G) differ, even though both pies can cause rabies. If one component of either pie is missing, the remaining components are not sufficient to result in the disease. In the rabies example above, if we take away the susceptibility of the individual via immunization, rabies will not occur even when all the other component causes remain. There are two implications of this model. First, the goal of wildlife population health research is not to elaborate every type and interactions of component causes (i.e., every slice of the pie). Rather, the pragmatic goal is to find the component cause(s) that when removed or modified, renders the circumstances insufficient to cause the disease or to alter the health outcome. Just as a disease outbreak is the result of a change in a rate-limiting component cause, disease control can be accomplished by manipulating a modifiable component cause(s).

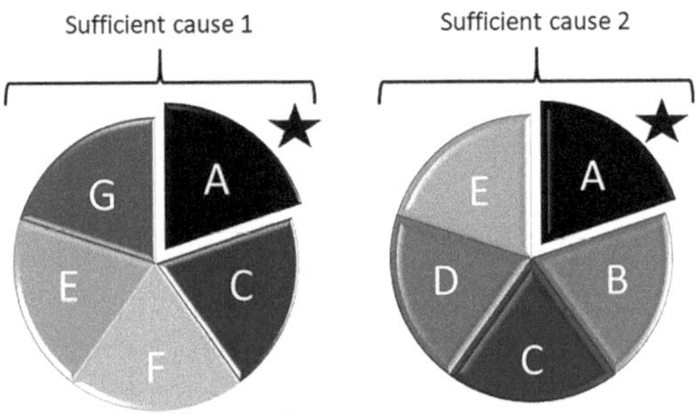

Fig. 1 The sufficient component cause model (based on Rothman and Greenland 2005). For the outcome to occur, all components of the sufficient cause must be present. Both sufficient causes can result in the same outcome. In both cases, "A" must be present and thus is a necessary component cause. Component causes B, C, D, E, F, and G are conditions or factors that are necessary to complete the sufficient cause and allow the outcome to occur. For hypothetical wildlife rabies outbreaks, A is the rabies virus, B is habitat changes that increase host densities and contact, C is susceptible hosts, D is an underfunded wildlife vaccination campaign that reduces immunization rates of susceptible wildlife hosts, E is viable exposure pathways between infected and susceptible hosts, F is dog importation policies that allow unvaccinated animals into a country, and G is cultural practices that enable a growing feral dog population. This is an illustrative rather than exhaustive list of possible component causes. Other component causes would affect rabies outbreaks

Second, while one slice of the pie might be necessary for the outcome to occur (i.e., the rabies virus in the preceding example), the contribution and variety of component causes that lead to the outcome can vary over time, place, and situation, because the amount and distribution of the component causes may also change. This complicates making generalizable declarations about the attributable contribution of a component cause that holds true across all circumstances. For the rabies example, the prevalence and influence of susceptible hosts will be substantially different in vaccinated versus unvaccinated populations, yet rabies can still occur in both populations. The multifactorial model of the determinants of health described in Chap. 1 corresponds with this multi-casual "sufficient component causes" model.

Determining the causes of injuries or diseases in an individual is usually a more straightforward process than determining the causes of population health. For example, it can be obvious that the fractured leg of an antelope was caused by the impact of the car you witnessed collide with the antelope. But, thinking about how to reduce the frequency and prevalence of automobile injuries in the antelope population requires thinking beyond the proximate cause of the acute injury to thinking about the social and ecological circumstances that reduce the probability and frequency of car–antelope interactions, ranging from influencing road designs, thinking about urban–wildlife interfaces, considering population dynamics of people and antelope and more. However, both causation at an individual level and at a population level is a complex accumulation of multiple, interacting, physiological, ecological, and social components.

2.2 Ecologic and Atomistic Fallacies

Wildlife health managers and researchers need to be aware of the atomistic and ecological fallacies. The atomistic fallacy refers to erroneous inference about causal relationships at the group or population level based on relationships observed in individuals. For example, chinook salmon surveillance based on postmortem examination of individual moribund fish captured at the water's surface associated plasmacytoid leukemia with high rates of individual mortality, and therefore, it was assumed that the disease was the cause of high rates (i.e., >90%) of population mortality. Subsequent investigation revealed that fish with plasmacytoid leukemia were more likely to be captured and tested than other moribund fish in the population and that the disease never accounted for more than approximately 10% of dead fish in the population when all dead fish were retrieved and studied (Stephen and Ribble 1995). These finding refuted the association of plasmacytoid leukemia with high population mortality rates.

The ecological fallacy is the reverse of the atomistic fallacy. It occurs when association made at the group level is used to make inferences on causation at the individual level. One example involves a study that demonstrated that Lyme disease incidence is negatively correlated with mammalian biodiversity, the abundance of fried chicken restaurants, and human obesity rates (Salkeld and Antolin 2020). While possible casual mechanisms for these associations could be proffered for how fried chicken consumption and obesity might affect individual exposures and susceptibilities, the authors showed that these were spurious associations arose due to the data examined being aggregated across large spatial scales. Examination of the data at small spatial subclusters or at the individual levels failed to support casual associations with fried chicken and obesity.

We can typically only measure change in disease or health at a population level in wildlife. We don't often get the opportunity to monitor individual wild animal health as we do with pets, zoo animals, or people, because we don't have access to them for longitudinal examinations. Too often we only access them at the end of the casual chain, when the animals are sick or dead, thus making it hard to explore their past histories and associations with potential risk factors or casual variables. Ecological bias is particularly important in wildlife management because exposure factors and disease determinants can seldom be measured at the individual animal level. For example, even if a population or ecosystem level study indicated more of an exposure opportunity at an ecosystem level, it does not mean that the individual animals with the disease were exposed to the hazard that is elevated in the ecosystem.

2.3 Health as a Cumulative Effect

The conventional approach to risk management has been to focus on identifying hazards, assessing risk from each identified hazard, and taking any necessary steps to control risk from each hazard separately. Population health, however, is the accumulation of causes at a population level. Disease (or the lack thereof) is a component cause of population health. The pragmatic reasons we think of disease as a stand-alone outcome is because disease is relatively easier to study than health, and we study it as a univariate outcome. While some people explore determinants that have a negative impact on health and others examine determinants that have a positive impact on health, a complete health model contains both sets of determinants combined in one causal health pie.

Thinking of health as the sufficient accumulation of a suite of determinants or thinking of disease as the sufficient accumulation of agents and circumstances can help make explicit a wider suit of wildlife health interventions and policy options (Wittrock et al. 2019). For example, Nuñez et al. (2020) proposed that introduction of rabies virus from Brazil to Uruguay likely prompted a cattle rabies outbreak, but it was habitat fragmentation that increased vampire bat population connectivity and environmental temperatures that determined the distribution of the outbreak. Controls targeting the virus were a plausible response to the outbreak, but these authors proposed that land-use planning that decreased grassland habitat fragmentation would reduce the risk of transmission to cattle and would be especially important in the context of climatic changes and increasing minimum temperatures in the winter.

Causal pathways that affect wildlife exposure, susceptibility, and resistance are structured by constraints and possibilities of biology, shaped by a species' evolutionary history, its ecologic context, individual histories, and by societal factors that determine the decisions that influence wildlife management. The challenge arises in combining different information and data collected at different times, means, and units of analysis. Population-level measures of health/disease are based on individual health/disease because we measure disease/health in individuals and then count them to get population-level statistics. This is often not the case for determinants of disease/health (e.g., climate, habitat connectivity) with are often measured at larger scales such as at the population or ecosystem level. Understanding population-level phenomena (such as rates of total mortality) should not be confused with mechanisms of disease causation (which occur within individuals) (Krieger 2008). But the connections between and contributions of disease/health in the individual to population health are important to understand to identify casual associations that can be modified to affect health outcomes.

3 How Are Causal Inferences Made?

Accepting the complexity and dynamics of a multi-casual world model does not excuse us from trying to help people answer the question, "what caused this?" It has been said that Leonardo Da Vinci wrote, while nature begins with causation and ends in experience, we must follow a contrary procedure, that is, begin with experience and with that discover the cause (as quoted in Ward 2009). Wildlife health researchers have used a variety of experiential approaches to understand causal relationships including laboratory experiments, epidemiological studies, theoretical models, experimental studies in artificial ecosystems and combinations of these approaches to better understand causal associations.

3.1 Experimental Studies

Experiments allow studying the relationship between an exposure(s) and an effect(s) under defined, controlled, and reproducible conditions. Experimental studies have been invaluable for screening and identifying etiologic agents, determining how attributes of the etiological agent influence variations in host outcomes, finding critical thresholds of exposure that leads to harms, understanding pathogenesis, and in developing and validating diagnostic methods (Brand 2013; Stallknecht 2007).

The unique challenges of meeting the needs of wildlife species in captivity (especially for large, wide roaming, and/or highly social species) plus growing public and political resistance against animal experimentation constrain the number, size, and variety of wildlife experiments that can be conducted. This, in turn, can impede the capacity for the experiment to produce reliable, valid, and generalizable findings. A prominent criticism of experimental studies has been their lack of ecological realism. Such criticism has encouraged researchers to try to mimic natural conditions while maintaining the reproducibility and controllability of laboratory studies through microcosm and mesocosm experiments. These experiments use artificially produced, simplified ecosystems that are used to simulate what might happen under natural conditions, but within controlled settings. Microcosm and mesocosm experiments have been used to study the environmental fate of etiologic agents (e.g., Avendaño-Herrera et al. 2006), explore how different stressors or hazards might interact (e.g., Baudrimont and De Montaudouin 2007), examine how pathogen effects other ecological interactions, like predation (e.g., DeBlieux and Hoverman 2019), and investigate how environmental conditions might alter the impacts of etiological agents on host outcomes (e.g., Hamilton et al. 2012).

Clinical trials are experimental studies undertaken in natural conditions that investigate new tests and treatments and evaluate their effects on health outcomes. Clinical trials, especially randomized controlled trials, try to control for most, if not all, sources of bias by means of randomization, subject and treatment allocation, blinding investigators to case or control classifications, and more. The strict expectations for selection, allocation, and monitoring in clinical trials can rarely be met for free-ranging wildlife. For example, unbiased randomized clinical trials are not possible unless all eligible animals in the target population have an equal probability of being included in the trial. Wildlife can be easily lost to follow-up and finding controls for cases can be impossible without access to and knowledge of all members of the target populations. The strict expectations for clinical trials can hamper the ability to generalize the results to populations outside of the highly controlled setting of the trial.

Clinical trials can be of two types; (1) trials designed to explore if and how an intervention works (these experiments are designed to control for all known biases and confounders) and (2) trials designed to test interventions in the full spectrum of everyday population settings to investigate whether an intervention actually works in real life (Patsopoulos 2011). The latter, known as pragmatic clinical trials, do not allow less attention to control of important biases but they do offer a human health model that could inform the design of wildlife clinical trials. Interpretation of field trials conducted for wildlife needs to account for the challenges in meeting clinical trial expectations and in the inherent difficulties of managing confounding and bias in populations that are not directly under our control (see Chap. 4).

3.2 Observational Studies

The experimental paradigm has been held for centuries as the "gold standard" of evidence. However, in his classic paper "Koch is Dead," Robert Hanson proposed that it is not until we examine the interactions of hosts, agents, and environments under the conditions in which wildlife live can we truly understand wildlife health and disease (Hanson 1988). Although observing how health outcomes manifest themselves under natural conditions should more accurately reflect the "real world" than experimental conditions, the inability to control the situation, find counterfactual situations to provide comparators, issues in accessing and measuring the prevalence or impacts of determinants and risk factors, logistic and economic limitations to sampling free-ranging animals, and ecological variations can all inhibit one's ability to consistently and accurately recognize cause–effects relationships in the "real world." Several possible explanations for an observed association need to be considered before one can infer cause–effect relationship from observational studies.

An observed association may be real or due to random error (chance), bias, or confounding. Using optimal study sample sizes and assessing the study results via statistical testing are the usual ways for dealing with random errors. Confounding is a distortion in the measure of association between a factor and an outcome that occurs when the primary factor of interest is mixed up with some other factor that is associated with the outcome. For example, when trying to study the spatial variation in chronic wasting disease in white-tailed deer (*Odocoileus virginianus*), researchers needed to account for the effects of age and sex because age and sex had an effect not only on the disease but also on spatial distribution of the deer (Osnas et al. 2009). Bias is a systematic error in a study that results in an incorrect estimate of the association between exposure and risk of disease. There are many causes of bias in observational studies that derive from how we decide to sample populations, the way we classify animal exposure or health status, the influence of investigator expectations for the data, and the amount of attention placed on some species versus others. Bias and confounding are affected by the study design or the methods used to obtain the study data (Dohoo 2014). The challenge with wildlife observational studies is that many of the expectations of veterinary epidemiology for study design, such as random sampling, cannot easily be achieved in wild populations (see Chap. 4).

3.3 Causal Criteria

Experiments, models, and field observations all give insight into cause–effect relationship, but each is notoriously inadequate to establish, with full confidence, that a cause–effect relationship exists. Criteria for identifying cause–effects relationships have plagued philosophers since antiquity. Many modern criteria can be traced back to John Stuart Mill's five canons (Ducheyne 2008) (Table 1), which reflect the way many of us intuitively make causal conclusions. The problem with Mill's canons, which were developed in the mid-1800s, is that they do not fit a multicausal worldview. Mill's approach assumes that we have a list of candidate causes to consider but does not tell us how to come up with that list. It also assumes that among the list of factors under consideration, only one factor is the unique cause of the effect.

Around the same time as Mill, the Henle–Koch postulates gained prominence as the standard for evaluating the causal relationship of an

Table 1 Two historically important criteria for causation. Mill's Canons and the Henle–Koch postulates (Ducheyne 2008; Evans 1976)

Mill's canons	1. Method of agreement—If two or more instances of the phenomenon under investigation have only one circumstance in common, the circumstance in which alone all the instances agree is the cause (or effect) of the given phenomenon 2. Method of difference—Where you have one situation that leads to an effect, and another which does not, and the only difference is the presence of a single factor in the first situation, you can infer this factor as the cause of the effect 3. Method of concomitant variation—If across a range of situations that lead to a certain effect, you find a certain property of the effect varying with variation in a factor common to those situations, then you can infer that factor as the cause 4. Method of residues—If you have a range of factors believed to be the causes of a range of effects, and you have reason to believe that all the factors, except one factor X, are causes for all the effects, except one, then you should infer that X is the cause of the remaining effect 5. The joint method—Applies both the method of agreement and the method of difference
Henle–Koch postulates	1. The pathogen or parasite occurs in every case of the disease in question and under circumstances which can account for the pathological changes and clinical course of the disease 2. It occurs in no other disease as a fortuitous and nonpathogenic agent 3. After being fully isolated from the body and repeatedly grown in pure culture, it can induce the disease anew

infectious agent to a clinical disease (Table 1). These postulates work well for diseases with single culturable agents for which there are experimental models for reinfection trials but are less well suited for multi-agent diseases, diseases with significant delays between exposure and outcome or for noninfectious diseases. Hanson (1988) argued that bringing a parasite/pathogen and host together is not enough. The circumstances under which this is done are equally important. What can be shown by the Henle–Koch postulates under controlled conditions may not reflect how the purported etiologic agent affects hosts in the face of environmental and host variation found under natural conditions. Furthermore, they do not account for asymptomatic infections, the impact of immunological process on diseases, latent infections, non-agent component causes, or the biological spectrum of disease (Evans 1993).

Hill's criteria were developed in the 1950–1960s and soon became the most used and taught criteria for causation in epidemiology (Table 2). Hill's criteria, while being highly influential and valuable for nominating casual variables, are like all causal criteria in that they are fallible. Even Hill recognized that some of his criteria may not be possible to meet even for truly causal relationships (e.g., plausibility and coherence are limited by the knowledge of the day). Hill also acknowledged that fulfillment of all criteria does not guarantee a casual relationship and failure to fulfill one or more cannot be used to rule out casual relationships. The only absolute requirement is temporality (i.e., cause precedes effect). In his classic paper, Evans reviewed the aforementioned and other developments in casual thinking in health science and built on Hill's criteria to develop, what he called, a unified concept for criteria for causation that were applicable to acute and chronic diseases (Table 3) (Evans 1976).

None of the authors of causal criteria believed their criteria were definitive. Evans (1976) warned investigators of several limitations in studies of causation for diseases including (1) the same pathologic or clinical state can be produced by different etiological agents; (2) causative agents may vary in different geographic areas, in different age groups, or with different patterns of host susceptibility; (3) some diseases require two or more agents or cofactors acting together to produce disease; (4) a single agent may produce different clinical and pathological responses in different settings; and (5) any cause or set of causes usually produces a biologic gradient of response which may include no observable or detectable reaction. Most authors writing about the concept of causation in health science

Table 2 Hill's criteria of causation (Fedak et al. 2015)

Strength of association	Strong associations between a putative cause and effect are more likely to be casual than weak associations
Consistency	Multiple studies using a variety of locations, populations, and methods showing a consistent association between two variables can indicate causal relations
Specificity	A cause leads to a single effect rather than multiple effects
Temporality	Cause must precede effect
Biological gradient	The presence of a dose–response relationship supports a causal association between an exposure and an effect
Plausibility	The hypothesized relationship between the cause and effect is biologically possible
Coherence	The proposed cause–effect relationship is coherent with what we know about the natural history and biology of the health outcome
Experiment	Evidence from experimental manipulation, particularly studies where disease risk declines following an intervention or cessation of exposure, leads to support for causal inference
Analogy	When one causal agent is known, the standards of evidence are lowered for a second causal agent that is similar in some way

Table 3 Evan's unified criteria for causation (Evans 1976)

Prevalence of the disease should be significantly higher in those exposed to the putative cause than in controls not so exposed
Exposure to the putative cause should be present more commonly in those with the disease than in controls without the disease when all risk factors are held constant
Incidence of the disease should be significantly higher in those exposed to the putative cause than in those not so exposed as shown in prospective studies
Temporally, the disease should follow exposure to the putative agent with a distribution of incubation periods on a bell-shaped curve
A spectrum of host responses should follow exposure to the putative agent along a logical biologic gradient from mild to severe
A measurable host response following exposure to the putative cause should regularly appear in those lacking this before exposure or should increase in magnitude if present before exposure
Experimental reproduction of the disease should occur in higher incidence in animals appropriately exposed to the putative cause than in those not exposed
Elimination or modification of the putative cause should decrease the incidence of the disease
Prevention or modification of the host's response on exposure to the putative cause should decrease or eliminate the disease
The whole thing should make biologic and epidemiologic sense

recognize that the world is a messy place and that a single set of causal criteria will never be able to capture all the various permutations of determinants, risk factors, or other variables that influence the relationship between health outcomes and their causes. Ultimately, all causal conclusions require some degree of judgment. However, criteria such as Evans' (Table 3) and Hill's (Table 2) provide an explicit way to structure studies and present evidence in a transparent and explicit manner and thus, hopefully build confidence in linking causes to their effects.

4 Pragmatism in the Face of Causal Uncertainty and Ambiguity

The goal of causal studies in population health is to produce rigorous information that is relevant to decision makers. Wildlife population health researchers and managers must often be pragmatic because cause–effect relationships can be uncertain or inexact and because context, the nature of the problem, and values influence how information and knowledge are translated into action. The biomedical, ecological, and social problems caused by a wildlife population health issue can rarely be solved by using a single scientific method. In practice wildlife population health research should be considered pragmatic research (as per Kaushik and Walsh 2019) in that researchers focus on the consequences of their research and on the research questions rather than on the "single best" method. Rather than striving to find a universal truth, pragmatic researchers emphasize work that is convincingly designed and interpreted, is supported by information from within and outside their investigation, and is trusted enough to help us understand what happened in the past or plan for what to do in the future (Stephen et al. 2016). Pragmatic research and practice should (1) produce good outputs as judged by peers and knowledge users; (2) have impacts affecting decisions and/or change outcomes, institutions, or understanding; and (3) be acceptable, efficient, and ethical (Stephen et al. 2016).

4.1 Dealing with Uncertainty

Uncertainty is expected due to variations in natural systems, lack of information or knowledge, scientific disagreement regarding cause–effect relationships, and/or inexperience in managing the issue. The degree of uncertainty will increase when considering longer time horizons and larger geographic areas. Uncertainty or disagreements about cause–effects relationships can trigger doubt in the advisability of an intervention or block actions. There are three possible uncertainty scenarios: (1) there is so much uncertainty that a confident, evidence-based, risk management decision cannot be made without further information; (2) there are uncertainties, but management options can plausibly reduce the risk to acceptable levels despite these uncertainties; and (3) the uncertainties are acceptable and within the normal scope of management practice.

Uncertainty is not an all or nothing phenomenon. Uncertainty has many different sources and decision makers are not typically uncertain about everything all at once (Bradley and Dreschler 2014). System uncertainty is related to the ecological or social components, relationships, and values in the system of interest and their effects on the decision makers mandate and responsibilities. Effect uncertainty is linked to the effects of changes in the system on outcomes. Response uncertainties are about which response options are available (feasible, acceptable, and effective) to generate desired changes and achieve the required outcomes. A variety of taxonomies of uncertainty have been developed in different fields (see, e.g., Walker 1990 Han et al. 2011; Bradley and Dreschler 2014). Figure 2 is a non-exhaustive attempt to integrate and present different taxonomies from a pragmatic wildlife health perspective.

Understanding the sources and types of uncertainty can help to refine and target subsequent inquiry. It is important to understand whose uncertainty one is addressing. The uncertainty preventing a politician to approve legislation to control a wildlife health problem can be different than the uncertainties of a local natural resource manager in permitting a wildlife health intervention which in turn can be different from the public stakeholder whose support is needed to allow a management action to take place. Observing uncertainty under a population health lens is not only a biomedical science matter alone but also requires participatory methods to understand the context under which decisions are being made and the needs of the decision makers.

One can make a decision in the face of uncertainty. This can be done unilaterally or as a participatory process involving those affecting and being affected by decisions or actions taken. A precautionary approach to this form of coping with uncertainty ideally errs on acting in a manner least likely to cause harm. Alternatively, a decision can be delayed until greater confidence emerges about the implications of acting in the presence of uncertainty, including the intended and unintended effects. This step will require negotiation amongst decision makers and stakeholders to gain a shared view about which uncertainties need to be resolved and how certain one needs to be before acting. Critical

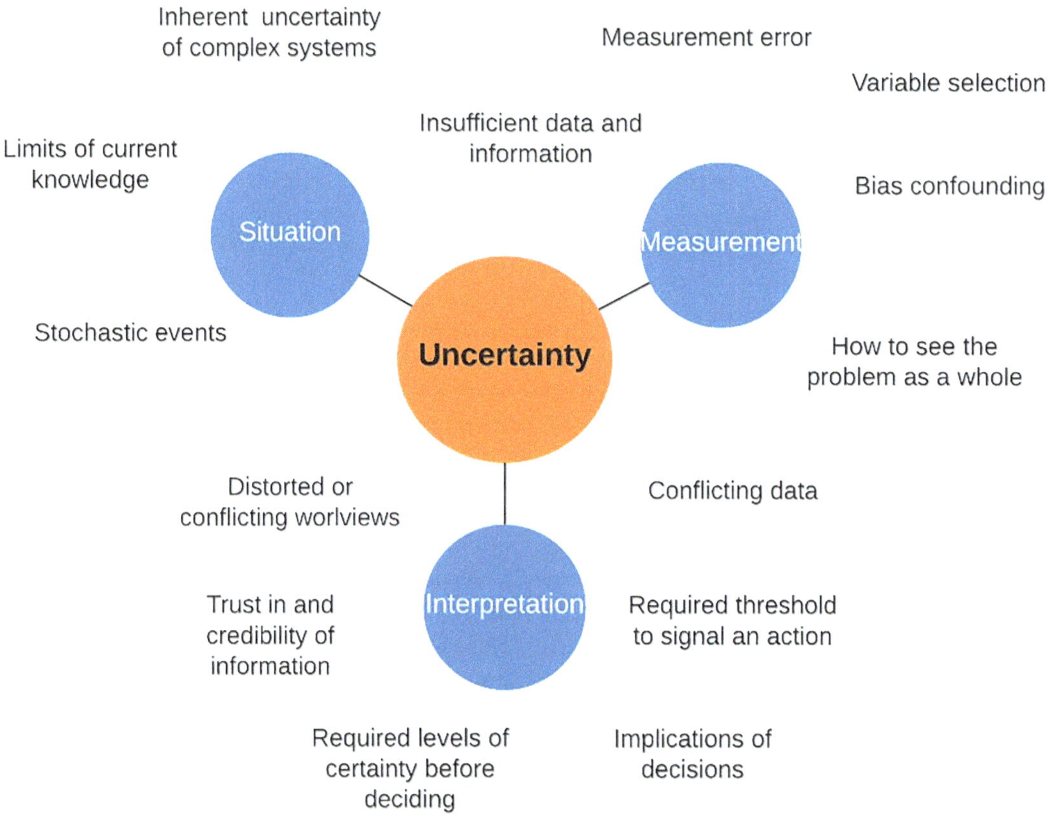

Fig. 2 Some general sources of uncertainty commonly confronted in wildlife population health research and practice

uncertainties can be reduced by collecting and analyzing missing information, mathematical modeling to better understand the implications of assumptions and uncertainties, and/or by developing expert consensus through formal participatory research processes.

4.2 Dealing with Ambiguity

Here the word ambiguity refers to something having more than one possible meaning. Ambiguity can come from uncertainty as well as from how different people assess and interpret the evidence available to them. For some people, evidence is restricted to knowledge produced in accord with the standards of a relevant academic discipline. For others, evidence can be knowledge, skills, and practices developed by and sustained between generations within a community. Others consider evidence as something they observe and document using their own senses. Political agendas, education, economic motivators, or cultural values influence people's willingness to use the various forms of evidence available to them. Along with taking steps to understand uncertainty and its implications for decisions, a wildlife health practitioner should also develop an understanding of why the situation is problematic and for whom. Each stakeholder and rights holder will be driven by a set of motivations, attitudes, knowledge, values, and barriers that shape their view of a wildlife health problem.

Managers have three options when faced with conflicting goals; (1) manage one goal and accept the collateral damage; (2) abandon management of their goal and accept its impacts; or (3) seek a strategy that allows multiple goals to be attained (Buckley and Han 2014). The mere acquisition of new information often does little to affect risk management goals and willingness to act (Gerrard et al. 1999), therefore, risk managers (be they government managers, politicians, citizens, or other decision makers) need to learn from each other to understand the social context before deciding which of these three options to take. One must recognize and respect the diverse and complex value systems operating in wildlife health problems if one hopes to inspire people to act in a way that addresses a suite of goals. Blending objective information with an understanding of peoples' emotions, values, and personal experiences is essential when trying to promote a shared interpretation of a situation. This includes reaching out to information and knowledge sources beyond the biological and biomedical communities that have predominated past wildlife health research and decision-making. Some degree of negotiation will be needed to create a shared vision that will help collaborators see how working toward collective interests will meet the interests of themselves or their organization. Collaborative assessment of information not only helps develop a shared understanding of the problem but also helps people see different aspects of the problem and, by exploring these differences, find solutions that go beyond their own perspective of the problem.

There are four preconditions to effective collaboration and cooperation (Thomson and Perry 2006; Singh and Kant 2008). First, rather than seeing the issue as someone else's problem, successful collaborators need to see their role and responsibilities extend beyond their interests. Second, collaborators need a shared goal of where they want to go and a hierarchy of achievable steps that, taken one at a time, will get them there. Third, collaborators need to understand and endorse a shared process for assessing available information, making decisions, and for moving what they know into action. Fourth, there needs to be trust. Trust can be built by being honest in negotiations, communicating purposefully and regularly, behaving in accordance with agreements, and not taking advantage of others or event when the opportunity is available.

5 Summary

Wildlife population health researchers and practitioners routinely need to confront the question of "what is causing this problem?" It is easy to get lost in the plethora of philosophical debates about the true nature of causation and be tempted by the quest to elucidate all component causes of a problem before acting. It is, therefore, important to differentiate the task of developing a theory of causation for a disease or other health outcomes from identifying the critical causal factors that influence if an intervention works or how the distribution of disease manifests in a natural setting. This chapter has provided some background information and proposed some ideas that could be used to allow casual inquiries to be helpful.

The first task in undertaking casual research in wildlife health is to distinguish the cause(s) of the problem from the cause(s) of the disease or deficiency facing the population. No single account of causation can fully capture the multiple perspectives and needs of the diversity of stakeholders interested in determining what is causing a wildlife health problem. One must recognize that the combinations of factors that lead to health outcomes in one setting or at one scale may not be the same, even for the same disease, in other settings or scales, that the cause of the disease in the individual is not the same as the cause of the distribution and impact of the problems at a population level, and that people will integrate and assess casual evidence in accordance with their perception of the problem. Rather than seeking a single test, experiment, or line of evidence to reveal causal relationships, a weight of evidence approach based on multiple lines of evidence is more typically needed. Clarifying what causal relationship you are examining (i.e., the cause of a pathological response vs. the cause of population effects vs. the cause of the social impacts) will help build a common vision that can serve as the foundation for selecting candidate component causes to prioritize. Formulating criteria for establishing causation in conjunction with those people your results intend to influence will not only provide a framework with which you can integrate and evaluate multiple forms of evidence but also will help build trust and transparency in any future causal claims you make.

References

Avendaño-Herrera R, Irgang R, Magariños B, Romalde JL, Toranzo AE (2006) Use of microcosms to determine the survival of the fish pathogen Tenacibaculum maritimum in seawater. Environ Microbiol 8(5): 921–928

Baudrimont M, De Montaudouin X (2007) Evidence of an altered protective effect of metallothioneins after cadmium exposure in the digenean parasite-infected cockle (Cerastoderma edule). Parasitology 134(2):237

Bradley R, Drechsler M (2014) Types of uncertainty. Erkenntnis 79(6):1225–1248

Brand CJ (2013) Wildlife mortality investigation and disease research: contributions of the USGS National Wildlife Health Center to endangered species management and recovery. Ecohealth 10(4):446–454

Buckley YM, Han Y (2014) Managing the side effects of invasion control. Science 344(6187):975–976

DeBlieux TS, Hoverman JT (2019) Parasite-induced vulnerability to predation in larval anurans. Dis Aquat Organ 135(3):241–250

Dohoo IR (2014) Bias—is it a problem, and what should we do? Prev Vet Med 113(3):331–337

Ducheyne S (2008) JS Mill's canons of induction: from true causes to provisional ones. History Philos Logic 29(4):361–376

Evans AS (1976) Causation and disease: the Henle-Koch postulates revisited. Yale J Biol Med 49(2):175

Evans AS (1993) Limitations of the Henle—Koch postulates. In: Causation and disease. Springer, Boston, pp 123–146

Fedak KM, Bernal A, Capshaw ZA, Gross S (2015) Applying the Bradford Hill criteria in the 21st century: how data integration has changed causal inference in molecular epidemiology. Emerg Themes Epidemiol 12(1):14

Gerrard M, Gibbons FX, Reis-Bergan M (1999) The effect of risk communication on risk perceptions: the significance of individual differences. JNCI Monogr 1999(25):94–100

Hamilton PT, Richardson JM, Anholt BR (2012) Daphnia in tadpole mesocosms: trophic links and interactions with Batrachochytrium dendrobatidis. Freshw Biol 57(4):676–683

Han PK, Klein WM, Arora NK (2011) Varieties of uncertainty in health care: a conceptual taxonomy. Med Decis Making 31(6):828–838

Hanson RP (1988) Koch is dead. J Wildl Dis 24(2): 193–200

Kaushik V, Walsh CA (2019) Pragmatism as a research paradigm and its implications for social work research. Soc Sci 8(9):255

Krieger N (2008) Proximal, distal, and the politics of causation: what's level got to do with it? Am J Public Health 98(2):221–230

Nuñez GB, Becker DJ, Lawrence RL, Plowright RK (2020) Synergistic effects of grassland fragmentation and temperature on bovine rabies emergence. Ecohealth 22:1–4

Osnas EE, Heisey DM, Rolley RE, Samuel MD (2009) Spatial and temporal patterns of chronic wasting disease: fine-scale mapping of a wildlife epidemic in Wisconsin. Ecol Appl 19(5):1311–1322

Patsopoulos NA (2011) A pragmatic view on pragmatic trials. Dialogues Clin Neurosci 13(2):217

Rothman KJ, Greenland S (2005) Causation and causal inference in epidemiology. Am J Public Health 95(S1): S144–S150

Salkeld DJ, Antolin MF (2020) Ecological fallacy and aggregated data: a case study of fried chicken restaurants, obesity and Lyme disease. Ecohealth 17(1):4–12

Singh MD, Kant R (2008) Knowledge management barriers: an interpretive structural modeling approach. Int J Manage Sci Eng Manage 3(2):141–150

Stallknecht DE (2007) Impediments to wildlife disease surveillance, research, and diagnostics. In: Wildlife and emerging zoonotic diseases: the biology, circumstances and consequences of cross-species transmission. Springer, Berlin, Heidelberg, pp 445–461

Stephen C, Burns T, Riviere-Cinnamond A (2016) Pragmatism (or realism) in research: is there an Ecohealth scope of practice? Ecohealth 13(2):230–233

Stephen C, Ribble CS (1995) An evaluation of surface moribund salmon as indicators of seapen disease status. Aquaculture 133(1):1–8

Thomson AM, Perry JL (2006) Collaboration processes: inside the black box. Public Adm Rev 66:20–32

Walker VR (1990) The siren songs of science: toward a taxonomy of scientific uncertainty for decisionmakers. Conn L Rev 23:567

Ward A (2009) Causal criteria and the problem of complex causation. Med Health Care Philos 12(3):333–343

Wittrock J, Duncan C, Stephen C (2019) A determinants of health conceptual model for fish and wildlife health. J Wildl Dis 55:285

Investigating Wildlife Disease as a Social Problem

Andrew Peters, Helen Masterman-Smith, and John Rafferty

Abstract

Social dimensions define whether wildlife disease is considered a problem, society's response to wildlife disease, and the available solution space for wildlife health problems. Wildlife disease can have interacting environmental, human health, economic and sociocultural impacts on society, but the way in which society constructs wildlife disease as a problem is shaped by a much broader sociocultural landscape. The sociological imagination, with lines of historical, cultural, structural, and critical inquiry, provides a framework for investigating the social dimensions of wildlife health. The investigation of wildlife disease as a social problem requires wildlife health professionals to work in transdisciplinary teams that include social scientists and to meaningfully engage community stakeholders in a process of cooperative inquiry. The development of effective, real-world solutions to a wildlife health problem demands comprehensive knowledge of the social dimensions of that problem, drawing on diverse social science disciplines, and the implementation of a translational approach to the research endeavor.

Keywords

Wildlife health sociology · Social dimensions · Social determinants · Sociological imagination · Translational science · Harm reduction model · Cooperative inquiry · Community-based · Participatory

Wildlife disease is both a scientific and a social phenomenon that often requires multipronged solutions. Disease has undoubtedly occurred in free-ranging animals since they first evolved hundreds of millions of years ago, as evidenced by the fossil record (Moodie 1917). There exist many wildlife disease events that are not likely associated with human activity and to which humans feel no need to respond.

As wildlife health professionals, we tend to internalize the process of deciding when a wildlife disease event demands investigation or intervention, and what that response is trying to achieve. Our decisions affect governments, non-government organizations, the public and First Nations peoples, among others. Given the crucial intersections between functioning ecosystems and the health and well-being of humans and other species (Humboldt-Dachroeden et al. 2020), our decisions must consider the social dimensions of wildlife disease events alongside etiological and ecological factors.

With a limited social science repertoire to draw upon, we often rely on our own social experiences, values, and worldviews when

A. Peters (✉) · H. Masterman-Smith · J. Rafferty
Charles Sturt University, Wagga Wagga, NSW, Australia
e-mail: apeters@csu.edu.au

C. Stephen (ed.), *Wildlife Population Health*,
https://doi.org/10.1007/978-3-030-90510-1_7

considering the social dimensions of wildlife disease events. Consciously or not, our thoughts and actions (or social agency) are interdependent with the social structures and cultures in which we are situated and, as such, are inescapably value laden. Many of us do not have the disciplinary expertise to make highly informed and self-reflective decisions about the social dimensions of wildlife disease events, including whether an event requires a solution at all and what form it might take.

1 When Is Wildlife Disease a Social Problem?

1.1 What Are We Really Calling a Problem and Why?

Animals have evolved, adapted, and become extinct for hundreds of millions of years. The existence of all contemporary living things, ourselves included, is contingent on that process, on the death of other animals and on the extinction of most animal lineages that came before us. The extinction or death of wildlife, or even the human species, is therefore not intrinsically a problem from an ecological perspective. Yet, species extinction and wildlife death are often publicly perceived to be problematic, for example, in terms of biodiversity loss, species justice, and human well-being (Cafaro 2015). The conventional criterion for determining if a wildlife disease event is problematic is whether it is directly or indirectly harmful to human society—for example, human health, social power, morality, and ethics—in other words, when it becomes a "social problem."

Not all social problems are responded to equally. Interpretations of, and solutions to, wildlife disease events are socially constructed, meaning that human perceptions of them are shaped by social structures and cultures, including the distribution and organization of social power in a given time and place (Cherry 2018; Yearley 2002). Influential sociologist, Peter Berger, explained that we inherit, often unquestioningly, diverse ways of seeing the world from the various social groups in which we are embedded (e.g., families, communities, professions, religions, nations, subcultures), or as he put it: "the many little workshops in which cliques of universe builders hammer out their models of the cosmos" (cited in Vera 2016). These socializing forces shape which socially problematic wildlife disease events will receive the necessary attention and resources.

Inequitable power structures and cultures are important parts of the complex social worlds in which wildlife disease events need to be understood. For example, environmental justice research demonstrates how social inequalities cause uneven distribution of environmental harm and exclude those most at risk of harm from decision-making structures (Mohai et al. 2009). These dynamics were vividly demonstrated by the socially inequitable impacts of the COVID-19 global pandemic (Cole et al. 2020).

The culture of any given society also reflects the contours of social inequalities. The dominant (or hegemonic) values, ideologies, worldviews, and customs of a society are the outcome of power struggles between subpopulations with competing interests (Van Krieken et al. 2006). Hence, social groups adhere to a diverse spectrum of norms or ideals regarding wildlife disease events. The response of society to the death or illness of wild animals depends on how those species are socially constructed (Straftford et al. 2000). Many aspects of wildlife health, including disease management activities such as culling programs (Mysterud and Rolandsen 2018), and even the protection of wildlife from the health impacts of industrial or agricultural development, or economic growth policies, for example, are controversial social questions. The structural and cultural factors that shape human relations with other animals and the rest of the environment unsettle simple answers to wildlife disease events (Dickens 1992; Peggs 2012).

So, when is a wildlife disease event a social problem? It is when powerful social actors (groups and institutions) construct it as a problem for particular parts of humanity. The basis upon which wildlife events are assessed aligns not only to objective facts of nature but also to subjective value systems produced by social forces within specific contexts. Importantly, the values that always underpin the social construction of wildlife disease as a social problem are heterogeneous, contested, and changeable. When we investigate a wildlife disease event and are interested in whether it is a social problem or not, we therefore need to understand the social landscape in which the event is situated to make sense of our findings.

1.2 A Harm Model for Understanding the Impacts of Wildlife Disease on Society

Wildlife disease can impact people, individually or collectively, directly, and indirectly. Obvious examples include direct economic loss associated with the emergence of disease in a commercially important species, as seen with pilchard herpesvirus (Whittington et al. 2008), human morbidity, and death due to viral disease following spillover from wildlife, as seen in the COVID-19 pandemic, or the loss of ecosystem services because of wildlife population changes due to disease, such as insect population control following mass mortality of insectivorous bats due to white nose syndrome (Blehert et al. 2009). Subtler examples include sociocultural harm associated with culturally significant species, such as the culling of reindeer to control chronic wasting disease emergence in Norway (Mysterud and Rolandsen 2018), the sociopolitical effects of the late nineteenth-century rinderpest epidemic in southern Africa (Phoofolo 1993), emotional or existential distress associated with environmental degradation (solastalgia) (Albrecht et al. 2007), or lost productivity because of ineffective control programs for bovine tuberculosis due to spillback from wildlife reservoirs (Tweedle and Livingston 1994).

Most wildlife disease events have more than one potential way of impacting people, and responses to wildlife disease can themselves have impacts. These impacts can perhaps be best conceptualized as interacting spheres of sociocultural, human health, economic and environmental harm (Fig. 1).

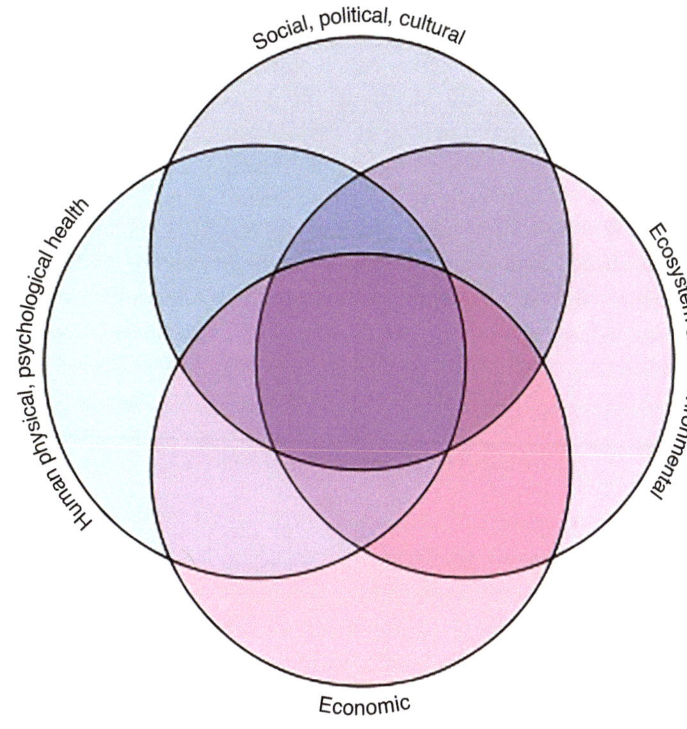

Fig. 1 The social impacts of wildlife disease events are distributed across spheres of sociocultural, human health, economic and environmental harm (Peters et al. 2019)

The social dimensions of wildlife disease events—their construction, causes, risks, impacts, harms, and solutions—are experienced differently across social groups, reflecting the uneven distribution of power in which they are embedded. In investigating and assessing the significance of wildlife disease events, and in considering the costs and benefits of any response to those events, it is, therefore, critical to understand where the event and potential response actions are positioned in terms of harm and impact for each stakeholder group.

1.3 Etiology and the Social Problem Are Not One and the Same

COVID-19 is a major societal problem, with substantial health, sociocultural, and economic impacts unevenly distributed within and across diverse human populations (Matthewman and Huppatz 2020). The etiology of COVID-19 is a virus. The social problem of COVID-19 is, however, something else. The problem differs for different groups in society. For some, the problem of COVID-19 may be evolving, counterintuitive, conflicting, and complex. It may be a problem of physical illness, loss and emotional hardship, unemployment, inequity, or loss of civil liberty. Biomedical solutions, including public health interventions and vaccinations, that target the etiology of COVID-19 may well solve the problem for many people, but for others, these solutions may exacerbate their experience of the problem (e.g., global inequity, unemployment, loss of civil liberty), even if temporarily. While understanding the etiology of a wildlife disease sheds light on appropriate scientific or clinical responses, it is only one part of developing solutions to the social phenomenon of wildlife disease.

Hendra virus spillover from wildlife reservoirs is considered a significant problem in eastern Australia. The problem of Hendra is not viral spillover from wildlife, it is the impact that it has on human health and on horses, an economically and socioculturally important species in Australia. This might seem like an unnecessary delineation, but it provides context for any response. Decades of viral and ecological research have characterized much of the etiology of Hendra virus spillover, including reservoir species and the ecological phenomena that are associated with spillover events (Field 2016). While this knowledge can inform solution development, any societal response to Hendra needs to align with the actual problem. Accurate spillover prediction is of little value without accompanying risk reduction treatments that may include social behavior change. The development of an effective Hendra vaccination in horses appears to have shifted the problem for some stakeholders from a low likelihood of severe health and sociocultural harm to one of high likelihood of low economic harm (Manyweathers et al. 2017). These social dimensions need to be considered and addressed in the development of solutions.

Chytridiomycosis, a disease caused by infection with the invasive fungal pathogen *Batrachochytrium dendrobatidis (Bd)*, is the cause of significant global declines and extinction of amphibian species (Scheele et al. 2019). The etiology of chytridiomycosis tells us very little about why, ultimately, it is constructed as a social problem. Global chytridiomycosis could be considered a social problem because it is a cause of biodiversity loss, which compromises the functional resilience of ecosystems that are critical to the well-being of human populations. It also impacts social groups who experience solastalgia, alienation from nature, ecological trauma, environmental injustice, or racism due to the anthropogenic decline or extinction of other species (Dickens 1992; Holifield et al. 2017; Galway et al. 2019; Panu 2020). For these social groups, attempts to conserve amphibian species ex situ may not be an effective response to their problem. Others again might hold values around the welfare or rights of individual wild animals and to them in situ conservation actions might not be relevant to their perception of the problem at all. Importantly, there are no universal social values legitimating one response to chytridiomycosis over all others, rather, whether solutions are effectively aligned with the problem is influenced by multiple social constructions and contexts.

2 Finding Effective Solutions Depends on Effectively Defining the Social Problem

2.1 The Challenge of Wildlife Health in the Anthropocene

Challenges to the health of wildlife are increasing predominantly because of anthropogenic factors (Daszak et al. 2001). The broad scientific explanation for the emergence of these challenges revolves around human disruption of historically relatively stable ecosystems. Ecological theory describes the interconnectedness and relationship between biological and physical components of the world. It infers that change propagates through ecosystems affecting those different components, influenced by the magnitude and distribution of change within a given ecosystem. Human activity in the Anthropocene has brought significant change to physical and biological components of almost all global ecosystems. It is hardly surprising in this context that challenges to wildlife health are emerging and likely accelerating in their emergence.

Likewise, the scale of many wildlife disease events appears to be increasing. White nose syndrome has caused millions of deaths of bats across North America, recent bushfires in Australia are thought to have killed billions of wild vertebrates alone, and chytridiomycosis has caused unprecedented amphibian decline and extinction associated with disease. These growing challenges have, in general, not been met with corresponding increases in centralized resourcing of solutions.

How, in this context, do we find solutions for problems relating to wildlife health? Biomedical solutions are undoubtedly part of the picture, but in general are highly system specific and resource intensive. While theory and technology can be transferred between systems (e.g., oral vaccination technology developed for rabies control can inform vaccine development for other wildlife infectious diseases), the different socioecological contexts of each wildlife health problem demand socially appropriate solutions alongside biomedical solutions.

Wildlife disease professionals might also consider that decisive solutions to serious human diseases have emanated far more often from the social realm, than the clinic or the laboratory (Germov 2018). Recognition of the pivotal role of social interventions in the elimination or minimization of many major human diseases led to the adoption of the social model of health by the World Health Organization in 1946 (Germov 2018). This model acknowledges that the absence of disease is an inadequate measure of health and well-being (see Chap. 1). Equally important are the social determinants of health, that is, the ways societies produce the more or less healthy conditions in which people live, work, and play (see Chap. 14). The One Health paradigm demonstrates the intersections between humans, ecosystems, and wildlife health. However, the One Health scholarship is dominated by scientific investigations, leaving our understanding of intersecting social determinants of health underdeveloped (Humboldt-Dachroeden et al. 2020). Finally, it is important to recognize that just as scientific data require theoretical analysis to establish meaning of results, so too social determinants data require theoretical interpretation from a social theory standpoint. In other words, understanding wildlife disease as a social problem requires attention to social data and social theory.

2.2 Discovering the Social Problem with Social Imagination

Wildlife health professionals at the investigatory interface for wildlife disease events are often confronted by an immediate problem: typically, the observation of sick or dead animals. These are not always the same as the social problem—the measure by which society would consider the event a problem.

Though most of us feel we have a reasonable grasp on the social world in which we live, systematic understanding of the complexities of society requires more than casual observations based on individual experiences and personal values. Peter Berger famously contended that "the first wisdom of sociology is this: things are not what they seem" (cited in Macionis and Plummer 2005). Sociologists systematically interrogate how societies work, including how individual or personal experiences and behaviors (micro) are linked to larger social forces and public practices (meso or macro) (Macionis and Plummer 2005).

In doing so, sociologists investigate the ways in which seemingly isolated personal matters are patterned to produce social trends, issues, and problems for ecosystems, humans, and other species. For example, from an individual standpoint, the decision to book a wildlife tourism holiday is a personal decision. A sociological standpoint reveals the causes and effects of millions of people making the same personal choices. Research indicates that people often interpret their personal participation in wildlife tourism as a socially and ecologically positive or benign practice overall, partly because its sociological dimensions are overlooked (Rizzolo 2017). For example, the commodification of wildlife can lead to habitat-destroying tourism infrastructure. The presence of wildlife tourists can disrupt natural eating and sleeping behaviors of animals. Wildlife tourism also contributes to harmful social constructions of human relations with other species. For example, wildlife tourism photography and souvenir collecting have fueled wildlife crimes (e.g., trafficking, poaching, illegal hunting). Tourism economies also affect human communities, often pricing low-income residents out of affordable housing. Although the social practice of wildlife tourism is also highly variable across time and place, many of these outcomes can escalate risks of wildlife disease events, the solutions to which lie as much within social structures and cultures as within clinical science (Tablado and D'Amico 2017; Odeniran et al. 2018; Ohmer et al. 2021).

Charles Wright Mills' seminal work, *The Sociological Imagination*, outlines an approach sociologists take to systematically understand the social world and solve social problems (Mills 2000). He argued that we all possess a sociological imagination, but many of us have not developed the analytical tools to deploy it effectively (cited in Macionis and Plummer 2005). Sociologists commonly utilize a four-dimensional template to analyze social problems based on Mills' concept of the sociological imagination. Drawing on Germov and Poole's version (Germov and Poole 2007), it consists of historical, cultural, structural, and critical lines of inquiry into any given topic (Table 1). The sociological imagination can be used to investigate the social dimensions of a wildlife disease event (Box 1).

Table 1 Lines of inquiry in the sociological imagination

Inquiry	Broad definition
Historical	Social changes or changes in social practices over time
Cultural	Social customs and traditions, value systems, ideologies, worldviews
Structural	Organizations and social institutions (e.g., governments, industry, media, family, education system, criminal justice system, legal system, healthcare system)
Critical	Distribution of social power, who benefits from status quo or social alternatives

Box 1 Example of the Use of the Sociological Imagination to Investigate the Social Dimensions of a Wildlife Disease Event

In a hypothetical example, a link is suspected between wildlife tourism and a wildlife disease event. Wildlife health professionals can usefully work alongside social science professionals to better understand the scope and causes of the problem and to forge solutions. Applying the sociological imagination approach to this topic, a sociological investigation might begin with the following types of questions.

Historical:

- How frequently has the wildlife disease event occurred over time?
- Have reporting practices of wildlife disease events changed over time?
- How has the scale, type, operations, tourism experiences, and social contexts of the relevant wildlife tourism precinct changed over time?

(continued)

- Do changes in the disease event and the tourism precinct coincide?

Cultural:

- Has the wildlife disease event also occurred at relevant wildlife tourism precincts elsewhere, including other countries?
- Which social customs, traditions, or values have influenced the scale, type, operations, and tourism experience of different precincts?
- Are dominant culture, values, and ideologies in which the precinct is located balanced with the interests of wildlife, ecosystems, and humans?
- Do diverse social groups and stakeholders have shared or competing worldviews about the commodification of wildlife?
- Do diverse social groups construct themselves as part of, or separate to, "nature" and "the environment?"

Structural:

- Does government policy appropriately regulate the wildlife tourism industry to monitor or prevent wildlife disease events?
- How effective are the legal mechanisms for preventing wildlife disease events within wildlife tourism context?
- Do labor processes and relations within wildlife tourism precincts influence the risk of a wildlife disease event or the social construction of wildlife and nature?
- How do media representations socially construct wildlife tourism; do they minimize or escalate risks of wildlife disease events?
- How effective are public education initiatives on the risks of a wildlife disease event associated with wildlife tourism?
- Do social inequalities (class, gender, cultural background, etc.), within communities dependent on the wildlife tourism industry, influence the risks of a wildlife disease event?

Critical:

- Who are the dominant decision makers on wildlife tourism matters?
- Which social groups most benefit from wildlife tourism?
- Which social groups are most harmed by a wildlife disease event?
- Do these two social groups have shared and/or competing interests?
- What ethical, justice, or equity considerations are at play?
- What alternatives are possible to prevent a recurrence of the wildlife disease event?

2.3 Opening the Solution Space

Solving a problem typically requires an in-depth understanding of the nature of that problem. Investigation of wildlife disease events focusing solely on scientific causation may open part of the potential solution space (the diversity of potential solutions that are available) but is both a risk and a lost opportunity.

Any approach to determining if and what kind of societal response to a wildlife disease event is warranted should investigate both etiology *and* the social dimensions of that event. The latter completes the solution space and is crucial to identifying an appropriate and effective response. Responses that are not aligned with an understanding of the problem as experienced by all relevant social groups risk unexpectedly shifting the problem to other types of impact or harm (e.g., from economic or human health to social or environmental harm) or unjustly burdening particular social groups with solving the problem. Prior knowledge of the social dimensions of a wildlife disease event facilitates the development of more sophisticated and more inclusive solutions that address the problem comprehensively. As discussed earlier, the development of a Hendra vaccination for horses not only effectively addresses part of the problem (health risk to horses and people) but also creates other problems for horse owners that may render the solution less effective than a more holistic approach to this problem (Manyweathers et al. 2017). Similarly, solutions to "One Health" problems (e.g., viral spillover from wildlife into people or domestic animals) should aim to reduce harm holistically, not merely shift the problem from human health or economic harm to one of environmental harm.

Considering the full social dimensions of a problem may open opportunities to engage with society, and decision-makers, more effectively in the development of solutions that are appropriate and sustainable. Understanding how wildlife disease events are a problem to different social stakeholders permits a "net benefit" approach to solutions, in which solutions or combinations of solutions offer benefit to more of society, encouraging better adoption, sociopolitical support, sustainability, and scalability.

3 Tools for Investigating the Social Problem of Wildlife Disease

3.1 Engagement of Stakeholders as Partners in the Investigative Process

Cooperative inquiry, in which researchers and relevant social groups and stakeholders collaborate in the creation of knowledge, contributes to more meaningful understanding of social dimensions by researchers, enhanced societal ownership of solutions, and greater potential for solutions that are contingent on social change (Heffner et al. 2003). The social science investigative toolkit includes methods for deeply engaging stakeholders, particularly those who may be negatively impacted by wildlife disease. For example, Participatory Action Research (PAR) and Community-Based Participatory Research (CBPR) specialize in stakeholder collaboration and power-sharing at all stages of an investigation, beginning with codesign partnerships (see Chaps. 5 and 21). These methodologies are commonly used in the fields of environmental justice and public health research. An important feature of these approaches is that they often hire members of "hard to reach" stakeholder groups as coinvestigators, outreach officers, or other kinds of intermediaries (Hacker 2013; Davis et al. 2020; Cossyleon and Spitz 2021; Milich et al. 2021).

3.2 Socio-Ecological Research Demands a Transdisciplinary Team

The effective creation of solutions-focused knowledge for wildlife disease events requires diverse disciplinary and theoretical expertise. While wildlife health professionals with biomedical or ecological disciplinary expertise may be able to contribute to components of a sociological investigation, application of the sociological imagination framework and deeper inquiry to characterize the social dimensions and determinants of wildlife health require great familiarity with sociological tools and theory. Social, biological, and ecological knowledge need to be integrated to effectively address the social problem associated with wildlife disease. Solution-focused investigation therefore needs to be transdisciplinary. Transdisciplinary teams go beyond siloed multidisciplinary participation in research by intertwining inquiry across all relevant disciplines to inform the creation of holistic knowledge regarding a problem and solutions.

Bennett et al. (2017) provide a useful overview of the social sciences relevant to conservation, all of which are equally relevant to many wildlife health problems. Social science in human health also has much overlap with wildlife health, especially in the area of risk communication (see Chap. 25). A broad summary of relevant social science disciplines is provided here (Table 2), however, it is important to recognize that each of these fields has diverse applications and it may be worth exploring to find whose expertise and experience best aligns with the social dimensions of wildlife health in the development of transdisciplinary investigatory teams. Wildlife health professionals should also consider the benefit of including disciplinary expertise which may seem peripheral to an investigation. For instance, decision-making around whether a response to a wildlife disease event occurs, and the form that the response takes, typically is made by authorities. These decisions usually weigh up expert opinion, diverse social perspectives and needs, policies and law, which can be perplexing and confounding to the recommendations developed by wildlife health professionals. Political scientists and experts in policy and law, who perhaps are thought of as being unnecessary in wildlife disease event investigations, can bring the relevant expertise to the investigatory process to ensure that knowledge and solutions consider and address downstream decision-making.

Working successfully in a transdisciplinary team requires the acceptance of different epistemological approaches and recognition of the value of different methods in the creation of social knowledge. Bennett et al. (2017) offer a concise and practical description of categories of methodology used in conservation social sciences that are relevant to wildlife health.

Where the social dimensions of a particular wildlife problem include the relationship between society and the environment, and/or ecosystem services, there is a need to contextualize the

Table 2 A selection of social science disciplines that may be relevant to investigating a wildlife health problem

Social science discipline	Relevance to wildlife health
Sociology	Contributes to knowledge of the social dimensions of wildlife health that are shaped by social interactions, relations, and organization. The field of environmental sociology is particularly relevant to wildlife health
Psychology	Contributes to knowledge regarding how people think, feel, and behave relating to wildlife and wildlife health problems. The fields of environmental and behavioral psychology are particularly relevant to wildlife health
Education	Contributes to knowledge about how people learn including how systems of learning interact with individual characteristics to affect learning. Education is instrumental to creating knowledge of wildlife health in society and to influence behavioral change
Communication	Contextualizes the environment in which communication takes place and contributes to the development of effective communication to solve wildlife health problems. Risk communication is a highly relevant field for wildlife health (see Chap. 25). The field of environmental communication also has considerable overlap with aspects of wildlife health communication
Integrated geography	Human–environment geography contributes to knowledge about the interactions between individual people, society, and the natural environment, which is relevant to understanding the way people perceive and are affected by wildlife health (see Chap. 14)
Anthropology	Contributes to knowledge of the relationship between humans, the environment, wildlife, and wildlife health, including over time, especially in the field of environmental anthropology
History	History is a key component of understanding the social dimensions of wildlife health, explaining the origin and form of contemporary relationships between people, the environment, wildlife, and health
Political science	Contributes to understanding relevant systems of governance and power relating to wildlife health. This can inform both the origin of wildlife health problems, aspects of their social dimensions, and the theory and practice of decision-making regarding wildlife health
Philosophy	Contributes to understanding of the fundamental ways in which individuals and societies experience and perceive wildlife health and wildlife health problems
Marketing	Contributes to the development of strategies and mechanisms to influence societal behavior relating to wildlife health
Policy and law	Contributes to knowledge of the legislative and policy landscape in which wildlife health problems occur, how decisions are made and to which solutions are confined

problem in terms of broader biological and ecological processes. This may require the participation of relevant biological science disciplines, including various fields of ecology. This is ancillary to investigating disease ecology to understand the drivers of wildlife disease and is rather about connecting wildlife health and disease with its impacts on human society through the environment.

3.3 A Translational Approach

Investigation of wildlife disease events by a transdisciplinary team in partnership with social stakeholders does not in itself provide a framework for creating effective, real-world solutions. The latter remains an enormous challenge in wildlife health. How can we structure our investigatory approach to effectively solve problems, rather than just effectively documenting the biological and social phenomena of wildlife health?

Human health offers a conceptually powerful framework in "translational science" for how research can contribute to real solutions. Translational science evolved in response to a disconnect that appeared in the twentieth century between the rapid growth in funding for biomedical research and the relatively paltry return on this investment in terms of real improvements in human health (Butler 2008). The gradual evolution of the translational approach has left diverse interpretations and models for its implementation but these converge on some broad, practical principles for solutions-focused research (Box 2). Peters et al. (2019) present a translational framework that incorporates these principles for solutions-focused wildlife health research.

Box 2 The Principles of Translational Research for Health

Translational approaches to health research are diverse but broadly converge on the following principles.

- The research program is focused on characterization of the nature of a problem and the creation of real world, effective solutions to that problem.
- Research engages transdisciplinary teams.
- End users and stakeholders are meaningfully engaged in all phases of research.
- Basic and applied research inform each other in a bidirectional continuum that nevertheless progresses to a final research phase focused on real-world deployment of solutions.
- Increased investment and focus on later phases of research that may be less innovative but more relevant to the application of solutions.
- Solution-focused research programs have independent governance, where priorities are not predominantly set by individual researchers or research teams.

4 Operationalizing Sociological Investigation of Wildlife Disease

It is not a cursory process to employ the principles and tools described here in investigating the social dimensions of wildlife health but, rather, is best achieved through long-term relationship building with diverse disciplinary experts, with community and with other stakeholders. However, with good intent, goodwill, and good leadership (see Chap. 22), the basis of a translational, transdisciplinary investigative approach that meaningfully engages relevant social stakeholders can be established for even urgent wildlife health challenges. The approach is scalable and can evolve, potentially initially involving collaboration with only a single colleague with relevant social science expertise and with direct, informal engagement with stakeholders on the ground. As the problem becomes clearer, further disciplinary expertise can be brought in and a more structured approach to stakeholder participation can be implemented. Once a wildlife health problem has emerged as a sociopolitical priority there is a need to establish robust, independent governance of research programs and to implement a translational framework to best achieve effective solutions out of the research endeavor.

References

Albrecht G, Sartore G-M, Connor L, Higginbotham N, Freeman S, Kelly B et al (2007) Solastalgia: the distress caused by environmental change. Australas Psychiatry 15(sup1):S95–SS8

Bennett NJ, Roth R, Klain SC, Chan K, Christie P, Clark DA et al (2017) Conservation social science: understanding and integrating human dimensions to improve conservation. Biol Conserv 205:93–108

Blehert DS, Hicks AC, Behr M, Meteyer CU, Berlowski-Zier BM, Buckles EL et al (2009) Bat white-nose syndrome: an emerging fungal pathogen? Science 323(5911):227

Butler D (2008) Translational research: crossing the valley of death. Nature 453:840–842

Cafaro P (2015) Three ways to think about the sixth mass extinction. Biol Conserv 192:387–393

Cherry E (2018) Birding, citizen science, and wildlife conservation in sociological perspective. Soc Animals 26(2):130–147

Cole HV, Anguelovski I, Baró F, García-Lamarca M, Kotsila P, Pérez del Pulgar C et al (2020) The COVID-19 pandemic: power and privilege, gentrification, and urban environmental justice in the global north. Cities Health 2020:1–5

Cossyleon JE, Spitz G (2021) Collaborating, then stepping back: doing public sociology through participatory action research. In: Hossfeld L, Kelly EB, Hossfeld C (eds) The Routledge international handbook of public sociology. Routledge, London, pp 69–78

Daszak P, Cunningham AA, Hyatt AD (2001) Anthropogenic environmental change and the emergence of infectious diseases in wildlife. Acta Trop 78(2): 103–116

Davis LF, Ramírez-Andreotta MD, Buxner SR (2020) Engaging diverse citizen scientists for environmental health: recommendations from participants and promotoras. Citizen Sci Theory Pract 5:1

Dickens P (1992) Society and nature: towards a green social theory. Temple University Press, Philadelphia

Field HE (2016) Hendra virus ecology and transmission. Curr Opin Virol 16:120–125

Galway LP, Beery T, Jones-Casey K, Tasala K (2019) Mapping the solastalgia literature: a scoping review study. Int J Environ Res Public Health 16(15):2662

Germov J (2018) Imagining health problems as social issues. In: Germov J (ed) Second opinion: an introduction to health sociology: an introduction to health sociology. Oxford University Press, Oxford, pp 2–23

Germov J, Poole M (2007) The sociological gaze: linking private lives to public issues. In: Germov J, Poole M (eds) Public sociology: an introduction to Australian society. Allen and Unwin, Crows Nest, NSW, pp 3–18

Hacker K (2013) Community-based participatory research. Sage, Los Angeles

Heffner GG, Zandee GL, Schwander L (2003) Listening to community voices: community-based research, a first step in partnership and outreach. J Higher Educ Outreach Engage 8(1):127–139

Holifield R, Chakraborty J, Walker G (2017) The Routledge handbook of environmental justice. Routledge, Abingdon

Humboldt-Dachroeden S, Rubin O, Frid-Nielsen SS (2020) The state of one health research across disciplines and sectors: a bibliometric analysis. One Health 10:100146

Macionis JJ, Plummer K (2005) Sociology: a global introduction, 3rd edn. Pearson, Harlow

Manyweathers J, Field H, Longnecker N, Agho K, Smith C, Taylor M (2017) "Why won't they just vaccinate?" horse owner risk perception and uptake of the Hendra virus vaccine. BMC Vet Res 13:1

Matthewman S, Huppatz K (2020) A sociology of Covid-19. J Sociol 56(4):675–683

Milich KM, Sorbello K, Kolinski L, Busobozi R, Kugonza M (2021) Case study of participatory action research for wildlife conservation. Conserv Sci Pract 3(2):e347

Mills CW (2000) The sociological imagination. Oxford University Press, Oxford

Mohai P, Pellow D, Roberts JT (2009) Environmental justice. Annu Rev Env Resour 34:405–430

Moodie RL (1917) Studies in paleopathology. I. General consideration of evidence of pathological conditions found among fossil animals. Ann Med Hist 1(4): 374–393

Mysterud A, Rolandsen CM (2018) A reindeer cull to prevent chronic wasting disease in Europe. Nat Ecol Evol 2(9):1343–1345

Odeniran PO, Ademola IO, Jegede HO (2018) A review of wildlife tourism and meta-analysis of parasitism in Africa's national parks and game reserves. Parasitol Res 117(8):2359–2378

Ohmer MEB, Alton LA, Cramp RL (2021) The role of conservation physiology in mitigating social-ecological traps in wildlife-provisioning tourism: a case study of feeding stingrays in the Cayman Islands. In: Madliger CL, Franklin CE, Love OP, Cooke SJ (eds) Conservation physiology: Applications for wildlife conservation and management. Oxford University Press, Oxford. https://doi.org/10.1093/oso/9780198843610.003.0010

Panu P (2020) Anxiety and the ecological crisis: an analysis of eco-anxiety and climate anxiety. Sustainability 12(19):7836

Peggs K (2012) Animals and sociology. Palgrave Macmillan, Houndsmill

Peters A, Carver S, Skerratt LF, Meredith A, Woods R (2019) A solutions-focused translational research framework for wildlife health. Bioscience 69(12): 1019–1027

Phoofolo P (1993) Epidemics and revolutions: the rinderpest epidemic in late nineteenth-century southern Africa. Past Present 138:112–143

Rizzolo JB (2017) Exploring the sociology of wildlife tourism, global risks, and crime. In: Gore ML (ed) Conservation criminology. Wiley, Chichester, pp 133–154

Scheele BC, Pasmans F, Skerratt LF, Berger L, Martel A, Beukema W et al (2019) Amphibian fungal panzootic causes catastrophic and ongoing loss of biodiversity. Science 363(6434):1459–1463

Stratford E, Mazur N, Lunney D, Bennett D (2000) Managing the koala problem: interdisciplinary perspectives. Conserv Biol 14(3):610–618

Tablado Z, D'Amico M (2017) Impacts of terrestrial animal tourism. In: Blumstein D, Geffroy B, Samia D, Bessa E (eds) Ecotourism's promise and peril. Springer, Cham, pp 97–115

Tweddle NE, Livingstone P (1994) Bovine tuberculosis control and eradication programs in Australia and New Zealand. Vet Microbiol 40(1):23–39

Van Krieken R, Habibis D, Smith P, Hutchins B, Haralambos M, Holden R (2006) Sociology: Themes and perspectives, 3rd edn. Pearson Longman, Sydney

Vera H (2016) Rebuilding a classic: The social construction of reality at 50. Theory Cult Soc 10(1):3–20

Whittington RJ, Crockford M, Jordan D, Jones B (2008) Herpesvirus that caused epizootic mortality in 1995 and 1998 in pilchard, *Sardinops sagax neopilchardus* (Steindachner), in Australia is now endemic. J Fish Dis 31(2):97–105

Yearley S (2002) The social construction of environmental problems: a theoretical review and some not-very-Herculean labors. In: Dunlap RE, Buttel FH, Dickens P, Gijswijt A (eds) Sociological theory and the environment: Classical foundations, contemporary insights. Rowman and Littlefield, Lanham, pp 274–285

Evidence-Based Decision-Making

Craig Stephen

Abstract

All forms of population health management are increasingly expected to base their decisions on evidence. But the reality is that few people make their decisions on scientific evidence alone. Wildlife population health practitioners need to distill evidence from research, context, and experience and use that evidence to inform and improve decisions. There are three general types of evidence needed for health management decisions; (1) evidence specific to the decision-making social, biological, and ecological context; (2) evidence extracted from other settings or situations; and (3) evidence pertaining to the values and expectations of the decision makers and those affected by the decision. Evidence-based decision-making is not merely using formal research findings to support your position. It is about having skills in weighing various types of evidence, facilitating appropriate use of evidence, combining various sources of evidence, and providing supplementary evidence appropriate to the local context. Wildlife population health practitioners need to embrace the reality of how decisions are made and develop collaborative and transparent processes to ensure a shared view that is tailored to the decision-making context on what is known about a problem and the implications of different action options.

Keywords

Evidence · Decision making · Wildlife · Health · PICOT · Standards

1 Introduction

Wildlife health management constantly requires decisions to be made. There are basically two parts to making a decision. First, a person must judge the situation needs a decision. Next, that person needs to make a choice between alternative options to fulfill the intent of the decision. Almost every wildlife health management decision occurs in the face of gaps, uncertainties and conflicts in knowledge, values, and expectations. Almost every decision is distinct because of the uniqueness of the social and ecological context of the problem at hand. Many of the decisions are made under pressure.

Many factors influence population health decisions including research outputs, community and political expectations and values, economic information, and the local context. Population health practitioners need to be adept at distilling the best available evidence from research, context, and experience and use that evidence to inform and improve decisions. The weight and influence of each factor depend on the skills,

C. Stephen (✉)
School of Population and Public Health, University of British Columbia, Vancouver, BC, Canada

C. Stephen (ed.), *Wildlife Population Health*,
https://doi.org/10.1007/978-3-030-90510-1_8

experience, and values of the decision maker. How then can society be assured that decision-making is being done in a consistent, logical, and fair way regardless of the circumstances? This chapter presents some core concepts, perspectives, and approaches to evidence-based decision-making that can help wildlife health professionals produce more transparent, reliable, and consistent decisions.

2 How Do We Really Make Decisions?

Do decision makers use evidence in an objective and balanced way or do they only select evidence that supports their view or distort evidence to suit the needs? How do political agendas, personal gains, or cultural values influence people's willingness to use the evidence available to them, without prejudice? Authors in conservation, public health, and management have all commented on the common failure of decisions makers to make rational decisions using evidence-based principles and approaches (e.g., Baba and HakemZadeh 2012; Pullin and Knight 2003; Brownson et al. 1999). They attribute this to the limited, faulty, and biased decision process most people naturally use. What people use as evidence to generate their decisions is a function of their experiences, education, training, and personal judgment (Baba and HakemZadeh 2012).

Some people believe that bad decisions can be avoided if proper information and incentives are given. By centralizing fact-based, politically neutral information that is equally and easily made available to everyone, better decisions should be made. Others believe that information is inherently incomplete, that preferences and experience put limits on rationality, and that the chaotic and complex nature of our environments precludes rational decision-making (Zhu 2007). The job of the population health manager is to negotiate these different perspectives to help people make decisions that maximize benefits, avoid unintended consequences, and result in fair and effective actions.

When people have not learned what decision to make through trial and error, they need to be able to extract relevant information, apply general values in specific settings, and integrate these pieces using some sort of decision rule (Parker and Fischhoff 2005). Most often, our decision rules are innate and tacit rather than explicit, transparent, and systematic. Limits in time, cognitive abilities at the moment of choice, and the decision-making environment lead people to knowingly or unknowingly use mental shortcuts (known as heuristics) when making a decision. Researchers have identified four prominent decision-making heuristics (Peters et al. 2006). The Affect Heuristic is the rapid, automatic reaction we have to the "goodness" or "badness" of a particular situation or decision. It is affected by our moods, personal experiences, social norms, and personal risk tolerance. In the Representativeness Heuristic, we estimate the likelihood of an event or object by comparing it to what we think of as the typical for these types of events or objects. In this case, people believe that two similar things or events are more closely correlated than dissimilar things. This heuristic is used commonly when using pattern recognition to make a diagnosis (i.e., this case looks like the textbook case and therefore its treatment should be like the textbook treatment). The Availability Heuristic describes the tendency to use quickly and easily remembered information to which you have more recently been exposed when making decisions about the future. The frequency of more vividly and easily remembered events will be overestimated, while the frequency of ordinary or difficult to recall events underestimated. The fourth heuristic, Anchoring and Adjustment, refers to where a person starts with an initial idea and adjusts their beliefs based off this starting point. For example, when discussing a budget, the decision maker may anchor her/his initial assessment on last year's budget and make decisions to change (adjust) relative to the initial anchor. These heuristics have a strong and significant influence on the assessment and judgment of decision options (McCaughey and Bruning 2010). When heuristics are aligned with significant experience, consensus, and evidence, they can guide efficient and reasonable decision-making. When poorly aligned, they can lead to errors in judgment (Kulkarni et al. 2019).

Population health practitioners need to acknowledge and work with these heuristics. The reality that these heuristics commonly affect the use of evidence needs to be balanced with the good intentions of those who seek to take ideology and politics out of the decision-making by bringing evidence into the decision-making process. Ensuring decision makers are given information that clarifies the similarity (or not) of one situation to a comparator, helping set accurate and reasonable anchors, and highlighting facts relevant to the decision maker (facts that often extend beyond scientific data) are all strategies that can help improve decisions made by those reliant on a heuristics approach to decision-making. Being alert to when these heuristics are influencing decisions can help to tactfully tailor conversations that promote more transparent and systematic decision-making processes.

3 What Is Considered Evidence in Evidence-Based Decision-Making?

Most people support the notion that wildlife health management decisions should be based on the explicit, consistent, and thoughtful use of evidence. Few would argue that it would be wrong to base actions on irrefutable evidence that has been rigorously and systematically assessed for relevance, reliability, and effectiveness before acting. But what do we mean by evidence? For some people, evidence is restricted to knowledge produced in accord with the standards of a relevant academic discipline. For others, evidence can be knowledge, skills, and practices developed by and sustained between generations within a community. Still others consider evidence as what they observe and document using their own senses. How law views evidence is not necessarily the same as an epidemiologist which may not be the same as an experimental parasitologist. A 2009 survey of over 200 human health care decisions makers found, "In spite of almost universal support in principle for using evidence in decision-making, there was little consensus among participants on what evidence is, what kind of evidence is most appropriate and how 'using evidence' can best be demonstrated" (Bowen et al. 2009). Different people with different experiences from different disciplines accept different types and standards of evidence upon which to make decisions. Chapter 6 highlighted how the evidence needed to prove cause and effect has changed over time and how debate remains about the relative reliability and utility of different criteria of causation.

There are two main questions to ask when setting out to use evidence-based decision-making. First, what types of evidence need to be considered? Second, how do I recognize good evidence? The answer to both questions is, "it depends." An important first step, therefore, is to explicitly establish the evidence needs for the specific decision-making context. Table 1 illustrates how population health management principles can help tailor the selection of evidence relevant to a situation. Given the lack of published wildlife population health management principles, these were extracted and modified from several sources including Wittrock et al. (2019), Bhattacharya and Bhatt (2017), Ibrahim et al. (2001), Radostits et al. (1994), and FAO (1991).

Table 1 helps us to see there are three general types of evidence needed: (1) evidence specific to the decision-making social and ecological context; (2) evidence extracted from other settings or situations; and (3) evidence pertaining to the values and expectations of the decision makers and those affected by the decision. It also helps us to see that evidence need not be restricted to peer-reviewed scientific outputs. Information on factors such as resource availability, political context, values, community experience, and available expertise should all be assembled, critically assessed, and presented in an integrated fashion.

The breadth of relevant information can be overwhelming, especially for complex problems involving multiple populations, conflicting information, and differing desired goals. A simple mnemonic—PICOT—serves as a framework

Table 1 Wildlife population health management principles as guidance for selecting relevant evidence for evidence-based decision-making

Principles for policies, programs, and practices	Guidance for selecting evidence for decision-making
Be outcome-based	Priority outcomes need to be identified through health needs assessments and input from affected communities to develop a shared vision of the preferred health outcomes, how they are measured and thresholds of acceptability to select evidence relevant to the management goal and those making and affected by the decisions
	There must be clarity on the population(s) being considered and the ecological or social expectations that determine the appropriate outcomes so the evidence selected can be tailored to that context
Be evidence-based	There must be an accepted standard for the nature of biological, ecological, and social evidence to be considered when making decisions and how to accommodate varying levels of reliability, validity, uncertainty, and representativeness of the evidence
	Evidence should come from both normative and empirical inquiries as required to support all component considerations of the decisions to be made
Emphasize all levels of prevention consistent with the population health model involving the determinants of health	Evidence collected should be attentive to how to prevent problems; manage long-term harm or premature losses, mitigate harmful events resulting in population impacts, reduce risk factors leading to harmful events, and prevent other populations from becoming at risk
	Evidence collected must be relevant to the variety of mechanisms to achieve outcomes including policies, plans, collaborations, and courses of action that may be required by law or expected by social norms
Be adaptable to heterogeneity resulting from different ecological, biological, and social conditions	There must be a balance between standardization of evidence used and customization of evidence to reflect the unique circumstances of the decision-making context
Be attentive to unanticipated or secondary consequences on subsections of the population(s) of concern and other ecological, biological, or social attributes impacted by health management programs	Evidence collected should consider the health status or access to health determinants for the entire population and those populations unintentionally or intentionally impacted by health management decisions

that can help focus evidence searches or production:

P = what is the Population of interest and what is the Problem of concern.
I = what actions and Interventions are of Interest.
C = what are the appropriate Comparators to this situation that can provide evidence.
O = what are the desired Outcomes.
T = what is the appropriate Time interval for decision-making (Melnyk et al. 2010).

Box 1 is a case study demonstrating the implications of inattention to PICOT.

Box 1 Inattention to the PICOT Mnemonic Prevents Program Evaluation—an Avian Influenza Surveillance Case Study

Case context: A national wild bird surveillance program delivered by a collaboration of multiple agencies faced funding cuts because of perceived irrelevance to risk management decision-making. The surveillance system was designed in 2005 at the height of interest and concern about wild birds being the source of highly pathogenic avian influenza viruses that put agriculture

(continued)

and public health at risk. Funders in 2016 began questioning how wild bird data was helpful.

Why inattention to PICOT prevented making the value proposition for the program.

P (population) = The funders had vaguely asked for "wild waterfowl" to be kept under surveillance without any guidance on which populations (species or geographic locations) or if live or dead birds were of greatest concern or if the problem of interest was documentation of the presence of the virus or potential exposure pathways to poultry. This led to inconsistent and highly variable surveillance approaches across the country.

I (interventions and interests) = There were no thresholds set for actions or interventions nor any means established to document the impact of viral detection on risk management or risk perception, precluding evaluation of the impact of the program on biosecurity decisions by poultry farmers or the implementation of other risk avoidance interventions.

C (comparators) = There were no compactors to determine if a trend in wild bird viral isolates reflected a true change in risk. This reduced the program to being deemed of value largely if a highly pathogenic strain could be detected because highly pathogenic strains were reportable by legislation and because the meaning of detection of low pathogenic isolates affected risk was unclear.

O (outcomes) = While the funder declared the outcome was to provide an early warning, they had no threshold established to determine if the findings were sufficiently early, who was to be warned, what factors they deemed were warning variables, or what criteria could confidently be used to declare that a situation was safe.

T (time) = The time delays between sample collection and viral strain confirmation were so long that affected waterfowl population had moved far from the collection location thus complicating conclusions about the risk status of the collection location.

Result: The lack of "PICOT clarity" in the program design phase left the agency responsible for program coordination without criteria to provide evidence that the program was working.

PICOT helps to show that more than one type of evidence is needed for decision-making. There is evidence derived from systematic and well-designed research, evidence from those experienced with similar circumstances, and evidence about the local context. This evidence can be formal, explicit, derived from research and scholarship and concerned with generalizability or can be informal, implicit, and derived primarily through experience. All these types of evidence need to be independently observed and verified and subject to open scrutiny (Melnyk et al. 2010).

It is important early in the decision-making process to establish a shared vision of the criteria for determining what constitutes best science, how conflicting evidence should be weighed and assessed, and the agreed-to standards of quality required to accept or reject evidence. Answering these questions will help to develop a shared understanding of the quality, meaning, reliability, or utility of the body of evidence upon which to make decisions. Imposing your criteria for selecting the best evidence without accounting for the need of decision makers can doom your efforts to fail. This is not to suggest that you compromise to a lesser quality of evidence to make decisions but instead, there needs to be a process of talking to and listening to each other to increase the likelihood that good quality evidence that is relevant to the needs of the decision makers is produced, thus increasing the efficiency and effectiveness of evidence-based decision making. Box 2 is a cautionary tale of the implications of failing to establish agreement on what constitutes good evidence.

Box 2 The Pitfalls of Inattention to Consensus on What Constitutes Evidence—A Case Study of Wild-Farmed Fish Conflicts

Case context: A multidisciplinary working group involving government agencies (from four departments), nongovernmental agencies, and community groups was convened to identify gaps, opportunities, and obstacles to setting and meeting farmed fish health regulations intended to prevent the transmission of infectious disease from farmed to wild fish.

Failure to establish rules of evidence doomed the process to failure.

The group did not undertake an initial process to establish agreement on the attributes, indicators, or thresholds that would be used to define a healthy population. In the absence of a shared vision, it was not possible to reach consensus on the objectives and outcomes of regulatory requirements that would lead to successful fish health management. There were no explicit criteria developed on what constituted best science, how conflicting evidence would be weighed and assessed, or the standards of quality required to accept or reject evidence. Evidence was presented via personal experience and interpretation rather than discussion or presentation of literature or data in the context of agreements on the use of evidence by the group. Social evidence was anecdotal and weighted inconsistently. No rules were developed to guide how to deal with uncertain or conflicting evidence. There were also no agreed to criteria for establishing causal relationships. These conditions limited the ability of the group to develop a shared understanding of the quality, meaning, reliability, or utility of the body of evidence upon which to make recommendations.

Result: No agreement could be made on the best practices for risk reduction largely due to conflicting interpretations of available evidence. Disagreements on the meaning of the available evidence were aired in the media, eroding confidence in the few items upon which the group agreed.

There are a variety of standards and hierarchies for assessing the quality of evidence. Most have been developed for use in decisions about people and cannot readily be used for free-ranging wildlife. For example, randomized clinical trials have for decades been viewed as a gold standard for making decisions about medical interventions. The expectations for randomized trial, however, can rarely be met in wildlife (see Chaps. 4 and 6). Similarly, wildlife health systematic reviews and meta-analyses can face challenges in finding enough studies of sufficient quality to allow their comparison and integration. Agreed standards for determining whether research evidence is appropriate and useful for a particular wildlife health decision and how it can be used have yet to be developed. Wildlife population health practitioners can still, however, systematically gather evidence from multiple sources for verification and quality assessment.

Evidence sources should be assessed for their validity, quality, and reliability (based on the most appropriate criteria for the nature of evidence being reviewed), whether the evidence supports or negates a proposition or conclusion, the size or strength of the reported effect, and the relevance to the specific hypothesis and situation (Salafsky et al. 2019). Baba and HakemZadeh (2012) build on these criteria when claiming that evidence is strong when (1) it fits the decision context; (2) its methods of production and analysis are transparent; (3) it includes more contextual information; (4) its method of production can be replicated; and (5) there is consensus about the evidence.

Often, practitioners and researchers will need to rely on evidence triangulation. This is the process of using more than one approach to researching a question to increase confidence in the findings through the confirmation using two or more independent measures (Heale and Forbes 2013). Triangulation can result in three outcomes: (1) the results may *converge*; (2) the results may be *complementary* to each other; and (3) the results may be *divergent* or *contradictory*. Convergence increases validity through verification, complementarity highlights different aspects of the issue under study. Divergence drives you to look for new and better explanations for the phenomenon under investigation.

4 The Process of Evidence-Based Decision-Making

Evidence-based decision-making is not merely using formal research findings and quantitative data to support your position. It is about having skills in weighing various types of evidence, facilitating appropriate use of evidence, combining various sources of evidence, and providing supplementary evidence appropriate to the local context (Bowen et al. 2009). Human medicine has led the development of evidence-based decision-making principles. Initially focused on decisions about the care of individuals, these principles have been expanded to public health, management, and conservation. Evidence-based medicine integrates the best research evidence with clinical expertise and patient values. Evidence-based medicine generally answers five questions (Porzsolt et al. 2003; Akobeng 2005).

Question 1—What is the question? The PICOT mnemonic can help organize the array of information associated with a decision situation to help formulate an answerable question to guide the evidence search. Because the goal is to inform a management decision as opposed to finding the most interesting scientific question, this step must be done in a collaborative manner involving both the evidence producers and evidence users.

Question 2—Where is the evidence? Evidence produced through experience and training needs to be complemented with evidence available in others' experience, the scientific literature, textbooks, and other external source. This step not only looks for the individual pieces of evidence but also sees the relationships and connections between the various pieces. Evidence searches aim to maximize the potential of retrieving information within a timespan suited to the pressing decision.

Question 3—How good is the evidence? Evidence should be appraised for its validity, importance, and applicability to the problem and population of concern. This process could validate and strengthen confidence in the decision, show that little evidence exists to support the decision, or that the available evidence is equivocal.

Question 4—Can the evidence be applied to the decision circumstance? This is the step in which an "evidence alliance" is sought. This occurs when there is high-quality information that fits with the values, resources, and needs of the circumstance and leads to a clear direction to address the shared decision goal. The population health practitioner is neither paternalistic nor passive but acts more as a guide and coach to help those making the decision and being affected by the decision come to a shared understanding of what the evidence can tell them in their situations.

Question 5—Was this helpful? Evidence-based decision-making includes looking for evidence that the decision made was implemented, helpful, equitable, effective, acceptable, maintained, and reached the intended knowledge users. Evaluation is an important part of the process.

While the five questions above seem logical, and on their surface achievable, this process of evidence-based decision-making is not without its critics. Proponents of evidence-based decision-making have focussed on making information more available, accessible, and attractive to decision makers. This assumed that the major barriers to decision makers' use of evidence are data availability, accessibility, and user capacity. This is, perhaps, a simplistic assumption. The problems with evidence-based decision-making in human health (see Straus and McAlister 2000) are amplified for wildlife. There is often a shortage of coherent, consistent evidence. While there is an information explosion going on, there is a fair share of evidence that will be irrelevant to your decision context, of poor quality, inconclusive, or inconsistent with other studies. Biological variation and the diversity of ecological and social situations confronting decision makers may make it hard to extrapolate available evidence to a certain context. Job pressure such as lack of time and resources, political pressures, or the lack of evidence in novel situations might prevent people from using evidence-based decision principles. Some people may not be trained in evidence extraction, appraisal, integration, and translation. All these challenges multiple when we shift from making relatively straightforward decisions to dealing with complex and wicked problems.

4.1 Wicked Problems and Decision-Making

Many wildlife health issues share characteristics with wicked problems such as each occurrence of a problem is unique in its own respect, the range of causal factors can be uncertain and can be very wide as can the suite of possible solutions, many people and organizations are involved or affected, problems are multifaceted, it takes time and resources to get to root issues driving the problem, and fixing one aspect of the problem can create new problems. In these cases, there is no single, elegant, rational, scientific finding that will point to the right decision to fix the problem. Rather, we need to assume that no one has *the* solution because this is a complex systems problem involving different biological and social subproblems caused by multiple factors. In these situations, the population health practitioner might switch from being the one to provide the solution, to the person who helps develop a collective view of the questions that need to be asked to allow for incremental improvements on aspects of the problem for which there is agreement and to promote a collective responsibility to act before we have perfect knowledge of the situation (Grint 2010).

It is important to understand the underlying assumptions people use when seeking solutions to complex problems. This helps to anticipate the consequences of the proposed solutions. Ulrich (2005) provided some helpful questions one may want to ask when reflecting on and discussing specific solution to a problem (Table 2). Dialogue among stakeholders and decision makers on these questions can provide a broader understanding of the proposed solution to a wildlife health problem.

Table 2 Probing questions to promote reflection and discussion of proposed interventions for wildlife health problems (adapted from Ulrich 2005)

Theme	Questions
Motivation	What are the purpose and intended consequences of the intervention? Who is the beneficiary of the intervention and whose interests are being served? What does success look like and who decides?
Power	Who is the decision maker? Who can make the desired changes and control the conditions for success? What conditions for success can/should the decision maker not control?
Knowledge	What counts as relevant knowledge? Who is seen as a competent provider of experience and expertise? Where do those involved seek some guarantee that improvement will be achieved?
Legitimacy	Who is seen as a legitimate stakeholder, and who argues for those stakeholders who cannot speak for themselves, including future generations and nonhuman nature? How is legitimacy determined? What different visions of "improvement" are considered, and how are different visions reconciled?

5 Summary

At their core, wildlife health management decisions are ultimately personal judgments involving practical reasoning of multiple types of biological and social evidence filtered through personal experience and beliefs. Population health practitioners can improve the quality of decisions made by:

1. Creating a shared vision of the decision-making goals,
2. Using processes that are aware of and inclusive of the current social realities of the decision-making situation,
3. Being attentive to how the decisions can be implemented in practice,
4. Enabling open, collaborative, and consistent collection, interpretation, and sharing of the necessary types of critically assessed evidence.

This chapter argues against the unfounded ideal that a single, or even set, of scientific studies will ensure decisions makers make the right choice. Instead, it advocates for population health practitioners to embrace the reality of how decisions are made and develop collaborative and transparent processes to ensure a shared view that is tailored to the decision-making context on what is known about a problem and the implications of different action options. More importantly, it advocates for openness to what we consider as evidence when promoting evidence-based wildlife health management.

References

Akobeng AK (2005) Principles of evidence based medicine. Arch Dis Child 90(8):837–840

Baba VV, HakemZadeh F (2012) Toward a theory of evidence based decision making. Manag Decis 50:832

Bhattacharya D, Bhatt J (2017) Seven foundational principles of population health policy. Popul Health Manag 20(5):383–388

Bowen S, Erickson T, Martens PJ, Crockett S (2009) More than "using research": the real challenges in promoting evidence-informed decision-making. Health Policy 4(3):87

Brownson RC, Gurney JG, Land GH (1999) Evidence-based decision making in public health. J Public Health Manag Pract 5:86–97

FAO (1991) Guidelines for strengthening animal health services in developing countries. http://www.fao.org/3/u2200e/u2200e00.htm#Contents

Grint K (2010) Wicked problems and clumsy solutions: the role of leadership. In: The new public leadership challenge. Palgrave Macmillan, London, pp 169–186

Heale R, Forbes D (2013) Understanding triangulation in research. Evid Based Nurs 16(4):98

Ibrahim MA, Savitz LA, Carey TS, Wagner EH (2001) Population-based health principles in medical and public health practice. J Public Health Manage Pract 7(3):75–81

Kulkarni SS, Dewitt B, Fischhoff B, Rosengart MR, Angus DC, Saul M, Yealy DM, Mohan D (2019) Defining the representativeness heuristic in trauma triage: a retrospective observational cohort study. PLoS One 14(2):e0212201

McCaughey D, Bruning NS (2010) Rationality versus reality: the challenges of evidence-based decision making for health policy makers. Implement Sci 5(1):39

Melnyk BM, Fineout-Overholt E, Stillwell SB, Williamson KM (2010) Evidence-based practice: step by step: the seven steps of evidence-based practice. Am J Nurs 110(1):51–53

Parker AM, Fischhoff B (2005) Decision-making competence: external validation through an individual-differences approach. J Behav Decis Mak 18(1):1–27

Peters E, McCaul KD, Stefanek M, Nelson W (2006) A heuristics approach to understanding cancer risk perception: contributions from judgment and decision-making research. Ann Behav Med 31(1):45–52

Porzsolt F, Ohletz A, Thim A, Gardner D, Ruatti H, Meier H, Schlotz-Gorton N, Schrott L (2003) Evidence-based decision making—the six step approach. BMJ Evid-Based Med 8(6):165–166

Pullin AS, Knight TM (2003) Support for decision making in conservation practice: an evidence-based approach. J Nat Conserv 11(2):83–90

Radostits OM, Leslie KE, Fetrow J (1994) Herd health: food animal production medicine. WB Saunders Company, New York

Salafsky N, Boshoven J, Burivalova Z, Dubois NS, Gomez A, Johnson A, Lee A, Margoluis R, Morrison J, Muir M, Pratt SC (2019) Defining and using evidence in conservation practice. Conserv Sci Pract 1(5):e27

Straus SE, McAlister FA (2000) Evidence-based medicine: a commentary on common criticisms. CMAJ 163(7):837–841

Ulrich W (2005) A brief introduction to critical systems heuristics (CSH). ECOSENSUS project site. https://i2s.anu.edu.au/wp-content/uploads/2014/02/ulrich_2005f.pdf

Wittrock J, Duncan C, Stephen C (2019) A determinants of health conceptual model for fish and wildlife health. J Wildl Dis 55:285

Zhu Z (2007) Decision making in the real world. Proceedings of the 51st Annual Meeting of the ISSS, Tokyo, Japan. https://journals.isss.org/index.php/proceedings51st/article/view/456

Wildlife Health Surveillance and Intelligence. Challenges and Opportunities

Craig Stephen and John Berezowski

Abstract

Surveillance systems must be designed with a clear purpose in mind. There are many impediments to designing and implementing a wildlife health surveillance and intelligence system that successfully fulfills all the expectations for "good" surveillance. Wildlife health surveillance practitioners need to be acutely aware of how the compromises and biases that arise from the practical constraints of limited resources and knowledge influence what they can say about the information their surveillance systems generate. There are growing expectations to expand the wildlife surveillance from early detection of harms to quickly minimize their impacts toward health intelligence that strives to protect and promote health by early actions in advance of harm. Health intelligence must be developed to ensure surveillance outcomes meet the needs of decision makers. A series of generic considerations and questions are provided to guide the design of surveillance and intelligence systems tailored to local needs that are helpful in wildlife health management despite the many challenges wildlife surveillance and intelligence programs face.

C. Stephen (✉)
School of Population and Public Health, University of British Columbia, Vancouver, BC, Canada

J. Berezowski
Scotland's Rural College, Aberdeen, UK

Keywords

Surveillance · Intelligence · Information systems · Decision support · Vulnerability · Wildlife

1 What Is Health Surveillance and Intelligence?

Surveillance, in the broadest sense, simply means to keep a close watch over something. In the health sciences, surveillance is the systematic, continuous collection and evaluation of pertinent data that are promptly assessed and shared with those who need to know to launch an effective response (Fig. 1). Fish and wildlife health surveillance should be designed with the goal of improving population health by providing timely and reliable signals that inform decisions and actions to protect, manage, and maintain health.

Surveillance is the primary source of timely information about changing health and disease risks that are used for health decision-making. It supplements what is known from experience and research with an understanding of the current dynamics of health issues or situations. Its timeliness makes health surveillance information a critical component of the total information, knowledge, and wisdom needed to make effective health decisions. Research and experiential knowledge are essential for effective decision-making, but they take time and do not provide

C. Stephen (ed.), *Wildlife Population Health*,
https://doi.org/10.1007/978-3-030-90510-1_9

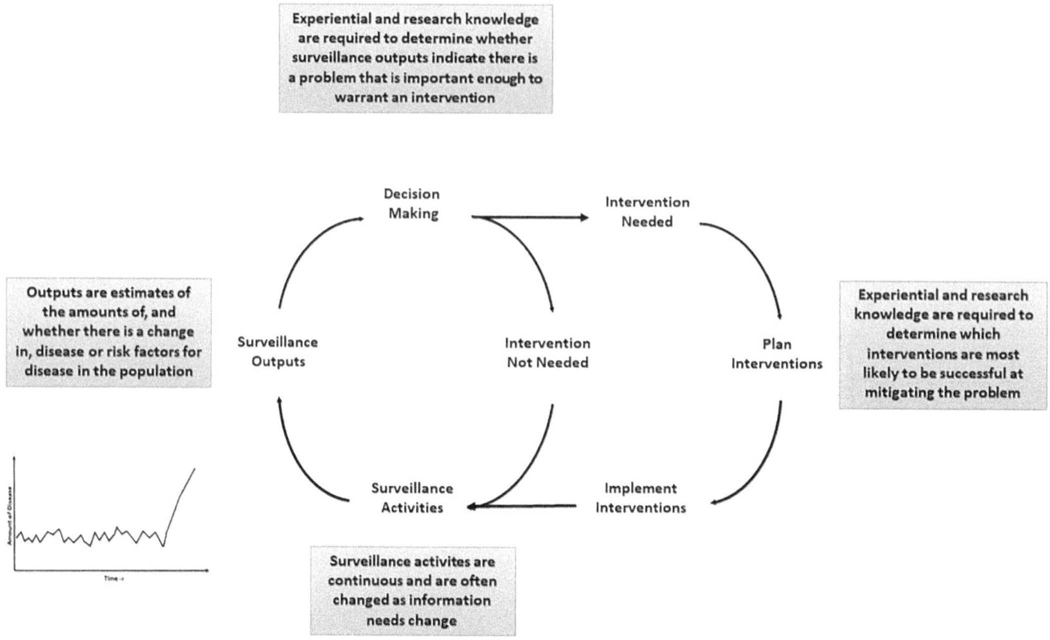

Fig. 1 Basic surveillance system structure and functions

the early warning needed to reduce the impact of rapidly changing health conditions. In contrast, surveillance provides continuous information about the current and changing disease/health situation in a population. It provides needed time for effective responses to be designed and implemented in a rapidly changing health situation.

Surveillance, in principle, is simple. It should generate information that tells health managers where the problem is, who or what is affected, if the problem is getting better or worse, and if interventions are making the desired change. Surveillance programs assess and characterize the burden and distribution of health events or risks, thus helping find priority problems, species, or areas. Surveillance can be used to measure the impact of interventions, identify emerging health conditions that may have a significant impact upon population health, identify risky settings or circumstances that need research or management, and evaluate the efficiency and effectiveness of wildlife health programs. Surveillance outputs should help decision makers target resources effectively to tackle the root causes of problems (Groseclose and Buckeridge 2017).

Surveillance is not a single event or an isolated process. It is an information system. Many pieces must be in place in this system to enable and sustain data collection, generation, and management, data quality monitoring, data analysis, interpretation of analytical results, and information sharing (Fig. 2). There must be adequate human resources to find and collect samples, analyze those samples with appropriate technology and methodologies, assess the resulting data to recognize important changes, and create and communicate outputs that influence actions. These human resources need to be supported with consistent and accessible infrastructure that must be created and maintained through appropriate legislation and fiscal resources. The cost of a large-scale, continuous surveillance system that targets the entire population at risk can exceed the resources of many wildlife health programs, thus forcing managers to make strategic decisions on how to best deploy their surveillance efforts.

Health intelligence is the process of creating knowledge products resulting from collecting and analyzing data, experience, and other learning in a way to make them understandable and usable for future decision-making (Jamot 2013) (Fig. 3).

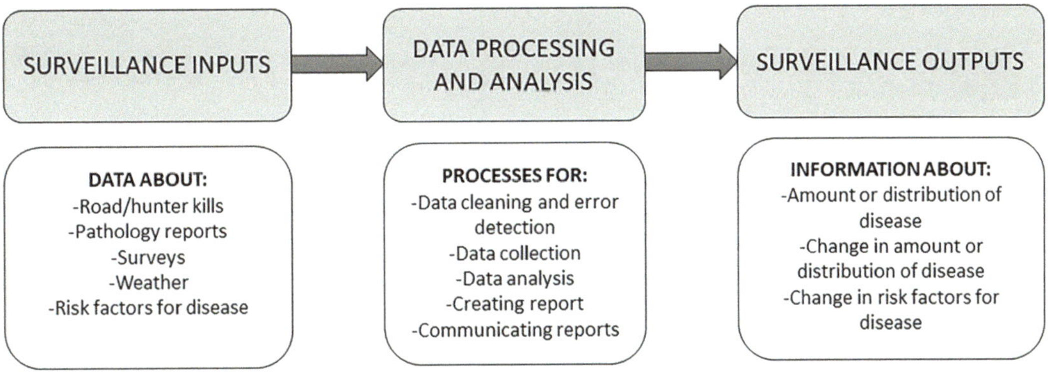

Fig. 2 Components of a wildlife surveillance information systems with example inputs, analyses, and outputs

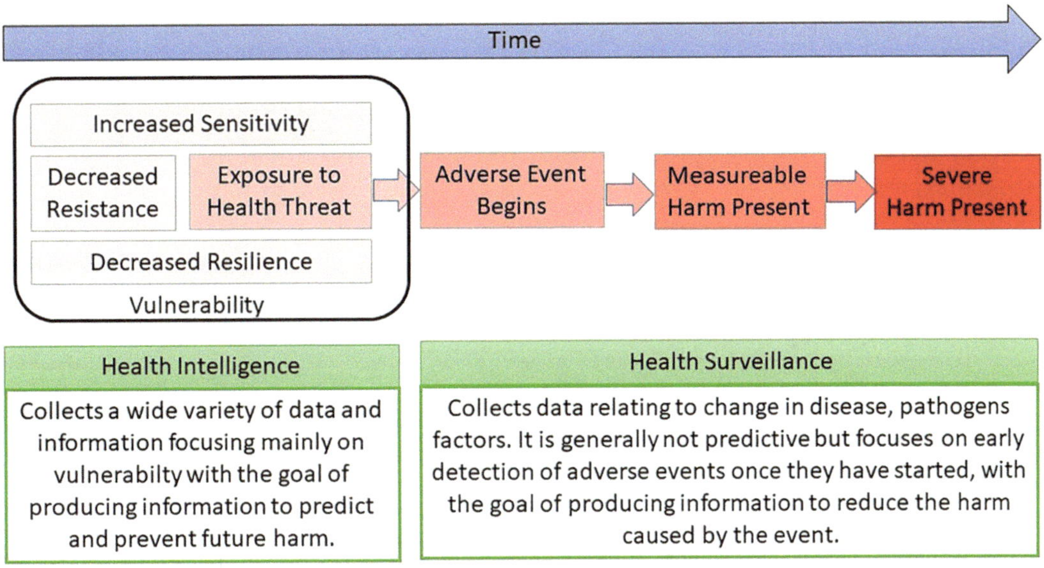

Fig. 3 Timeline of the occurrence of harm in a population (shades of red) and the locations where health surveillance and health intelligence collect data and produce information

The goal is to protect and promote health by early actions in advance of harm. This contrasts with the usual wildlife surveillance goal of early detection of harms to quickly minimize their effects. The health intelligence process begins by accessing and collecting observations of health outcomes, threats, or other indicators of vulnerability. Vulnerability is the product of the likelihood of sufficient exposure to a threat, sensitivity of a population to harm from that threat, and the capacity of the population to resistant, cope, and recover from those harms (Fig. 4). Analysis, organization, and presentation of these data help reveal information about trends or patterns in vulnerable places, populations, or situations. Placing this information into the context of the prevailing state of knowledge and future probabilities and scenarios adds understanding of the importance of intelligence information (also known as intelligence knowledge). Expert understanding and judgment of the context in which the knowledge is generated helps to provide an understanding of the significance of the intelligence knowledge.

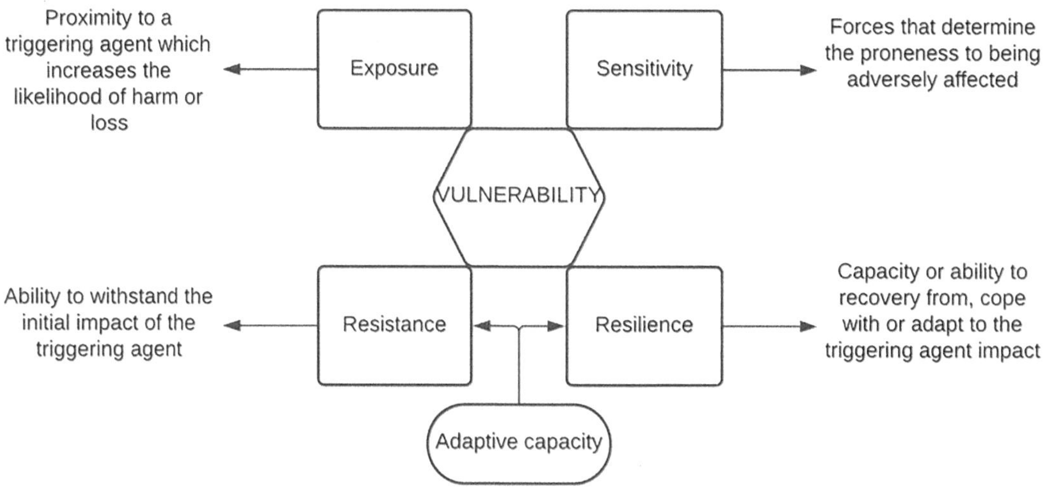

Fig. 4 The contributors to vulnerability

Decision support arises when the intelligence considers real-world opportunities, constraints, perspectives, and priorities for action, be these social and/or ecological (Berezowski et al. 2020). Intelligence products, therefore, consist of information and knowledge that have been refined to provide an understanding that meets the needs of decision makers. To serve this function, a health intelligence system needs to be able to find actionable and meaningful signals of change, provide insight into future risk and trajectories, and characterize possible opportunities for intervention. Intelligence, therefore, relies on information from multiple sources to provide a stream of information that can be inspected to support decisions about prevention, surveillance, or responses (Han and Drake 2016).

Health intelligence creates knowledge by building on surveillance of health outcomes and/or risk factors with reconnaissance of the populations of interest and the collection, evaluation, analysis, and interpretation of social and environmental information that are valuable for strategic planning. Health intelligence draws from a range of different sources (from the determinants of health through to population trends, to ecosystem functions, to social capacities for management) rather than an in-depth analysis of individual pieces of pathological or epidemiological data. Health intelligence aims to integrate research, routine data, and experience to support evidence-based decision-making (see Chap. 8) to improve health outcomes. Intelligence outputs help decisions makers act in the most appropriate way, resulting in the least harm and greatest benefit to as much of the population and as many of the stakeholders as possible (Berezowski et al. 2020).

The design and implementation of the intelligence portion of surveillance and intelligence systems often get less explicit attention and investment than the surveillance arm. Often, intelligence functions are done informally or implicitly, rather than as an explicit and regular part of a program. This is due in part to the historic focus of wildlife surveillance on documenting disease and risk patterns rather than on its fundamental role in decision support. Conversations on the design of surveillance systems tend to be very technically focused on the necessary set of activities for population sampling, sample testing, and epidemiological analysis to generate a signal. Health intelligence requires transdisciplinary teams that include a range of scientific and practical specialties as well as representatives of different parts of society who can share their knowledge to identify surveillance signals that are important and support effective and efficient decisions in terms of effects on wildlife, society, and the environment (Berezowski et al. 2020). Care and attention to designing this side of the surveillance and intelligence system warrant equal time in system design planning.

2 Designing a Surveillance and Intelligence System

We said earlier that surveillance, in principle, is simple. Surveillance and intelligence systems are, in practice, complex, involving different purposes, stakeholders, components, infrastructure, processes, policies, and regulations that are assembled and deployed differently depending on the surveillance context. This section offers some questions that can help design surveillance and intelligence systems that are adaptable to different settings, species, and issues. These questions provide the "simple" foundation upon which the "complex" system can be built.

2.1 Why Are We Doing This?

Surveillance and intelligence are purpose oriented. Their essential purpose is to create information and knowledge for health decision-making leading to action. The specific purposes vary with the situation of concern. All elements of system design and operations should be informed by its purpose and objectives. They set the scope of the system, informing issues such as who should participate and which methods are most appropriate. Explicit objectives provide a mechanism for evaluating the successes or shortcomings of a surveillance and intelligence system.

Different wildlife health problems require different surveillance purposes, objectives, information systems, and actions. Surveillance systems designers need to be aware of the epidemiological situation, what decisions need information and understanding, and what needs to be collected, analyzed, assessed, and communicated to aid in making those decisions (Tables 1 and 2). For example, a system intended to establish the

Table 1 Examples of how the epidemiological situations and desired actions influence the purpose of a surveillance system (adapted from Berezowski et al. 2020; Häsler et al. 2011)

Epidemiological situation	Desired action	Surveillance implications
A known etiologic agent/hazard of importance is not currently present in the population	Rapid response is required when the agent/hazard is introduced to eliminate the agent/hazard from the population or environment	Activities are targeted to finding cases conforming with the accepted case definition as quickly as possible
A new pathogen is discovered but its impacts are not known	Need to detect harms linked to this pathogen that warrant intervention	Activities need to be broad enough to detect any harms linked to the presence of this new pathogen
Occurrence of an epidemic of an important disease that escapes the current control activities	Insights are required on the most effective and efficient ways to bring the disease under control including finding high-risk populations or places and evidence on the effects of interventions	A variety of information (e.g., incidence, geographic distribution, species affected, changes in response to intervention) is needed to develop a new control strategy
Disease control activities are ongoing	Decisions need to be made about whether to continue with the current control activities or to consider new ones	Information about the changing amount and distribution of disease in the population must be produced with enough reliability and precision to detect if the disease is above or below the control target
A spreading threat exists elsewhere but has yet to be detected in the populations of concern	Opportunities to reduce vulnerability to the threat must be identified to prompt action before harms occur	Information is needed on changes in factors that influence the exposure, sensitivity, or adaptive capacities of the population(s) of concern

Table 2 An introductory taxonomy of some different types of health surveillance (adapted from Hoinville 2013)

Surveillance type	Purpose
General surveillance	Use general tests or signals (e.g., clinical examination or gross pathology) to detect a variety of threats or outcomes that might reveal a change in population health warranting further investigation or a timely impact assessment to determine if action is required E.g., A diagnostic laboratory receiving dead wildlife from the public for assessment
Early warning surveillance	Detect early signals of changing risk using health, social, or ecological indicators and/or disease occurrences in defined populations to increase the likelihood of timely detection of a new or unexpected threat E.g., virus detection in corvids as a warning for changing public health risk from West Nile virus in North America
Hazard specific surveillance (also known as targeted surveillance)	Focus on one or more predefined diseases or hazards and use agreed-upon case definitions to point towards the presence or absence of a threat and/or track their changing epidemiology over space and time among one or more hosts E.g. surveillance of dead waterfowl for highly pathogenic avian influenza
Syndromic surveillance	Uses health-related information that might precede a formal diagnosis to indicate a possible change in population health that deserves further investigation. This is not usually focused on a particular hazard so it can be used to detect a variety of diseases or etiologies including new (emerging) diseases E.g., surveillance for dead bats outside of a hibernaculum
Risk factor–based surveillance	Surveillance that targets: (1) populations exposed to factors that may predispose it to disease or other impacts; (2) subpopulations where host risk factors, the disease or hazards are most likely to be found; or (3) populations where the consequences of disease or exposure to the risk factor could be severe E.g., Peri-agricultural wild rodent surveillance to detect spill-over of antimicrobial resistance from domestic animals into the environment
Sentinel surveillance	Collection of information from the same selected sites or groups of animals that act as a proxy for the larger population to estimate patterns of disease occurrence in the larger population E.g., surveillance for deformities in groundfish in chemically contaminated environments to detect biological impacts of industrial pollutants

absence of a disease in a wildlife population to support claims of disease freedom will use different sampling protocols and case definitions than will a surveillance program trying to establish if an industrial pollutant is creating biological harms or a system intended to find the first incursion of a pathogen. To be effective, surveillance must create information about changing conditions at a rate that is appropriate for the decisions that need to be made. Making decisions for slowly changing processes (such as impacts of industrial pollutants) may require information at a much slower rate than fast-moving processes (such as an emerging infectious disease). Without clarity on why surveillance is being undertaken and what decisions it intends to inform, one cannot make decisions on priorities or methods, nor can one communicate to partners and stakeholders the value of the surveillance system.

Surveillance and intelligence objectives must be tempered with practical realities and pressures. Table 3 summarizes some considerations to make when asking if a desired surveillance system is possible within the existing operational circumstances. Often, the ideal surveillance system needs to be modified to be imperfect yet helpful within the preexisting opportunities and constraints to its implementation. For instance, while the purpose of an early warning system might be to find the index case of an outbreak in

Table 3 Themes of operational considerations and comments on their implication for wildlife health surveillance system design

Themes	Comments
Impacts	The issue under surveillance must be important enough to warrant management. Conservation, economic, social, and public perception criteria can determine an issue's importance as can scientific evidence and expert opinion. Documenting importance can help to establish the value proposition for investment in the surveillance system
Actions	What decisions or actions need to be made to manage the problem? In the absence of criteria or thresholds for action or in the presence of uncertainties about what to do, short-term surveillance or a series of surveys can help characterize the problem, the outcomes of which can inform risk assessments and research
Legislation	Agencies or organizations conducting surveillance must have the legal authority to work with the animals or samples being tracked, such as having the necessary permits to handle wildlife or their products
Resources	Surveillance activities will be constrained by the variety, availability, and sustainability of expertise, infrastructure, and finances
Species status	Some species may be prioritized because the surveillance outcomes can influence decisions critical to their persistence (e.g., species at risk), to retention of their social value (e.g., harvested food safety), or to their ecological role (e.g., keystone species)
Decision maker and benefactors needs	The technical elements and purpose of a surveillance system will need to balance scientifically interesting goals with the knowledge needs of decision makers and the utility of the outputs for people being affected by the decisions
Logistics	Surveillance goals might be limited by challenges in accessing animals, samples, or other data. For example, wildlife in remote locations may not be routinely and representatively sampled because of the costs of finding the animal and obtaining samples over a vast landmass with a sparse human population with negligible transportation infrastructure
Level of certainty	The types of information collected and the approach to assessing surveillance data can be influenced by the willingness of decision makers to accept uncertainty in surveillance outputs
Public priorities	Wildlife health management is highly influenced by public values and popular opinion. For example, some wildlife capture, handling, and collection methods may be deemed unethical or socially unacceptable. Public investment priorities can bias efforts to charismatic or economically important species

time to respond, the information chain from case detection to sample submission to testing, analyzing, and reporting may be too slow, given a mismatch between the epidemiology of the disease, dynamics of the population and resources available. For example, the earliest warning for highly pathogenic avian influenza in migratory waterfowl might be 6 weeks due to delays in samples reaching testing laboratories, requirements for samples to be confirmed by a reference lab, and concerns about communicating results in a manner that does not influence international trade. Within that 6-week window, the affected subpopulation might have moved many hundreds of kilometers away from their sample collection site and be unavailable for further testing or interventions (given that some waterfowl can fly on average 80 km/h and often benefit from 50 km/h tailwinds over their 8 h/day flights) (Miller et al. 2005). While true early warning may not be achievable, this system can still reveal important seasonal trends, heighten poultry farmer biosecurity concerns, and reveal much about the ecology of the pathogen. As another example, formula to estimate sample sizes to establish freedom from the disease usually need information on the size of the population and the sensitivity and specificity of the tests used (information often unknown for many species) as well as assume representative sampling of the population (a situation rarely attainable for free-ranging wildlife). Similarly, sample sizes intended to find rare diseases can require numbers that exceed the human and financial resources available to collect and test samples. Proving the true absence of disease or developing systems to find very rare outcomes with the same rigor and confidence as for domestic animals may not be possible, but ongoing watchfulness of a population with consistent biases and limitations can contribute to evidence-based arguments about the presence of a hazard in an environment or population. These challenges should not dissuade surveillance developers from undertaking wildlife surveillance. Rather, it serves to challenge them to think of ways to gather ancillary information to triangulate their surveillance data toward a common understanding and to account for the biases and limitations created by real-world constraints when assessing and communicating surveillance outputs and information.

2.2 What Are We Going to Keep Under Surveillance?

2.2.1 Case Definitions

Once the priorities, purposes, and objectives have been established, the next step is to select the metrics (measurable quantities associated with the health problem under surveillance) that allow decision makers to estimate the size of the problem, monitor its spread, and measure the effects of an intervention (Nsubuga et al. 2006). Many wildlife surveillance systems track issues that effect a range of conservation, economic sectors, human and ecological communities, and geographic concerns. As such, there can potentially be an exceptionally large number of surveillance metrics, even for a single issue.

Surveillance systems aim to identify surveillance signals that are significant changes in one or more of the metrics under surveillance. Surveillance signals are important as they can indicate an ongoing change in the health status of the population or system under surveillance. Signals can come in many forms, from changing death rates to detection of increases in specific disease risk factors to changes in disease patterns. There must be agreement about what metrics (health outcomes or risk factors) to monitor to generate a signal. The metrics under observation should be meaningful, understandable, and support the system's objectives. To see trends in surveillance metrics, the metrics must be monitored consistently over time.

A case definition is used to provide standardized criteria for case reporting to increase reporting accuracy and consistency. They were originally developed for reporting cases of the disease but have since been generalized to provide standards for a variety of metrics. Case definitions aim for uniformity, simplicity, and brevity (Wharton et al. 1990) to reduce the likelihood of misclassification biases in surveillance data. A shared case definition enhances the ability to compare surveillance outputs over time and between different surveillance systems. Case definitions should be defined in precise, unambiguous terms that describe clearly and exactly what is being measured. They should give a clear idea of the data required and the population among which the cases are being sought. A case definition can include not just information from diagnostic test results but also information on the affected animals, populations, and/or situations and risk factors.

Surveillance case definitions are not meant to be used for clinical or pathological diagnosis. These require additional information as well as the experience of the diagnostician. Differences between surveillance and clinical case definitions can create apparent conflicts in that a surveillance system might not consider a clinical or pathological diagnosis to be a case for reporting purposes or vice versa for diagnostic purposes. Clinical or pathological diagnoses are intended to inform case management, whereas surveillance case definitions are intended to produce a clear and consistent accounting of a disease (or other problems) within a population management context. Their criteria for classifying each can be different, as they are built to suit their purpose.

Disease case definitions may include confirmed, clinical, or suspect classifications. None of these classifications is better or more accurate than others per se. Rather they serve different purposes. A broad suspect case definition will increase the number of signals (as well as the number of false alarms) but will reduce the likelihood of missing true positive animals. This approach can be useful if the goal is to launch rapid emergency responses to a high-risk disease, where the "costs" of acting before diagnostic confirmations are less than the costs of missing a case and losing the opportunity to intervene early. A confirmed case can use criteria that increase confidence in classifying a case as a true positive and therefore reduce the number of signals but increases the probability of having undetected positive cases that fail to meet all the case definition criteria (Guberti et al. 2014). This is an important consideration if, for example, the objective is to establish freedom from disease in a region or there is a high cost of a false positive (such as a trade embargo).

The biological variation in the diversity, frequency, and intensity of presentation of diagnostic signs across a population ensures that not all cases will match the case definition. The positive- and negative-predictive value of an animal or sample that is classified as a case (as defined by a specific case definition) influences the frequency and type of case misclassifications. The predictive value of a case definition is affected not only by the sensitivity and specificity of the diagnostic test or criteria used but also by the prevalence of the disease (or problem or risk factor) in the population and the number of animals sampled (Martin et al. 1992). The prevalence of the disease in the population is especially important as the probability that an animal meeting a disease case definition truly has the disease can change from very low when the population prevalence is low to very high when the population prevalence is high. Unfortunately, well-validated and field-tested diagnostic criteria are rare for many wild populations as are accurate disease prevalence estimates. This creates important gaps in knowledge about the predictive value of case definitions. Surveillance systems need to be aware of the implications and likelihoods of misclassification biases when analyzing and reporting their results.

2.2.2 Target Populations

A second aspect of "what are we going to track" refers to the animals being kept under surveillance. Careful consideration of the sample to be collected is critically important because results generated from the sample can be either a true reflection of what is happening in the population or a false reflection of what is happening in the population resulting from a bias in the sample. An ideal surveillance system should be based on unbiased (representative) samples of the entire population of concern (target population). This would require random sampling of the entire target population or sampling that truly represents the distribution of subpopulations across the landscape. Both require accurate information about the size and geographical distribution of all members of the target population, which is rarely, if ever, achievable for free-ranging fish and wildlife. Often practical constraints necessitate the selection of a convenience subset from the target population to act as a surrogate for the target population. When such non-probability sampling of subsets is used, representativeness can only be achieved if risk factors relevant to the issues under surveillance can be measured and weighted to show the relative differences in risk and proportion between the sample and the target population (OIE 2019). Such information is rarely available for free-ranging animals. Fish and wildlife surveillance systems, therefore, are usually reliant on biased population samples.

Decisions need to be made about which animals will be sampled and the biases those decisions impart to the surveillance information outputs. Hunted animals or animals conveniently found dead, for example, often make up a significant source of samples for wildlife surveillance. Hunters' preferences for healthy-looking animals or animals of a specific age or sex preclude generalizations of surveillance findings from this sample to other sexes or ages. The use of animals found dead as a sample source can not only impact the quality of samples for testing purposes but also can prevent the detection of animals that remain asymptomatic or have health impacts other than mortality. Relying on the public or natural resource officers to provide samples can bias systems away from cryptic species and towards larger, more charismatic species. Decisions may need to be made about whether a vector or intermediate host might be more informative than tracking a primary host or if environmental samples for the detection of a hazard better suit the purpose of the surveillance system. Selecting what will be sampled and how a case will be defined must be done within the context of the purpose and objectives of the surveillance program and the implications of biases or uncertainties in the results.

2.2.3 Preemptive Signals

There are five prerequisites to inspire action: (1) awareness that a problem exists; (2) a sense that the problem matters; (3) understanding of the causes; (4) the capability to intervene or influence. and (5) the political will to deal with the problem. Disease or hazard-based outcome surveillance makes significant contributions to some of these prerequisites, but their dependence on tracking the "bad" outcomes (e.g., death, disease, deformity) reduces their contributions to getting ahead of the curve. These "bad" metrics measure outcomes resulting from a causal chain of events that has already occurred. In other words, they measure the outcome of past activities and therefore are not suitable for prediction or prevention and preparedness planning. There are other metrics that warrant watching to develop intelligence systems that are helpful in protecting wildlife health. Health intelligence that provides additional information to characterize that the problem matters (e.g., social impacts and population ecology effects), understand what causes a population to be more prone or likely to be harmed, and track changing social capacity and priorities for action expands the contributions of surveillance to action-inspiring decisions by expanding the metrics kept under surveillance. Risk factor surveillance, vulnerability intelligence, and population health monitoring are all examples of approaches to expanding the knowledge created by surveillance and intelligence systems.

Risk factor surveillance tracks changes in the determinants of disease rather than disease outcomes or etiologic agents. Risk factor surveillance has been developed for tracing preventable or modifiable behavioral or biological risk factors for human chronic diseases (WHO 2005). These approaches benefit from significant knowledge of the types and importance of risk factors that serve as actionable signals for population health interventions. Due to less attention on risk factor analysis in wildlife and disconnects between surveillance and management of risk factors found beyond the host level, risk factor surveillance is rarer for wildlife. In animal health, more emphasis has been placed on risk-based surveillance, where risk assessments are used to refine the targeting of surveillance efforts to subpopulations or situations at a higher risk, thus increasing the efficiency of surveillance efforts (Stärk et al. 2006). Risk-based surveillance uses information about the probability of hazard occurrence and the magnitude of the biological and/or economic consequence of hazards to plan, design, and/or interpret the results obtained from surveillance systems.

Vulnerability intelligence is an emerging but unexploited concept in wildlife health. It can be used to develop a shared understanding of the state and determinants of vulnerability and identify options for action to reduce vulnerability. Knowledge gaps plus the reality that local circumstances will affect the most influential contributors to the vulnerability of a situation, population, or place preclude prescription of a specific, standard set of recommendations on the best variables to monitor for assessing wildlife health vulnerability and their associated action thresholds. Selection of the most useful variables to be assessed and tracked (from a management support perspective) will be context-specific and will require a combination of data, experience, and judgment. Collective processes, such as Delphi exercise, should be undertaken to seek commonalities in understanding and perspective amongst experts and decision makers to identify a set of vulnerability indicators, processes for their collection and assessment in a manner adaptable to local situations, and to create guidance on action thresholds. First principles of population health (e.g., Karpati et al. 2002; Gowan et al. 2014) can be used to argue that vulnerability intelligence must focus on the determinants of health (see Chap. 1) rather than only on the presence of hazards or adverse outcomes like death and disease. Based on emerging models of wildlife health (e.g., Wittrock et al. 2019), it seems reasonable to conclude that determinants of vulnerability need to gather information on (1) the biological endowment of the population (e.g., diseases and stressors); (2) the animal's social environment (e.g., extent of competition and demographics); (3) the quality and abundance of the needs for daily living (e.g., food supplies and habitat availability); (4) their abiotic environment (e.g., extreme weather variables); (5) sources of direct mortality (e.g., harvest), and (6) changing human expectations (e.g., social attitude and management policies).

2.3 Who Needs to Be Involved?

The answer to who needs to be involved is affected by two questions, "who needs to know and how will you gather, analyze and assess your surveillance data and samples?" All health surveillance systems involve different stakeholders, components, infrastructure, processes, policies, and regulations. Differences between surveillance systems relate to the populations or species that should be monitored to create information, the nature, number, and diversity of domains that should collaborate, and the stakeholders that will benefit from the decisions informed by the surveillance system.

Many surveillance systems rely on citizens, natural resource workers, or other field personnel to collect and submit samples or data routinely and safely. In other cases, the surveillance system might employ its own staff or others to purposively go out searching for data or samples. As the types of surveillance and intelligence data desired by a system expands from hazards or diseases to information on the population and social systems affecting wildlife management, the suite of people who must be made aware of the surveillance system, its purpose, its operating procedures, and its expectations grow.

Technical experts (e.g., pathologists and epidemiologists) are essential but not the sole participants in the surveillance system. The flow of data and information from source to the decision maker needs a suite of participants ranging from citizens or resource management staff in the field who find animals or samples, logistics personnel to get the samples to the lab, communications staff to share results, decision makers and policy makers who turn the intelligence understanding into action and interests groups, stakeholders and other community members who will be affected by the decisions and interventions launched as a result of the surveillance outputs. Each of these groups contributes their own inputs and processes to the detection and assessment of surveillance signals. Failing to engage the necessary people along this information chain creates the potential for gaps in the detection of surveillance signals and skews their assessment toward a limited set of needs. The diversity of participants who are included in the design, implementation, and use of a surveillance and intelligence system influences assessment objectives, the selection and application of acceptable and useful methods and tools, and how outputs are translated into action.

2.4 How Will We Recognize a Signal?

The first step in signal recognition is to establish a case definition for the metric that is meaningful, interpretable, and understandable to those people whose decisions or actions you intend to influence. The next step needs to find agreement on the degree of change in the metric under surveillance that would result in an actionable signal (such as a change in the spatial distribution of cases beyond a given boundary, the emergence of a new problem in a population or change in the frequency of case detection). There is no gold standard or consensus on methods for identifying and calculating wildlife surveillance thresholds.

The goal of surveillance is to find a change. Therefore, the first two questions to ask are, what type of change is important to find and how much of a change warrants action? To answer the first question, one must return to the purpose and objectives of the surveillance system. The answer to the second question can be more challenging.

For human or livestock populations there may be sufficiently reliable population sampling over a long enough time to determine statistically defined changes in the case count, prevalence, or incidence to signal an unacceptable change. For example, in public health, an increase of two standard deviations in the 3-month rolling average of the number of cases of meningococcal meningitis in a community could be used as a signal of the need to launch an outbreak investigation. Years of experience and research plus ongoing feedback and evaluation of the surveillance system give confidence in this action threshold. Rarely is there enough information to establish such quantitative thresholds for wildlife. Surveillance systems intended to detect and contain the introduction of a disease might use a single case as an action threshold, whereas surveillance systems supporting programs to contain the impacts of an endemic infection might use thresholds that allow for a certain amount of circulating disease that produces tolerable harm.

The simplest way to establish a threshold is to use existing literature and expert opinion to develop consensus among decision makers about the level of disease or hazards or risk factors that are concerning for them. This will require an adaptive management approach that modifies the thresholds as experience is gained about the reliability and utility of the current threshold. Thresholds might also be based on the implications of misclassification. Trade regulation might also establish action thresholds. Research and experience that show cause–effect relationships between the amount of disease in a population and population impacts can also provide the basis for an action threshold.

2.5 How Will We Know If the Surveillance and Intelligence System Is Good?

Criteria for evaluating surveillance systems were first popularized for public health. Many of the evaluation criteria emphasize the following attributes: usefulness, simplicity, flexibility, data quality, acceptability, sensitivity, predictive value positive, representativeness, timeliness, and stability (German et al. 2001). The systems should be both technically and economically efficient (Hoinville et al. 2013). Technical efficiency is driven by the sensitivity, specificity, and timeliness of the system. Sensitivity is affected by the diagnostic test or case definition sensitivity and the coverage of the surveillance system both of which affect the probability of finding and correctly classifying a case and a population. Specificity reflects the likelihood that a system will create false alarms. This too is affected by the case definition and test performance as well as by thresholds set for alarms. Economic efficiency considers the social return on investment from the surveillance system.

Unrepresentative access to the populations at risk, biased case ascertainment, the lack of adequately validated diagnostic tests, inaccurate or missing denominator data, lack of standard case definitions and diagnostic protocols, regulatory restrictions, ecological complexities, and fiscal constraints are but some of the factors that prevent the direct application of evaluation standards from domestic animals or public health to wildlife surveillance. Despite the growing demand for wildlife surveillance, there is little guidance on the necessary performance standards of wildlife health surveillance program. The lack of performance standards complicates attempts to show that investments and activities are meeting the expectations of public or private investors in wildlife disease surveillance services (Stephen et al. 2019).

3 Summary

There are many impediments to designing and implementing a wildlife health surveillance and intelligence system that fulfills all the expectations for a health surveillance system. Most of these expectations were developed for human health or livestock health surveillance. Failure to meet these expectations does not mean wildlife health surveillance is not useful. It does mean that wildlife health surveillance practitioners need to be acutely aware of how the compromises and biases that arise from the practical constraints of limited resources and knowledge influence what they can say about the information they generate. It means that the assessment and analysis phase will benefit from using additional information to supplement surveillance findings to increase confidence in trends described or signals made. It means that surveillance systems need to be designed with a clear purpose in mind and that the purpose must be developed with good situational awareness of the capacity for surveillance outcomes to meet decisions and actions support expectations.

References

Berezowski J, Carmo LP, Stephen C (2020) Building health surveillance for decision support at the animal, human, environment nexus. In: Stephen C (ed) Animals, health and society: health promotion, harm reduction and health equity in a one health world. CRC Press, Boca Raton

German RR, Horan JM, Lee LM, Milstein B, Pertowski CA (2001) Updated guidelines for evaluating public health surveillance systems; recommendations from the guidelines working group. https://www.cdc.gov/mmwr/preview/mmwrhtml/rr5013a1.htm

Gowan ME, Kirk RC, Sloan JA (2014) Building resiliency: a cross-sectional study examining relationships among health-related quality of life, well-being, and disaster preparedness. Health Qual Life Outcomes 12(1):85

Groseclose SL, Buckeridge DL (2017) Public health surveillance systems: recent advances in their use and evaluation. Annu Rev Public Health 38:57–79

Guberti V, Stancampiano L, Ferrari N (2014) Surveillance, monitoring and survey of wildlife diseases: a public health and conservation approach. Hystrix Italian J Mammal 25(1):3–8

Han BA, Drake JM (2016) Future directions in analytics for infectious disease intelligence: toward an integrated warning system for emerging pathogens. EMBO Rep 17(6):785–789

Häsler B, Howe KS, Stärk KD (2011) Conceptualising the technical relationship of animal disease surveillance to intervention and mitigation as a basis for economic analysis. BMC Health Serv Res 11(1):1–0

Hoinville L (2013) Animal health surveillance terminology. Final report from pre-ICAHS workshop . https://www.fp7-risksur.eu/sites/default/files/partner_logos/icahs-workshop-2011_surveillance_tewrminology_report_V1.2.pdf

Hoinville LJ, Alban L, Drewe JA, Gibbens JC, Gustafson L, Häsler B, Saegerman C, Salman M, Stärk KD (2013) Proposed terms and concepts for describing and evaluating animal-health surveillance systems. Prev Vet Med 112(1–2):1–2

Jamot E (2013) Developing a framework for assessing wildlife health: a model for wildlife health intelligence for use in the Sri Lanka wildlife health Centre doctoral dissertation, UM2. https://agritrop.cirad.fr/573529/1/document_573529.pdf

Karpati A, Galea S, Awerbuch T, Levins R (2002) Variability and vulnerability at the ecological level: implications for understanding the social determinants of health. Amer J Pub Health 92(11):1768–1772

Martin SW, Shoukri M, Thorburn MA (1992) Evaluating the health status of herds based on tests applied to individuals. Prev Vet Med 14(1–2):33–43

Miller MR, Takekawa JY, Fleskes JP, Orthmeyer DL, Casazza ML, Haukos DA, Perry WM (2005 Dec) Flight speeds of northern pintails during migration determined using satellite telemetry. Wilson J Ornithol 117(4):364–374

Nsubuga P, White ME, Thacker SB, Anderson MA, Blount SB, Broome CV, Chiller TM, Espitia V, Imtiaz R, Sosin D, Stroup DF (2006) Public health surveillance: a tool for targeting and monitoring interventions. Dis Contr Prior Develop Countr 2:997–1018

OIE (2019) Animal health surveillance. In: Terrestrial animal health code. World Organisation for Animal Health, Paris. https://www.oie.int/fileadmin/Home/eng/Health_standards/tahc/current/chapitre_surveillance_general.pdf

Stärk KD, Regula G, Hernandez J, Knopf L, Fuchs K, Morris RS, Davies P (2006) Concepts for risk-based surveillance in the field of veterinary medicine and veterinary public health: review of current approaches. BMC Health Serv Res 6(1):20

Stephen C, Zimmer P, Lee M (2019) Is there a due diligence standard for wildlife disease surveillance? A Canadian case study. Can Vet J 60(8):841–847

Wharton M, Chorba TL, Vogt RL, Morse DL, Buehler JW (1990) Case definitions for public health surveillance. https://www.cdc.gov/mmwr/preview/mmwrhtml/00025629.htm

Wittrock J, Duncan C, Stephen C (2019) A determinants of health conceptual model for fish and wildlife health. J Wildl Dis 55(2):285–297

World Health Organization (2005) WHO STEPS surveillance manual: the WHO STEPwise approach to chronic disease risk factor surveillance. World Health Organization, Geneva

Health Protection and Promotion for Disease Management in Free-Ranging Wildlife Populations

Colin M. Gillin

Abstract

The strategies of wildlife disease management are built on the goals of prevention, surveillance, and management of causes of disease found in, or diseases associated with, wildlife. However, many health protection and disease control options are not always available, affordable, feasible, or socially acceptable to use in wild animals. This chapter highlights successful actions and interventions, challenges, and limitations related to wildlife disease management and reviews important disease examples that highlight these concepts. Insights are discussed on methods to prevent disease proactively and preemptively through building population resilience or protecting the determinants of health.

Keywords

Prevention · Control · Wildlife disease · Preparedness · Intervention · Strategy · Management

1 Introduction

To protect and promote health in free-ranging wildlife populations, managers combine methods based on science with strategies practiced and proven through decades of managing disease. The formal management of disease in wildlife populations is more than a century in practice (Westmore 1918; Allen 1924; O'Roke 1928). Management tools range from field studied interventions to laboratory methods proven in controlled settings that are applied on free-ranging wildlife experiencing uncontrolled influences in natural habitats.

The tools used and outcomes of disease control in a wildlife population depend on the unique epidemiological characteristics of the etiologic agent, the host wildlife species, and the environmental factors influencing the disease (Wobeser 2006) including human influences. Understanding the dynamics and control of wildlife diseases requires addressing issues and interactions at the interface of population biology, animal behavior, community ecology, epidemiology, and statistics (Hayden 2008) as well as sociology, political science, and human behavior (O'Brien et al. 2011). When first confronted with a new disease problem, managers are often pressed to decide if eradication or control is achievable and if not, why? Often the reasons are related to challenges in changing human behavior and attitudes, political and special interest influences, available funding, and the scale of the outbreak or capacity of agencies to respond. The limited number of success stories in the control or eradication of pathogens and the diseases they impart are evidence of the challenges and complexity of

C. M. Gillin (✉)
State Wildlife Veterinarian, Corvallis, OR, USA
e-mail: colin.m.gillin@odfw.oregon.gov

C. Stephen (ed.), *Wildlife Population Health*,
https://doi.org/10.1007/978-3-030-90510-1_10

wildlife disease control (Bernard et al. 2020; Decker et al. 2006; Deem et al. 2001).

Understanding disease epidemiology to minimize impacts on human, wild, and domestic animal health is important. But understanding the relationship of human behavior and how we facilitate or manage the spread of disease in wildlife will undoubtedly become a central objective to affecting disease outcomes and their impact on people, domestic animals, wildlife, and ecosystems.

Much can be learned by formally and informally evaluating if health management actions worked or missed the management objectives, thus gaining insight into how to better respond with new solutions and strategies to promote and protect health in wildlife populations. This chapter reviews some of the strategies and tools used to prevent or mitigate disease in free-ranging wildlife, discusses their utility and limitations, and proposes new approaches that may provide health protection and promotion in free-ranging wildlife populations.

2 Strategies and Tools of Free-Ranging Wildlife Disease Management

The fundamental disease management strategies used in wildlife populations are planning, prevention, surveillance, and management. Disease prevention and management rely on manipulating the disease agent, the host, and/or the environment (Wobeser 2006).

2.1 Planning and Preparation

Effective planning and preparation are desired objectives before a disease occurs. However, it is often difficult to predict the type and level of preparation required because the disease classification, risk factors, and severity of the disease can vary with different circumstances or diseases as can availability of resources, funding, and public support. When agencies are less than prepared for disease response, managers too often employ a reactive "crisis management" strategy typical in many wildlife disease outbreaks (Voyles et al. 2014). This approach may involve a basic plan to respond after a disease is detected and established in a wildlife population (Grant et al. 2017), a strategy that may be too late to effectively manage the disease.

Coordinated planning and cooperative agreements are key to planning and securing prevention strategies and funding for effective response and increased chance of success. Wildlife disease management is dependent on strategies implemented by wildlife agencies with public input and assistance, often from those who engage in wildlife use, such as hunting (Gillin and Fischer 2018).

2.2 Prevention

Preventing the spread and establishment of a disease to a new area or a naive population is the single most effective and efficient action a wildlife management agency do to can protect wildlife population health. Prevention of disease establishment is a cost-effective action compared to the personnel effort, resources required, and cost of managing an established disease. Initial steps in disease prevention involve identifying the risks for introduction and transmission of new hazards, mitigating specific risks via public education, providing recommendations, and forming regulatory instruments to avoid disease introduction (Fischer and Davidson 2005).

A disease that is established on the landscape will challenge agencies both fiscally and in acquiring support and momentum for actions aimed at reducing disease prevalence and distribution. Prevalence rates may already be high when the disease is first detected because of the difficulty of detection in the early stages of disease establishment and nuances in sampling related to the slow growth rate of prevalence (Miller and Fischer 2016). Agencies may spend many times the annual prevention expenditures attempting to eradicate an established and spreading pathogen.

Wobeser (2002) identified several effective proactive management actions to prevent the establishment of disease in wildlife populations. These involve modifying human behavior by altering activities that increase risk such as banning live animal translocations and reintroductions and stopping supplemental feeding and baiting of wildlife. Other actions that reduce risk proactively include effective vaccines to immunize populations or erecting or managing barriers to prevent animals and their diseases from moving into new areas (Sutmoller 2002). Regulatory and legislative tools can be used to restrict the movement of potentially infected animals and animal parts, thus protecting uninfected populations. Carcass disposal protocols of potentially infected carcasses can be implemented as well as banning the use or transportation of reproductive (e.g., semen and embryos) and other tissues to restrict the human-associated pathways for disease spread into new areas.

2.3 Surveillance

Surveillance is an important component of both prevention and mitigation measures. Surveillance in free-ranging wildlife is often conducted during predisease emergence and implemented through testing and sampling in a statistically meaningful manner using validated diagnostic tests to search for disease-causing agents and evaluate and verify the current health or disease status of a population. It is an effective tool for the early detection of pathogens and can be used to reduce the risk of disease establishment after pathogen introduction. Surveillance can also be used to identify pathogens in source populations prior to translocations as part of restoration programs. Chapter 9 provides details on the design and challenges of wildlife health surveillance.

2.4 Containing Disease

Removing an introduced disease from a population is the ideal disease management goal. However, it is rarely attainable. Controlling the course of the disease is the next best option. Managers have options to contain disease in space or time and/or sustain infection rates at tolerable prevalence rates. Initial disease management objectives are focused on aggressive attempts to remove infected animals while determining the prevalence and distribution of the outbreak.

As a pathogen becomes more established and spreads through the population by natural animal movements or human activities, reducing transmission or spread of the disease becomes an objective that may be hampered by animal movements (seasonal movements, immigration, and emigration). Therefore, population density reduction is a potentially effective option to decrease contact between animals and reduce transmission as animal movements occur. Herd or population-level interventions more often involve manipulation of environmental conditions or decreasing transmission opportunities through density reduction via harvest, an effective method used currently and historically to reduce prevalence (Mohler 1926). Other reasons to reduce host numbers are that high population density can impact habitat quality and quantity and provide conditions for rapid pathogen transmission and geographic spread (Gortázar et al. 2006).

Other interventions might include relocation of rare or sensitive populations to avoid infection or dispersal of animals (e.g., uninfected animals hazed away from a diseased population or outbreak area) and habitat manipulations to prevent, attract, or maintain wildlife use of an area (Friend et al. 1999).

Disease control interventions generally involve the use of several integrated strategies (Gortazar et al. 2015). For example, population reduction may be combined with arthropod vector control to reduce transmission of vector-borne disease. Other innovative strategies involve fertility control to reduce populations and direct contact transmission or predator protection to encourage the removal of diseased animals (Killian et al. 2007; Wild et al. 2011). However, the principal intervention method used by managers on game animals has been via population reduction.

There is an increasing call among wildlife managers for better prevention, detection, and response to emerging diseases (Kuiken et al. 2005; Voyles et al. 2014) with an emphasis on effective surveillance such that, in the event of a disease incursion, it is discovered early, managed, and removed from the population. To attain this goal during a disease outbreak, access to infected animals and having the ability to quickly remove them from the population is a requirement.

3 Lessons Learned: Case Studies of Wildlife Disease Control

3.1 Chronic Wasting Disease in North America

Chronic wasting disease (CWD) (Williams and Young 1980) is a devastating disease that continues to spread across North American landscapes. CWD is a transmissible spongiform encephalopathy or prion disease affecting species in the family Cervidae. It is a fatal, degenerative neurological disease of deer, elk, moose, and reindeer and likely other related species, subspecies, and genetic crosses. The causative prion is persistent and infective in the environment. At high prevalence, CWD has been shown to negatively impact cervid populations (DeVivo et al. 2017; Edmunds et al. 2016). Geographic spread may occur through natural migration or dispersal of cervids as well as through human movement of infected animals or materials. CWD's lengthy preclinical incubation period in multiple host species, shedding of prions in saliva, urine, and feces of asymptomatic animals, and lengthy persistence of prions in the environment, make the disease especially difficult to detect early and manage in free-ranging animals.

The strategies and management tools implemented to prevent, detect, and manage CWD have been exhaustive and expensive. In 2018, the Association of Fish and Wildlife Agencies developed and adopted best management practices (BMPs) for wildlife management authorities as guidance on strategies and technologies for CWD prevention, surveillance, and management (Gillin and Mawdsley 2018). The BMPs cover dozens of management topics and strategies including restricting movement of live cervids and carcasses, banning of baiting, feeding, and urine-based lures, and using adaptive management practices, along with validated testing and other management tools to plan, prevent, and mitigate CWD spread and limit economic impacts.

Aggressive preventative actions taken by states and provinces are the only effective measures to combat this disease once detected. Along with restricting or banning every conceivable route or activity to exclude the pathogen from entry into healthy North American cervid populations, effective surveillance has been shown to potentially identify the disease early before it becomes established. In 2005, the state of New York detected and removed what appeared to be their index case of CWD and successfully halted the establishment of the disease in the state. This success has not been repeated or maintained, to date, in other states or provinces.

Curtailing the geographic spread of CWD requires impeding (1) natural migration—which is an impractical and biologically unsound strategy and (2) human-assisted movement of animals or parts—which is a strategy unpopular among farmed cervid owners or decision makers balancing political will against sound science. However, managing to decrease CWD prevalence by reducing transmission and geographic spread through selective population reduction has been an effective adaptive management strategy in some areas (Miller et al. 2020).

Interjurisdictional movement of live farmed cervids, which is allowed by many North American states and provincial governments, facilitated anthropogenic disease spread (Makua et al. 2020; Miller and Williams 2004) despite U.-S. Department of Agriculture and Canadian Food Inspection Agency programs to promote or certify low-risk herds. Hunters have moved infected carcasses and carcass parts interjurisdictionally due, in part, to a lack of knowledge of the regulations, limited law enforcement capacity, and a lack of regulatory consistency, coordination, and communication between jurisdictions. Although a dynamic array of best practices provides direction for protecting and promoting health in North American cervids, the slow spread of CWD by natural animal movements, or the rapid spread by human-facilitated movements, present biological, social, and policy challenges. The long-term management of CWD will require public education and the development of regulations to limit animal movement by humans and keep disease prevalence low through harvest strategies to reduce the likelihood of spread of CWD to new areas via natural animal movement.

3.2 Bovine Tuberculosis: The Tale of Two States

Bovine tuberculosis (bTB) is a cattle disease that can infect multiple mammalian hosts, including humans, causing morbidity often seen as thoracic granulomas, with lung, pleura, and lymph node abscesses in affected animals. It also leads to severe agricultural economic losses from herd depopulations and trade restrictions, leading to cultural and political ramifications (O'Brien et al. 2006).

A bTB eradication program that began in 1917 led to the last known bTB-infected cattle herd in the state of Michigan being depopulated in 1974. However, the following year, a bTB-positive wild white-tailed deer was harvested by a hunter with the infection believed to have originated from livestock spillover (Schmitt et al. 1997). Over the next two decades, bTB was not detected until a hunter-harvested positive deer was identified in 1994 in the northeastern portion of the Lower Peninsula of Michigan. Intensive surveillance found additional infected wild deer in subsequent years and detection of bTB in domestic cattle herds in the affected region soon followed. It was concluded that deer had become a reservoir for this domestic cattle disease.

Schmitt et al. (2002) concluded that the persistence of the disease in deer was most likely due to high deer densities and concentrating animals with baiting and feeding to maintain high densities for hunting purposes on private land. These human-associated risk factors facilitated nose-to-nose contact between infected and uninfected animals and are most likely responsible for establishing self-sustaining bTB in free-ranging Michigan deer. Beginning in the 1930s, a series of actions contributed to disease establishment including limited commercial and public hunting following large land acquisitions by hunt clubs that allowed deer densities to exceed habitat carrying capacity (O'Brien et al. 2006). Supplemental feeding was implemented by hunt club landowners who also raised cattle on these same properties. By the 1990s, supplemental feeding was a common practice as was fencing to limit migration. Live deer relocations also occurred between properties for genetic enhancement.

Disease management strategies included (1) banning baiting and feeding of deer to reduce contact infection in counties where the disease was found and (2) deer herd reductions through unlimited antlerless hunting permits to decrease densities in the endemic area. These measures decreased bTB prevalence by half (Schmitt et al. 2002); however, eradication of the disease has not been accomplished to date. This is partially due to public opposition to further reductions of deer population densities (Dorn and Mertig 2005) and the illegal and continued use of bait on private property (Gwizdz 2004) which concentrates deer and increases transmission potential. Diminished cooperation and compliance by landowners and stakeholders have limited the effectiveness of hunting to reduce herd densities and the use of baiting restrictions as bTB intervention tools, although they are known to provide the greatest potential for effective control strategies or disease eradication (O'Brien et al. 2011). Considering this limitation, managers felt improving hunter cooperation and compliance and implementing publicly acceptable control policies provided the most effective means of controlling bTB in wild deer and preventing spillover into cattle herds (O'Brien et al. 2006).

The state of Minnesota also experienced a bTB outbreak in wild deer and domestic cattle, but circumstances led to a different and favorable outcome. The disease had not been detected in the state for 35 years but reemerged in 2005 in a beef cow in northwestern Minnesota (Carstensen and DonCarlos 2011). Before year's end, four additional infected cattle herds were identified and depopulated in the region with cattle owners indemnified. As part of the surveillance plan, the Minnesota Department of Natural Resource (MNDNR) sampled 474 hunter-harvested deer for bTB near the infected farms and found one bTB-infected deer near the index cattle herd (Carstensen and DonCarlos 2011). Landowners of bTB-infected cattle farms were issued deer shooting permits to cull deer for testing; however, no new cases were detected. In 2006, a one-time statewide sampling effort tested 1554 cattle herds and 4000 harvested deer. Cases of bTB were found in two more cattle herds and five more deer in the region of the index herd. A bTB Management Zone was established.

Wildlife managers implemented an aggressive disease management campaign with the goal of eradicating bTB in Minnesota deer (Carstensen et al. 2011). Critical to this response was the continued banning of recreational feeding of wild cervids in the affected area, a practice prohibited since 1991. Deer densities were reduced in the bTB Core Area using government agency-sponsored sharpshooters. The core area focused disease management efforts on deer and was part of a larger bTB Management Zone delineating affected surveillance areas. Aggressive deer removal continued through 2010 by ground and aerial sharpshooting, resulting in the detection of 14 bTB-positive deer. Hunters took advantage of liberalized hunting seasons involving early and late season hunts and reduced-cost permits within the bTB Management Zone, which also became a special deer management unit. The MNDNR allowed landowners with farms inside the bTB Management Area to harvest deer on their private lands beyond the period of regular deer harvest seasons.

Management of bTB in livestock also was aggressive. The Minnesota State Legislature funded a voluntary cattle buy-out program of 46 farms, effectively removing most cattle from the bTB Management Zone. Herd owners were indemnified for existing cattle as well as for future calf crops. Farms that continued cattle operations inside the zone were required to fence feed storage and winter-feeding sites with state purchased fencing to restrict wild cervid access (Carstensen and DonCarlos 2011).

Surveillance using hunter harvest and sharpshooters continued. The disease has not been detected in wild deer in the state since November 2009 (Carstensen and DonCarlos 2011). The strategies utilized to successfully eradicate bTB in free-ranging white-tailed deer and spillover into cattle included aggressively reducing transmission potential among deer by reducing deer densities, prohibiting supplemental feeding, and mitigating risks at the cattle-wildlife interface.

In comparing the Michigan and Minnesota outcomes, several contributing factors limited Michigan's success, despite both states implementing similar intervention strategies. One major factor contributing to Michigan's less desirable outcome was the apparent establishment of bTB in wild deer years before it was first detected. Identifying a disease outbreak early demonstrates why surveillance is so critical in any wildlife disease management program. This advantage was realized by Minnesota where infected deer were removed before widespread disease establishment. Infected deer were only identified in proximity to infected farms and shared the same strain of bTB as cattle. The assumption that deer were likely recent spillover hosts was ascertained because positive deer were not detected beyond the area near the index farm (Carstensen and DonCarlos 2011).

Additional factors contributed to Minnesota's success. The affected area was much smaller than Michigan's and deer density was much lower (Carstensen and DonCarlos 2011). Also, only 10% of Michigan's affected area was public land compared to 60% in Minnesota's affected area (Carstensen and DonCarlos 2011) affording greater access for aggressive disease management activities using agency sharpshooters and hunters. Michigan has not been able to eliminate baiting and feeding of deer through regulation, where Minnesota experienced much better compliance. Minnesota experienced positive legislative and agency support from many stakeholder groups (public, animal agriculture, hunters, etc.) for aggressive intervention strategies and made available dedicated funding for disease management activities (Carstensen and DonCarlos 2011).

3.3 New World Screwworm in North America

The eradication of the economically devastating flesh-eating screwworm, a species of blowfly, *Cochliomyia hominivorax* (Coquerel), from the Florida Keys, United States in 2017, is a disease eradication success story (Skoda et al. 2018). Female blowflies lay eggs once during their lifetime, usually at the edge of an open wound. The hatching larvae then feed on living tissue causing extensive cavitated lesions in the skin and underlying tissues, which can lead to the death of the host.

In North and Central America, screwworm has been eradicated using the Sterile Insect Technique (SIT). The United States and Panama are cooperative partners in this control program designed to prevent the spread of screwworms from infested South American countries (Concha et al. 2020). The program maintains a permanent barrier through the release of millions of sterile male and female flies at the border between Panama and Colombia.

In 2016, the identification of new world screwworm fly larvae samples from infested Key deer (*Odocoileus virginianus clavium*) in the Florida Keys was followed by an immediate interagency response to collect blowfly larvae from infested animals to develop a sterile fly colony for release. Approximately 188 million sterile flies were released to mate with wild flies involved in the outbreak, resulting in infertile eggs. A long-acting parasite treatment was also applied prophylactically to many deer. Eradication of the parasite was declared on March 23, 2017. The population of approximately 1000 Key deer lost 135 infected deer to euthanasia during the outbreak and an estimated equal number died but were not detected.

The success of this eradication intervention was unique for many reasons including the benefit of a relatively non-migratory closed population host of limited distribution. Also, a beneficial prophylactic treatment was applied to part of the population. Foremost, the delivery of an ingenious intervention technique involving the prescriptive release of modified sterile screwworm flies that significantly reduced the reproduction of new generations of screwworms created a biological barrier that led to the local eradication of the fly. Infected countries of South America and several island nations (Cuba, the Dominican Republic, Haiti, Jamaica, and Trinidad and Tobago) continue to serve as a challenge to screwworm elimination; however, implementation of SIT programs is an effective tool to eradicate this parasite (Whitworth 2006).

4 Additional Options for Disease Control

Wildlife disease management has benefited from adapting techniques and tools developed for domestic animal disease control. Managers have used vaccination of free-ranging wildlife populations as a prevention practice to limit the spread of a small number of diseases in terrestrial wildlife when treating focal outbreaks (Slate et al. 2009). For example, vaccinating highly susceptible prairie dogs (*Cynomys* spp.), a critical prey species of the endangered black-footed ferret (*Mustela nigripes*), has shown promise as a conservation tool in endemic plague regions (Abbott et al. 2012; Salkeld 2017). Vaccination also has very successfully limited the spread of the rabies virus in wildlife in defined areas (Slate et al. 2009). However, fiscal, political, and logistical challenges can limit the effectiveness of vaccinating or treating free-ranging animals for reasons such as the insufficient capacity to access a statistically appropriate proportion of the population to attain a measurable therapeutic result at a population scale (Schreiner et al. 2020).

Vector control strategies have been used to limit human infections with wildlife-associated zoonotic disease agents carried by mosquitos, such as West Nile and Zika viruses, and tick-vectored diseases such as Lyme disease and ehrlichiosis. However, this strategy is relatively limited for broad regional application as evidenced by vector-borne zoonotic disease persistence due in part to land-use changes and social factors (Kilpatrick and Randolph 2012).

Strategies involving the removal of infected and diseased wildlife carcasses may be used to reduce mortality rates by reducing sources for pathogen transmission. A successful example of this involves the removal of waterfowl carcasses from bodies of water during avian cholera or botulism outbreaks. This can reduce the severity of the outbreak. However, infected birds may survive and continue to serve as a source of *Pasteurella* bacteria or botulinum toxin, and spores from *Clostridium botulinum* can persist in the environment.

Quarantine of free-ranging wildlife is another intervention strategy that has been used in rare circumstances (Nishi et al. 2002). Unless the population is small and geographically isolated, corralling or restraining wildlife at the herd or population level for quarantine, treatment, or testing is rarely feasible.

5 Limitations and Challenges

The effectiveness of disease management strategies depends not only on the scale and timing of interventions but also on factors that relate to our ability to effectively respond at the population level. White-nose syndrome (WNS), caused by the introduced fungal pathogen, *Pseudogymnoascus destructans*, has caused widespread declines in bat populations in North America and threatens several species with extinction as it spreads continent-wide (Drees et al. 2017; Warnecke et al. 2012). Few tools exist presently to reduce WNS impacts on affected bat populations on a regional scale (FWS 2011), but local strategies have been attempted including cave closures, decontamination protocols, and cave or bat treatments.

Most common ubiquitous wildlife diseases (e.g., sylvatic plague, tularemia, toxoplasmosis, and many other parasitic infections) have no effective or practical disease intervention strategy other than local prevention measures. For emerging diseases, such as Treponeme-Associated Hoof Disease in elk (Clegg et al. 2015), there is no effective environmental, animal, or population treatment nor disease mitigation strategy other than to limit human-associated animal movements. For other diseases, prevention and response interventions that show promise in controlled settings cannot be economically or logistically implemented in free-ranging animals or they face public opposition. For example, selective slaughter and culling to decrease animal density by managers, hunters, or sharpshooters has shown to be effective but increasingly may not be socially acceptable as a long-term management strategy and may be met by opposition from stakeholders. For this reason, proactive efforts to reduce risk factors, reduce population vulnerability, and enhance population resilience need to form the foundations of wildlife health programs.

Measures to prevent and combat disease establishment, spread, and impacts using the best practice intervention methods are expected by the public. However, in many jurisdictions, statutes, administrative rules, and other regulatory instruments often oblige wildlife management agencies to protect and sustain public use of wildlife (Fischman 2003; Freyfogle and Goble 2019). In such circumstances, disease control practices cannot supersede the mandate to manage and sustain populations for their economic and cultural benefit to the general public and hunters (O'Brien et al. 2006, 2011).

Managing any disease in wildlife likely will be compromised if human-associated risk factors are permitted and/or access to land and animals to conduct management activities is impeded. Furthermore, disease management efforts are only as effective as the management agency's planning, available personnel and resources, and funding to implement an effective response. Unfortunately, funding is very often a common factor limiting the feasibility and effectiveness of management activities.

Management response efforts may require coordination across jurisdictions or continents.

Cross jurisdictional response adds to the challenge of consistent disease management and messaging as responding agency management capacity may vary between jurisdictions as can agency funding, resources, and mandates (Bernard and Grant 2019). The lack of a single regulatory framework or management protocol used to respond to emerging diseases in free-ranging wildlife adds to the dysfunction of effective control or eradication of important diseases (Langwig et al. 2015).

Many of the important diseases involving wildlife have specific attributes that are conducive to their persistence and spread. The complexity of infection and transmission presents major challenges for response and management when a pathogen can infect multiple host species or maintain transmission opportunities using the environment as a reservoir (Turner et al. 2016). Anthrax, botulism, highly pathogenic avian influenza, and CWD are examples of diseases with these characteristics. Pathogens that are quickly and easily spread over ecosystems and landscapes via multiple routes are also very problematic to contain. This can occur via pathogen transport in the animal host through natural movements, by humans moving the host and pathogen, via a parasitic or insect vector, or in the feces of scavengers.

The presence of a disease in a species that is hunted can lead to a decrease in hunter participation (Needham et al. 2004). In jurisdictions relying on hunting-associated revenues, diseases, or efforts to control them, that reduce hunting can directly affect wildlife program funding used to prevent and manage disease (Vaske et al. 2004). Ethical, fiscal, and ecological issues will all influence how disease risks are communicated and controlled.

6 Preemptive and Proactive Strategies

The future of managing wildlife health will undoubtedly need to include new ways to address disease threats. We have learned to react with logical strategies, tested methods, aggressive response, and persistence to prevent, manage, and occasionally eradicate disease. Agencies are starting to more regularly incorporate human dimensions evaluations to understand the socio-cultural and ecological relationships in the management of wildlife (Decker et al. 2012) and use transparent public and media communications as part of a comprehensive response effort (see Chap. 25). This is a major point of importance as public tolerance of specific management actions can affect the success of agency control efforts (Carstensen et al. 2011). Regardless of the agency's will, funding, and strategy, managing wildlife disease is a variable undertaking with population-level disease eradication a rare event. Proactive strategies to prevent disease and bolster sustainability and resiliency in wildlife populations will require productive advances in research, with new methodologies developed to affect the factors that sustain health and mitigate the factors that lead to conditions that facilitate the emergence and/or impacts of diseases.

Research has made great strides in wildlife disease prevention through novel discoveries in diagnostic and surveillance technology. Many of these methodologies have led to earlier diagnosis and more effective response and control strategies. Examples include new detection of volatile organic compounds in air or animal breath to identify disease at the molecular level (Bayn et al. 2013). Trained detection dogs have also been used to identify disease from airborne molecules (Angle et al. 2016). Other remote field detection techniques have been developed based on organismal DNA found in the environment (eDNA) and used as a surveillance tool for parasite detections and other wildlife pathogens in aquatic environments (Huver et al. 2015; Sieber et al. 2020). Real-time quaking-induced conversion (RT-QuIC) assays amplify minute amounts of CWD prion protein (Cooper et al. 2019) to a detectable level (Henderson et al. 2015a) to identify the protein in a variety of tissues, secretions, and excretions from live cervids (Henderson et al. 2015b).

New methods and tools used in the prevention, surveillance, and management of disease in free-ranging wildlife populations will improve wildlife manager's ability to plan and react more effectively while applying adaptive management strategies to control and eradicate pathogens from populations. However, these tools alone will not affect the source of the ecological instability making diseases more prevalent or impactful and will not build sustainability or resiliency in wildlife populations to mitigate threats to health. This level of planning requires additional foresight and an understanding of connections between biotic and abiotic systems, humans and their values, and the root causes driving morbidity, mortality, and pathogen transmission in wildlife populations.

Disease prevention and management strategies will always be required to respond to approaching threats or established and emerging disease outbreaks. But to provide better prevention, managers will need to affect factors, events, or circumstances that indirectly cascade into environmental, ecological, and biological "disharmony" and ultimately lead to the emergence of disease in vulnerable populations of wildlife.

Protecting and promoting health in free-ranging wildlife is a goal of management (Wittrock et al. 2019). Population health is driven not only biologically and ecologically but also socially and economically (Stephen 2013, 2014). A public health concept known as the determinants of health (Eyles and Furgal 2002; Cieza et al. 2006) has been applied to wildlife populations and used in modeling the factors that determine or influence health (Wittrock et al. 2019; see Chap. 1). This approach allows managers to identify issues that may reduce the resiliency of a population ahead of a disease outbreak, potentially reducing the level of control interventions because of the population's ability to adapt. For example, in modeling caribou and salmon, Wittrock et al. (2019) could identify numerous factors influencing health that were not limited to the occurrence of disease. The authors used this proactive and multifactorial modeling approach as a planning tool for potential strategies and actions. These types of prevention concepts are key to building a foundation of resiliency and sustainability in wildlife prior to the need for reactive actions.

Another way to look ahead of outbreaks has been proposed in relation to pandemic disease and wildlife trafficking. Felbab-Brown (2021) proposed reducing the human interface with wildlife and wild habitats and focusing on the elimination of probable pathogen transmission points where spillover from wildlife to humans can occur. The conservation of natural habitats is an important driver for maintaining health and resiliency in wildlife populations. Felbab-Brown (2021) points out that habitat conservation may require drastic changes in how humans encroach upon and utilize the land for food production, housing and communities, and recreation. This concept is not necessarily focused on restricting humans from outdoors or wild places but rather works to bring people into the outdoors and in doing so, recognize that wildlife and their environments are worth saving. Environmental education and conservation engagement by all stakeholders is prevention based on a fundamental restructuring of how humans treat nature.

7 Summary

Sustainable and resilient healthy wildlife populations in a world of increasing pressures on wildlife and habitats from the ever-expanding influence of humans will require mitigating the effects of this influence through integrated planning at the ecosystem, if not global, level. Management decisions should incorporate the concepts of health protection and promotion within long-term wildlife population and habitat strategic planning, with goals to mitigate disease emergence and establishment proactively and preemptively before the threat is even apparent.

References

Abbott RC, Osorio JE, Bunck CM, Rocke TE (2012) Sylvatic plague vaccine: a new tool for conservation of threatened and endangered species. Ecohealth 9: 243–250. https://doi.org/10.1007/s10393-012-0783-5

Allen AA (1924) The grouse disease. Bul Amer Game Prot Assoc 1924:12–14

Angle TC, Thomas P, Waggoner PL, Fischer TD, Rogers BP, Galik PK, Maxwell HS (2016) Real-time detection of a virus using detection dogs. Front Veter Sci 2:79

Bayn A, Nol P, Tisch U, Rhyan J, Ellis CK, Haick H (2013) Detection of volatile organic compounds in brucella abortus-seropositive bison. Anal Chem 85(22):11146–11152. https://doi.org/10.1021/ac403134f

Bernard RF, Grant EHC (2019) Identifying common decision problem elements for the management of emerging fungal diseases. Soc Nat Resour 32:1040–1055

Bernard R F, Reichard JD, Coleman JTH, Blackwood JC, Verant ML, Segers JL, Lorch JM, White JP, Moore MS, Russell AL, Katz RA, Lindner, Toomey RS, Turner GG, Frick WF, Vonhof MJ, Willis CKR, Grant EHC. Identifying research needs to inform white-nose syndrome management decisions. Conservation Science and Practice. 2020; 2 8, e220. https://doi.org/10.1111/csp2.220

Carstensen M, DonCarlos MW (2011) Preventing the establishment of a wildlife disease reservoir: a case study of bovine tuberculosis in wild deer in Minnesota, USA. Veter Med Int 2011:413240. https://doi.org/10.4061/2011/413240

Carstensen M, O'Brien DJ, Schmitt SM (2011) Public acceptance as a determinant of management strategies for bovine tuberculosis in free-ranging U.S. wildlife. Vet Microbiol 151(1–2):200–204

Cieza AC, Oberhauser J, Bickenbach RN, Jones TB, Ustun N, Kostanjsek JN, Morris JN, Chatterji S (2006) Health is not just the absence of disease. Int J Epidemiol 45:586–587

Clegg SR, Mansfield KG, Newbrook K, Sullivan LE, Blowey RW, Carter SD, Evans NJ (2015) Isolation of digital dermatitis treponemes from hoof lesions in wild north American elk (*Cervus elaphus*) in Washington state, USA. J Clin Microbiol 53:88–94

Concha C, Yan Y, Arp A, Quilarque E, Sagel A, Pérez de León A, McMillan WO, Skoda S, Scott MJ (2020) An early female lethal system of the New World screwworm, *Cochliomyia hominivorax*, for biotechnology-enhanced SIT. BMC Genet 21:143. https://doi.org/10.1186/s12863-020-00948-x

Cooper SK, Hoover CE, Henderson DM, Haley NJ, Mathiason CK, Hoover EA (2019) Detection of CWD in cervids by RT-QuIC assay of third eyelids. PLoS One 14(8):e0221654. https://doi.org/10.1371/journal.pone.0221654

Decker DJ, Riley SJ, Siemer WF (2012) Human dimensions of wildlife management. John Hopkins Univ. Press, Baltimore, p 304

Decker DJ, Wild MA, Riley SJ, Siemer WF, Miller MIM, Leong KM, Powers JG, Rhyan JC (2006) Wildlife disease management: a manager's model. Hum Dimens Wildl 11(3):151–158. https://doi.org/10.1080/10871200600669908

Deem SL, Karesh WB, Weisman W (2001) Putting theory into practice: wildlife health in conservation. Conserv Biol 15:1224–1233

DeVivo MT, Edmunds DR, Kauffman MJ, Schumaker BA, Binfet J, Kreeger TJ, Richards BJ, Schatzl HM, Cornish TE (2017) Endemic chronic wasting disease causes mule deer population decline in Wyoming. PLoS One 12(10):e0186512. https://doi.org/10.1371/journal.pone.0186512

Dorn ML, Mertig AG (2005) Bovine tuberculosis in Michigan: stakeholder attitudes and implications for eradication efforts. Wildl Soc Bull 33:539–552. https://doi.org/10.2193/0091-7648(2005)33[539:BTIMSA]2.0.CO;2

Drees KP, Lorch JM, Puechmaille SJ, Parise KL, Wibbelt G, Hoyt JR, Sun K, Jargalsaikhan A, Dalannast M, Palmer JM, Lindner DL, Kilpatrick AM, Pearson T, Keim PS, Blehert DS, Foster JT (2017) Phylogenetics of a fungal invasion: origins and widespread dispersal of white-nose syndrome. MBio 8:e01941. https://doi.org/10.1128/mBio.01941-17

Edmunds DR, Kauffman MJ, Schumaker BA, Lindzey FG, Cook WE, Kreeger TJ, Googan RG, Cornish TE (2016) Chronic wasting disease drives population decline of white-tailed deer. PLoS One 11(8):e0161127. https://doi.org/10.1371/journal.pone.0161127

Eyles J, Furgal C (2002) Indicators in environmental health: identifying and selecting common sets. Can J Public Health 93:S62–S67

Felbab-Brown V (2021) Preventing pandemics through biodiversity conservation and smart wildlife trade regulation. Editorial brief in brookings blueprints for american renewal & prosperity. https://www.brookings.edu/research/preventing-pandemics-through-biodiversity-conservation-and-smart-wildlife-trade-regulation/

Fischer JR, Davidson WR (2005) Reducing risk factors for disease problems involving wildlife. In Proceedings, North American wildlife and natural resources conference, Washington, DC, pp 289–309

Fischman RL (2003) The national wildlife refuges: coordinating a conservation system through law. Island Press, Washington, p 278

Freyfogle ET, Goble DD, Wildermuth TA (2019) Wildlife law: a primer, 2nd edn. Island Press, Washington, p 339

Friend M, Franson JC, Ciganovich EA (1999) Field manual of wildlife diseases: general field procedures and diseases of birds. U.S. Department of the Interior, U.-S. Geological Survey, p 426

FWS US (2011) A national plan for assisting states, federal agencies, and tribes in managing white-nose syndrome in bats. https://pubs.er.usgs.gov/publication/70039214

Gillin CM, Fischer JR (2018) State management of wildlife disease. In: Ryder TR (ed) State wildlife conservation and management. John Hopkins University Press, New York, pp 176–193

Gillin CM, Mawdsley JR (2018) AFWA technical report on best management practices for surveillance, management and control of chronic wasting disease. Association of Fish and Wildlife Agencies, Washington, p 111

Gortázar C, Acevedo P, Ruiz-Fons F, Vicente J (2006) Disease risks and overabundance of game species. Eur

J Wildl Res 52:81–87. https://doi.org/10.1007/s10344-005-0022-2
Gortázar C, Diez-Delgado I, Barasona JA, Vicente J, De La Fuente J, Boadella M (2015) The wild side of disease control at the wildlife-livestock-human interface: a review. Front Vet Sci 1:27. https://doi.org/10.3389/fvets.2014.00027
Grant EHC, Muths E, Katz RA, Canessa S, Adams MJ, Ballard JR, Berger L, Briggs CJ, Coleman JTH, Gray MJ, Harris MC, Harris RN, Hossack B, Huyvaert KP, Kolby J, Lips KR, Lovich RE, McCallum HI, Mendelson JR III, Nanjappa P, Olson DH, Powers JG, Richgels KLD, Russell RE, Schmidt BR, Spitzen-van der Sluijs A, Watry MK, Woodhams DC, White CL (2017) Using decision analysis to support proactive management of emerging infectious wildlife diseases. Front Ecol Environ 15(4):214–221
Gwizdz B (2004) NRC needs to have courage with bait ban. The Grand Rapids Press, Grand Rapids, Michigan
Hayden E (2008) Biological tools revamp disease classification. Nature 453:709. https://doi.org/10.1038/453709a
Henderson DM, Davenport KA, Haley NJ, Denkers ND, Mathiason CK, Hoover EA (2015a) Quantitative assessment of prion infectivity in tissues and body fluids by real-time quaking-induced conversion. J Gen Virol 96(Pt 1):210–219. https://doi.org/10.1099/vir.0.069906-0
Henderson DM, Denkers ND, Hoover CE, Garbino N, Mathiason CK, Hoover EA (2015b) Longitudinal detection of prion shedding in saliva and urine by chronic wasting disease-infected deer by real-time quaking-induced conversion. J Virol 89(18):9338–9347
Huver JR, Koprivnikar J, Johnson PTJ, Whyard S (2015) Development and application of an eDNA method to detect and quantify a pathogenic parasite in aquatic ecosystems. Ecol Appl 25:991–1002. https://doi.org/10.1890/14-1530.1
Killian G, Fagerstone K, Kreeger T, Miller L, Rhyan J (2007) In: Nolte DL, Arjo WM, Stalman DH (eds) Management strategies for addressing wildlife disease transmission: the case for fertility control, pp 265–271
Kilpatrick AM, Randolph SE (2012) Drivers, dynamics, and control of emerging vector-borne zoonotic diseases. Lancet 380(9857):1946–1955
Kuiken T, Leighton FA, Fouchier RAM, LeDuc JW, Peiris JSM, Schudel A, Stohr K, Osterhaus ADME (2005) Pathogen surveillance in animals. Science 309:1680–1681
Langwig KE, Voyles J, Wilber MQ, Frick WF, Murray KA, Bolker BM, Kilpatrick AM (2015) Context-dependent conservation responses to emerging wildlife diseases. Front Ecol Environ 13:195–202
Makau DN, Vander Waal K, Kincheloe J, Wells SJ (2020) Implications of farmed-cervid movements on the transmission of chronic wasting disease. Prev Vet Med 182: 105088
Miller MW, Fischer JR (2016) The first five (or more) decades of chronic wasting disease: lessons for the five decades to come. Trans N Am Wildl Nat Resour Conf 81:110–120
Miller MW, Runge JP, Holland AA, Eckert MD (2020) Hunting pressure modulates prion infection risk in mule deer herds. J Wildl Dis 56(4):781–790. https://doi.org/10.7589/JWD-D-20-00054
Miller MW, Williams ES (2004) Chronic wasting disease of cervids. In: Harris DA (ed) Mad cow disease and related spongiform encephalopathies. Current topics in microbiology and immunology. Springer, Berlin, Heidelberg, p 284. https://doi.org/10.1007/978-3-662-08441-0_8
Mohler JR (1926) Foot and mouth disease, with special reference to the outbreaks in California in 1924 and Texas in 1924 and 1925. US Dept of Agric Dept 400:82
Needham MD, Vaske JJ, Manfredo MJ (2004) Hunters' behavior and acceptance of management actions related to chronic wasting disease in eight states. Hum Dimens Wildl 9:211–231
Nishi JS, Elkin BT, Ellsworth TR (2002) The Hook Lake wood bison recovery project. Ann N Y Acad Sci 969: 229–235. https://doi.org/10.1111/j.1749-6632.2002.tb04384.x
O'Brien DJ, Schmitt SM, Fitzgerald SD, Berry DE (2011) Management of bovine tuberculosis in Michigan wildlife: current status and near-term prospects. Vet Microbiol 151:179–187
O'Brien DJ, Schmitt SM, Fitzgerald SD, Berry DE, Hickling GJ (2006) Managing the wildlife reservoir of *Mycobacterium bovis*: the Michigan, USA, experience. Vet Microbiol 112:313–323
O'Roke EC (1928) Intestinal parasites of wild ducks and geese. Calif Fish Game 14(4):286–296
Salkeld DJ (2017) Vaccines for conservation: plague, prairie dogs and black-footed ferrets as a case study. Ecohealth 14:432–437. https://doi.org/10.1007/s10393-017-1273-6
Schmitt S, O'Brien D, Bruning-Fann C, Fitzgerald S (2002) Bovine tuberculosis in Michigan wildlife and livestock. Ann N Y Acad Sci 969:262–268. https://doi.org/10.1111/j.1749-6632.2002.tb04390.x
Schmitt SM, Fitzgerald SD, Cooley TM, Bruning-Fann CS, Sullivan L, Berry D, Carlson T, Minnis RB, Payeur JB, Sikarski J (1997) Bovine tuberculosis in free-ranging white-tailed deer from Michigan. J Wildl Dis 33:749–758
Schreiner CL, Nuismer SL, Basinski AJ (2020) When to vaccinate a fluctuating wildlife population: is timing everything? J Appl Ecol 57:307–319. https://doi.org/10.1111/1365-2664.13539
Sieber N, Hartikainen H, Vorburger C (2020) Validation of an eDNA-based method for the detection of wildlife pathogens in water. Dis Aquat Organ 141:171. https://doi.org/10.3354/dao03524
Skoda SR, Phillips PL, Welch JB (2018) Screwworm (Diptera: Calliphoridae) in the United States: response to and elimination of the 2016–2017 outbreak in Florida. J Med Entomol 55(4):777–786. https://doi.org/10.1093/jme/tjy049

Slate D, Algeo TP, Nelson KM, Chipman RB, Donovan D, Blanton JD, Niezgoda M, Rupprecht CE (2009) Oral rabies vaccination in North America: opportunities, complexities, and challenges. PLoS Negl Trop Dis 3: e549. https://doi.org/10.1371/journal.pntd.0000549
Stephen C (2013) Toward a new definition of animal health: lessons from the Cohen commission and the SPS agreement. J Public Sector Manage 43:1
Stephen C (2014) Toward a modernized definition of wildlife health. J Wildl Dis 50(3):427–430
Sutmoller P (2002) The fencing issue relative to the control of foot-and-mouth disease. In: Gibbs EPJ, Bokma BH (eds) Domestic animal/wildlife interface: issue for disease control, conservation, sustainable food production, and emerging diseases, vol 969. Annals of the New York Academy of Sciences, New York, pp 191–200
Turner W, Kausrud K, Beyer W, Easterday WR, Barandongo ZR, Blaschke E, Cloete CC, Lazak J, Van Ert MN, Ganz HH, Turnbull PCB, Chr N (2016) Lethal exposure: an integrated approach to pathogen transmission via environmental reservoirs. Sci Rep 6: 27311. https://doi.org/10.1038/srep27311
Vaske JJ, Timmons NR, Beaman J, Petchenik J (2004) Chronic wasting disease in Wisconsin: hunter behavior, perceived risk and agency trust. Hum Dimens Wildl 9:193–209
Voyles J, Kilpatrick AM, Collins JP, Fisher MC, Frick WF, McCallum H, Willis CKR, Blehert DS, Murray KA, Puschendorf R, Rosenblum EB, Bolker BM, Cheng TL, Langwig KE, Lindner DL, Toothman M, Wilber MQ, Briggs CJ (2014) Moving beyond too little, too late: managing emerging infectious diseases in wild populations requires international policy and partnerships. Ecohealth 12:404–407
Warnecke L, Turner JM, Bollinger TK, Lorch JM, Misra V, Cryan PM, Wibbelt G, Blehert DS, Willis CKR (2012) Inoculation of bats with European *Geomyces destructans* supports the novel pathogen hypothesis for the origin of white-nose syndrome. Proc Natl Acad Sci USA 109:6999–7003
Westmore A (1918) Duck sickness in Utah. U. S. Dept. Agr., Bul. No. 672. Biol. Survey, Washington
Whitworth T (2006) Keys to the genera and species of blow flies (Diptera: Calliphoridae) of America north of Mexico. Proc Entomol Soc Wash 108(3):689–725
Wild M, Hobbs NT, Graham MS, Miller MW (2011) The role of predation in disease control: a comparison of selective and nonselective removal on prion disease dynamics in deer. J Wildl Dis 47(1):78–93. https://doi.org/10.7589/0090-3558-47.1.78
Williams ES, Young S (1980) Chronic wasting disease of captive mule deer: a spongiform encephalopathy. J Wildl Dis 16:89–98
Wittrock J, Duncan C, Stephen C (2019) A determinants of health conceptual model for fish and wildlife health. J Wildl Dis 55(2):285–297
Wobeser G (2002) Disease management strategies for wildlife. Revue Scientifique et Technique de l'Office Int des Epizooties 21:159–178
Wobeser GA (2006) Essentials of disease in wild animals. Blackwell, Ames, p 243

Part III

Confronting Twenty-First Century Challenges to Wildlife Population Health

From Amazon Floods and Australian Wildfires to Human Spills and Explosions: What Disasters Mean to Wildlife

Christa Gallagher and Heather Fenton

Abstract

Natural and anthropogenic disasters have been impacting populations of people and animals and our shared environments for ages. Natural disasters are an important component of many ecosystems with many species relying upon natural fires, floods, storms, and other environmental changes to complete life cycles, reproduce, and colonize new geographic regions. However, in the face of global climate change and industrialization, the frequency and severity of disasters are increasing and are impacting global biodiversity in a manner that justifies consideration, planning, and response by wildlife managers to minimize the impacts of disasters. Preparedness and response approaches utilized in disasters are traditionally implemented with a public health focus. Here we propose the incorporation of wildlife and ecosystem health and sustainability considerations to disasters within the entire disaster management cycle and recommend the utilization of the One Health approach and Incident Command System in conjunction with resiliency strategies to deal with present day and emerging disasters. Specific examples relevant to wildlife health and biodiversity conservation in the face of natural and anthropogenic disasters are examined in this chapter.

Keywords

Disasters · Wildlife · One Health · Incident command system · Climate change · Disaster management · Risk reduction · Resiliency · Wildfires · Flooding · Drought

1 Introduction to Disasters

Disasters have existed for millennia and continue to plague societies and ecosystems to the present day. Disasters can be localized events or more widespread global phenomena. The number of natural disasters has more than quadrupled worldwide in the last 50 years. In 2020, there were 389 natural disasters recorded globally that killed 15,080 people and affected 98.4 million others (Centre for Research on the Epidemiology of Disasters (CRED) 2021). Anthropogenic (man-made) disasters are also similarly trending upwards over recent decades. The ongoing COVID-19 pandemic, with its unprecedented, catastrophic consequences had, by 2020s end, claimed almost two million lives, resulted in more than 80 million confirmed cases, triggered worldwide economic crisis with trillions of dollars lost, and caused extreme socioeconomic disruption, with substantial worsening of poverty and inequality in many low- and middle-income

C. Gallagher (✉) · H. Fenton
Ross University School of Veterinary Medicine, Basseterre, Saint Kitts and Nevis
e-mail: cgallagher@rossvet.edu.kn

C. Stephen (ed.), *Wildlife Population Health*,
https://doi.org/10.1007/978-3-030-90510-1_11

areas of the globe (World Bank 2020; World Health Organization 2020). Although perhaps too early to evaluate the effects of the COVID-19 pandemic impacts on wildlife and conservation, some experts warn that the net impacts will be negative, as the pandemic is expected to contribute to unfavorable perceptions of certain wildlife species (which could further threaten fragile populations such as the Chinese pangolin (*Manis pentadactyla*)), divert funding from conservation activities, and increase wildlife harvesting, poaching, and damage to ecosystems (Lindsey et al. 2020; Neupane 2020). No matter the origin of any disaster, these continuous adverse events have created a great sense of urgency to prepare, mitigate, respond, and build resilience to them. Disasters are expected to increase in severity and frequency due to planetary climate change, as well as increased human activities including population growth, land-use change, urbanization, industrialization, and social inequality (Arnold 2002). These anthropogenic drivers are interacting and amplifying in dynamic and unanticipated ways, which are predicted to have devastating harmful effects on all populations of humans and animals and their vital ecosystems.

According to the United Nations Office for Disaster Risk Reduction (UNDRR), a disaster is "a serious disruption of the functioning of a community or a society at any scale due to hazardous events interacting with conditions of exposure, vulnerability and capacity, leading to one or more of the following: human, material, economic and environmental losses and impacts ... The effect may test or exceed the capacity of a community or society to cope using its own resources" (UNDRR 2020). Disasters are largely defined by their humanitarian statistics and events are so-named when at least one of the following criteria is met: ten or more people are reported killed; 100 or more people are reported affected; a state of emergency is declared; or there is a call for international assistance (Centre for Research on the Epidemiology of Disasters 2020). Animals, including free-ranging species, are not mentioned in the definition of a disaster or its statistical criteria. When animals are depicted as casualties of disasters, the focus is most often placed on domesticated animals, companion animals, and livestock species, while wild animals may be disregarded or ignored except by special interest groups. However, all types of animals can be impacted by the profound and overwhelming effects of these events because they share the same vulnerabilities as humans. Unlike domesticated animals, wildlife tends to not be directly cared for or supported by human interventions, although there are a number of exceptions with food provisioning and rehabilitation (Cox and Gaston 2018; Taylor-Brown et al. 2019). Any disaster could cause significant morbidity and mortality of wildlife and call for human aid and assistance to prevent severe population losses, in addition to responding to prevailing public sentiment. Distribution of resources may be skewed toward reactive or immediate responses rather than long-term measures designed to maximize overall conservation outcomes. The legal jurisdiction, legislation, and responsibility for wildlife conservation and management are often shared by multiple levels of government that can complicate a coordinated response.

2 Profiling Disasters

Disasters have historically been characterized by their cause, onset, frequency, and impact. Natural hazards typically comprise the largest proportion of disasters (about 70%) (Rondeau et al. 2020) and include geophysical, meteorological, hydrological, climatological, extraterrestrial (as with aster-/meteoroids), or biological phenomena (Centre for Research on the Epidemiology of Disasters 2009). Naturally caused disasters include hurricanes, floods, wildfires, earthquakes, and landslides, or in the case of biological disasters, infectious disease as seen in the COVID-19 pandemic, which arguably has anthropogenic components. Wildlife plays a key role in the transmission of most emerging zoonotic diseases by serving as reservoirs, sources, and conduits for the transmission of communicable diseases across interfaces with other wildlife species, livestock, the environment, and ultimately humans (Kruse et al. 2004). Infectious pathogens originating in wildlife include examples such as West Nile virus (WNv), Ebola virus, human immunodeficiency virus (HIV), rabies, and many others that have substantially impacted human health, agriculture, and wildlife-based economies (Bengis et al. 2004). Anthropological disasters, which may be intentional or unintentional, are caused by human action, inaction, or error. They are typically classified as either technological (i.e., transportation accidents, engineering failures such as structural collapse, explosions, and environmental disasters associated with a spill of a toxic substance) or sociological disasters (i.e., war, acts of terrorism, and civil unrest). Although the origins of disasters are described as distinct categories, there is sometimes overlap between natural and unnatural causes, as seen with intentionally set wildfires or incidents of bio-and agroterrorism.

A disaster's onset may be variable in terms of time scale and may be described as slow and insidious as in a drought, or conversely, a sudden incident such as an earthquake or volcanic eruption (Fig. 1). Their occurrences may fall somewhere in between these divergent time scales and have more variable timing like what might be expected with infectious disease or invasive species perturbation. Disasters may be: (1) frequent and expected to occur at certain intervals, as seen with tropical storms; (2) infrequent, such as volcanic eruptions; and (3) variable frequency, as with incursions of infectious disease in people and animals.

Disasters are largely described and recognized by the severity of their impacts. Although not historically well documented outside of specific incidents localized to a specific region or location, wildlife involvement in disasters represents a growing body of literature. As the health and well-being of wild animals are linked to their environment, significant damage to the natural environment can have immediate to long-lasting impact, usually depicted as primary, secondary, and tertiary effects (as illustrated for oil spills in Fig. 2). The severity of impact of a major natural disturbance and resultant population decline is primarily dependent on the intensity (strength) and type of disturbance, and the specific vulnerability of a given species or ecosystem to that disturbance (Iwasaki and Noda 2018). A triggering event is common to all disasters that cause devastating damage to wild animals and their natural environments. A full description of the spectrum of disasters that could potentially impact wildlife goes beyond the scope of this

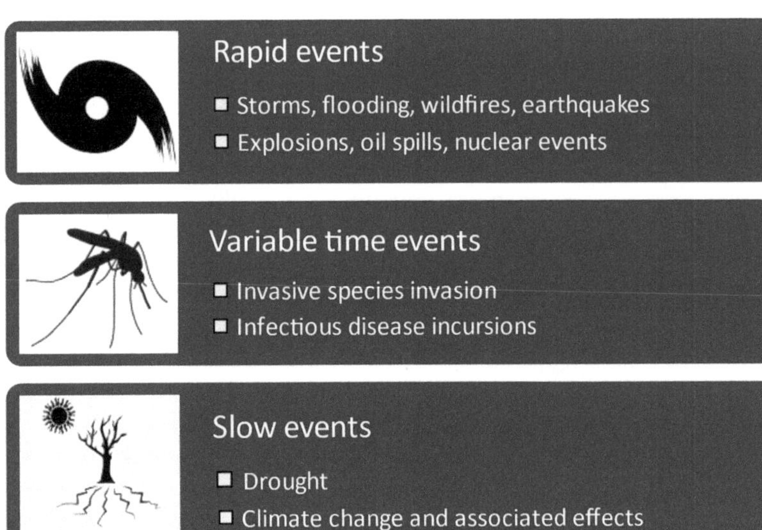

Fig. 1 Time onset of disasters affecting wildlife

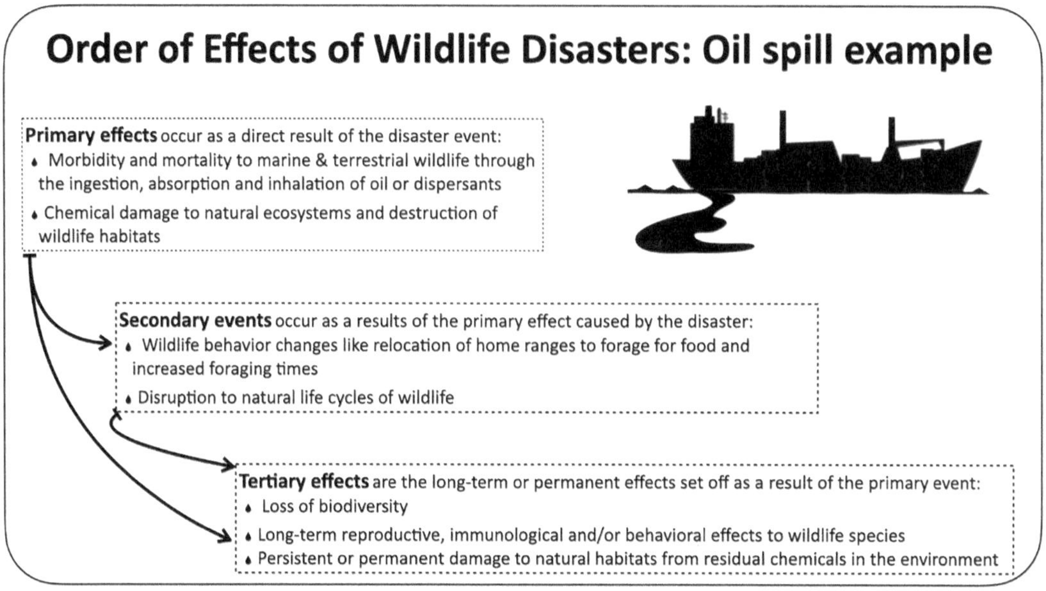

Fig. 2 Order of effects of wildlife disasters: oil spill example

chapter. However, some key examples are discussed along with their impacts and implications for future management considerations.

3 Disaster Management

Wildlife exists within complex socioecological systems (Daszak 2005; Petrosillo et al. 2015), implying that a systems-based approach to management would best address the complexities of disasters that impact wildlife. The disaster management cycle is used in most disaster response situations to organize, mobilize, prioritize, and manage resources and responsibilities in a crisis (Fig. 3). The four phases of the disaster management cycle including mitigation, planning, response, and recovery collectively aim for the prevention or reduction in injury, losses of life, and property, which could be reframed as direct mortality, morbidity, and availability of adequate habitat and food resources for wildlife. Additionally, disaster management measures aim to reduce other second- and third-order effects of adverse events, and provide for prompt and proper assistance to disaster victims and achieving the most rapid and successful recovery possible. Although unique actions may occur during each stage, they are not always distinct and are considered continuous. Although typically considered for humanitarian disasters, all four phases should be considered for events involving and affecting wildlife, which should be reflected in plans and actions for optimal results. Notably, the response phase is critical to reduce morbidity and mortality of affected animals (as well as people considering zoonotic risks and human–wildlife interactions in disasters), and is highly recognized and monitored by the public during a crisis (Lunney and Moon 2012). However, the most positive outcomes for people and animals are ones that stem from the other three disaster management phases that call for deliberate, advanced planning and preparation pre-event, and continuous longer term efforts that may contribute to restructuring for a better future (Heath and Linnabary 2015).

Planning for potential crises should be managed through an "all hazards" operational approach which: (1) identifies the full range of threats and risks that are most probable and likely to cause disruption and/or damage in a given area/region and (2) plans for threats through consideration of the aspects that are common to all of the

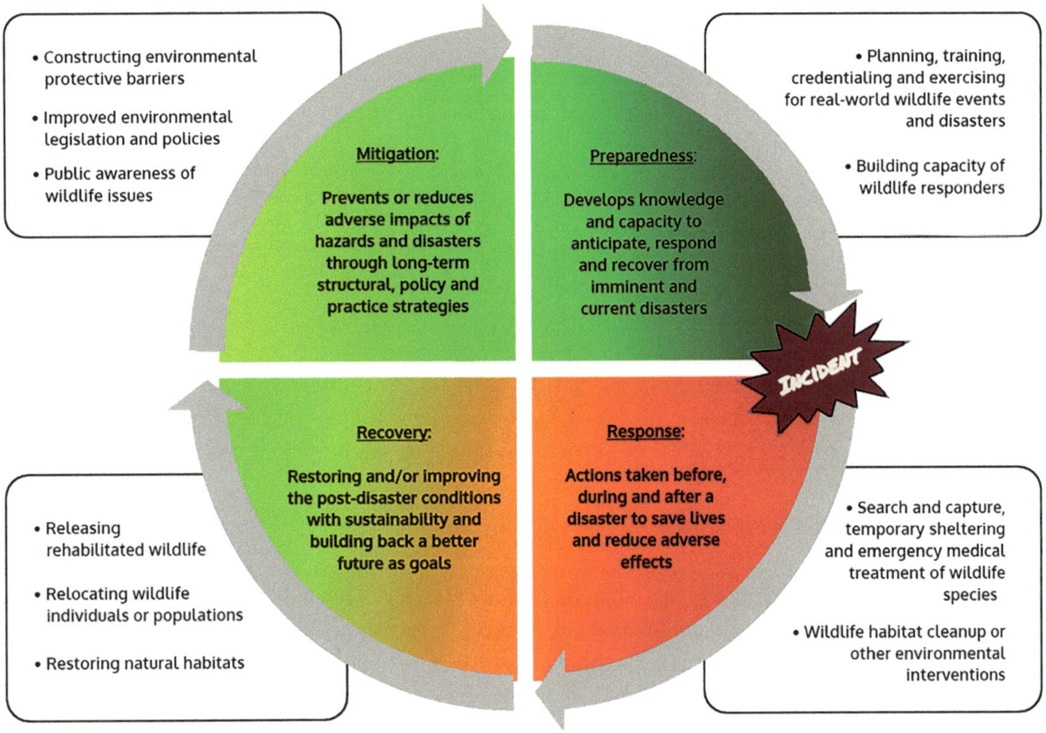

Fig. 3 The disaster management cycle with wildlife examples

hazards, not on a hazard-by-hazard basis (World Health Organization 2021). This preferred approach puts the focus on the consequence of the hazard rather than its cause. No matter if the hazard was fire, flooding, or man-made, each would require personnel and resources needed for the tactical response involved in the rescue and medical treatment of wildlife in addition to conservation planning that acknowledges the potential impact of disasters on free-ranging wildlife populations. Recent experience dictates that emergency planners should also plan for multiple concurrent disasters, as well as sequential disasters (Marjanishvili 2012; Ruiter et al. 2020). For example, any disaster that occurred globally in 2020 was compounded by the COVID-19 pandemic, causing cascading effects that added complexity to the existing crisis and any response to it (Sohrabizadeh et al. 2021).

The Disaster Risk Equation summarizes key attributes that lead to the generation of risks from hazards (Fig. 4) (International Federation of Red Cross and Red Crescent Societies 2020). A caveat to recognize from this equation is that while the hazard itself may not be reduced, or minimally so, by human interventions vulnerability to the hazard and our capacity to manage it is modifiable. Both should be continuously addressed through disaster management, policymaking, funding sustainability, and individual and societal resilience strategies.

4 Key Considerations in Approaching Disasters That Involve Wildlife

4.1 One Health Approach

The One Health approach should be considered with disasters involving wildlife. One Health is "a collaborative, multisectoral and transdisciplinary approach—working at local, regional, national and global levels—to achieve optimal health and

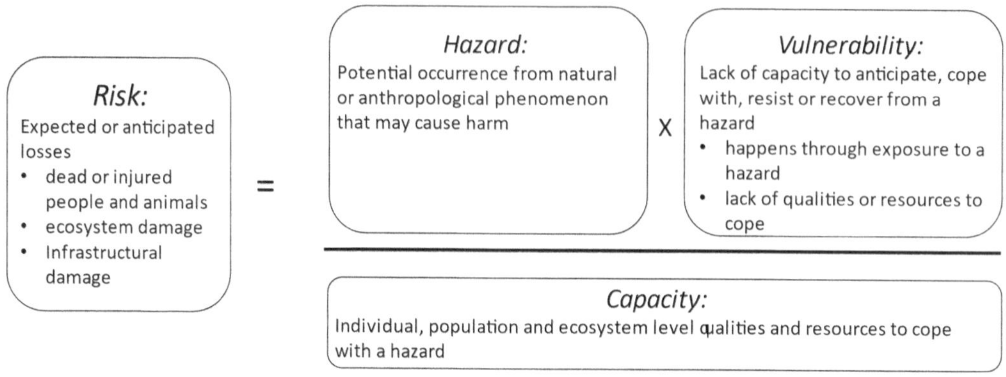

Text for image adapted from:
International Federation of Red Cross and Red Crescent Societies. (2020). *Risk Reduction: Disaster Preparedness training Programme.* https://www.ifrc.org/Global/Risk.pdf

Fig. 4 The disaster risk equation for human and animal populations

well-being outcomes recognizing the interconnections between people, animals, plants and their shared environment" (One Health Commission 2021). Recognizing that human, animal, and environmental involvement is common to all disasters necessitates the use of this approach. Furthermore, there are dynamic complex relationships that exist among these entities, such that if any of these domains are negatively affected, they could exert a detrimental effect on the remaining domains. Conversely, positive influences can lead to positive outcomes for all domains. Due to this interwoven relationship, all domains should be considered collectively and synergistically. As applied to disasters, One Health disaster management is a comprehensive systems-thinking approach that incorporates human, animal, and environment domains into the continuous and overlapping disaster management phases. Doing so ensures the consideration and response to the immediate and delayed impacts of a disaster from unique, but interconnected perspectives. Stakeholders from each domain will be represented, so it is inclusive, and minimizes the more traditional "siloed" approach, to produce more cohesive mitigation and response to these adverse events. The relationship between relevant stakeholders will be cemented by pre-event engagement for planning and participation in joint training and exercises.

Importantly, no single agency or organization holds the responsibility or has the necessary capacity to deal with a wildlife disaster. Using this One Health approach to a wildlife disaster could bring together emergency management, government regulatory agencies, wildlife health professionals, wildlife nongovernmental organizations (NGO), environmental experts, academia, traditional first responders, and public volunteers (Fig. 5). Community involvement is crucial as embedded members are most directly affected and have the most to lose in a crisis, in addition to providing crucial local and traditional knowledge. Many communities interact with wildlife either through consumptive or nonconsumptive practices with the value of wildlife difficult to quantitatively measure (Chardonnet et al. 2002). However, there is increasing recognition of the importance of wildlife in terms of sociocultural well-being of Indigenous people, consumption of wild game resources, and overall benefits of nature with reference to the interconnection between environmental and human health (Buttke et al. 2015; Decker et al. 2016). All of these factors have driven justification for consideration of wildlife in disasters as important community assets held within the public trust that should be included in disaster preparation, response, and planning.

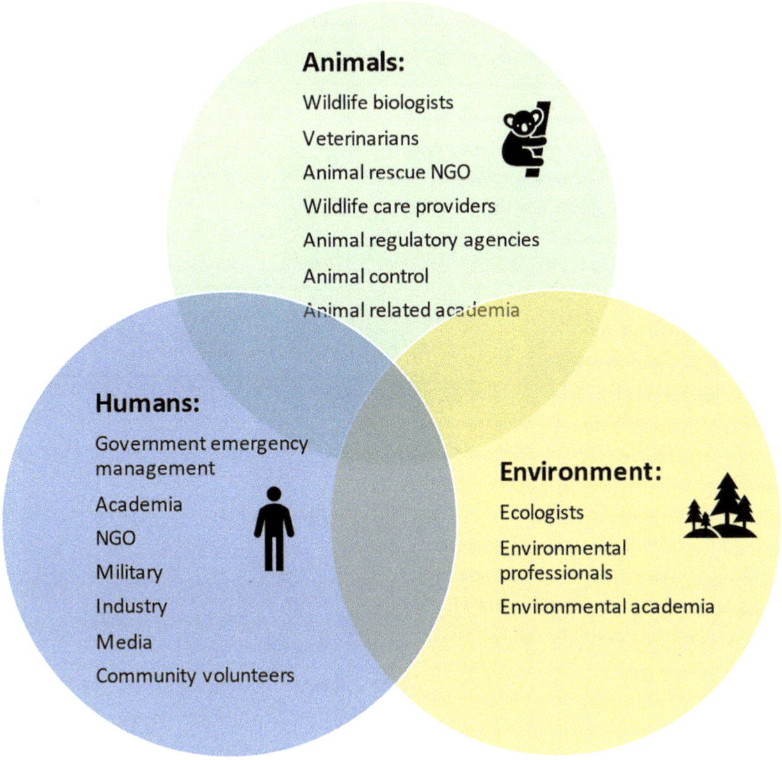

Fig. 5 Schematic of wildlife disaster practitioners in each of the One Health domains

4.2 Incident Command System

To make certain that the human, animal, and environmental health needs are met symbiotically in disasters, all managing agencies should consider the use of an established organizational system. One prominent example is the United States' National Incident Management System with its Incident Command System (ICS). Used and recognized internationally, ICS is a "standardized approach to the command, control, and coordination of on-scene incident management that provides a common hierarchy within which personnel from multiple organizations can be effective" (Federal Emergency Management Agency 2018) (Fig. 6). ICS provides an authoritative structure for integrated tactical response, which directs information and communication flow, analysis, and implementation of response measures across a wide variety of relevant agencies and partners.

This structure addresses the complexity across a spectrum of emergency events that vary in size and intensity, to reach common desired objectives. Using this system, civil organizations, NGOs, and the public can, through pre-event planning, be safely incorporated into the response to an event and utilized to the maximum extent possible. ICS has been successfully used in all types of disasters including natural disasters, terrorist acts, and chemical spills. It has been used in the animal realm with infectious disease and invasive species events (Burgiel 2020) and repeatedly with oil spills. The United States provides two good examples of ICS application: (1) The U.S. Department of Agriculture's Animal and Plant Health Inspection Service has used ICS responding to multiple invasive species incursions (i.e., Asian longhorned beetle [*Anaplophora glabipennis*], fruit fly [*Rhagolletis cerasi* (Linnaeus], and the New World screwworm [*Cochliomyia hominivorax*]) and it is central to their response framework (Veterinary Services, U.S. Department of Agriculture 2017; Burgiel 2020; United States Department of

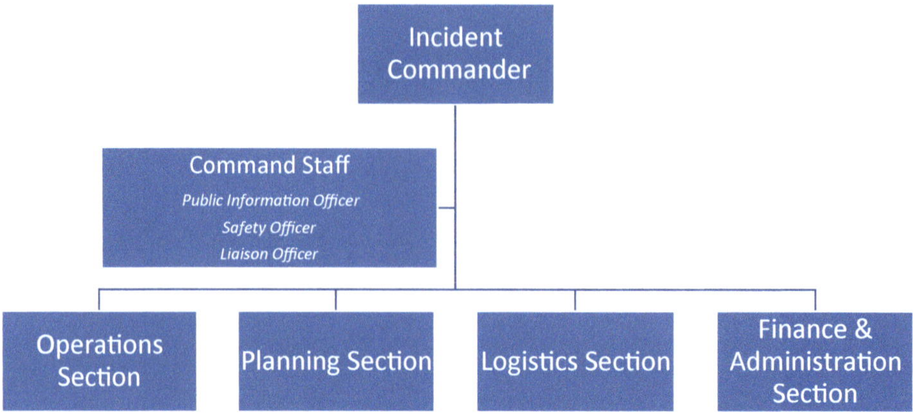

Fig. 6 Standard Incident Command System (ICS) organizational structure (Federal Emergency Management Agency 2018)

Agriculture 2020); (2) the Oiled Wildlife Care Network (Cho 2018) uses ICS to rescue oiled wild animal species in California. International organizations such as Focus Wildlife (Focus Wildlife International, Ltd. 2021) and International Fund for Animal Welfare (International Fund for Animal Welfare 2021) also promote and use the ICS framework for their work in wildlife emergency response. Many wildlife agencies incorporate the ICS framework for public emergency response to events such as wildfires and dangerous animal encounters, so agency managers should be familiar with the process and framework.

4.3 Broad Risk Reduction

The future approach to disasters is the adoption of early and purposeful mitigation of these events to improve societal and ecosystem resilience. To accomplish this contemporary disaster management now goes beyond traditional hazard identification (and associated risk) and aims for broader risk reduction. Disaster risk reduction reduces existing disaster risk from hazards and manages residual risk, as well as prevents new risk and impact for individuals and populations. The United Nations endorsed the Sendai Framework for Disaster Risk Reduction 2015–2030, and it was adopted globally in 2015 to strengthen economic, social, health, and environmental resilience (United Nations Office for Disaster Risk Reduction 2020). The concept of seeking to achieve resilience in disasters is an important one and should be integrated into wildlife emergencies. The United Nations Office for Disaster Risk Reduction defines resilience as: "the ability of a system, community or society exposed to hazards to resist, absorb, accommodate, adapt to, transform and recover from the effects of a hazard in a timely and efficient manner, including through the preservation and restoration of its essential basic structures and functions through risk management" (United Nations Office for Disaster Risk Reduction 2020). This definition emphasizes the need for societies and ecosystems to not only react to disturbances but also respond by resisting damage in addition to enabling rapid recovery from any perturbation with minimal negative impact. The practice of seeking resilient habitat and ecosystems for wildlife would call for resolve and cooperation from all involved stakeholders, but it could make a huge difference for their long-term survival and sustainability of biodiversity. Supporting and actively promoting an "all of society" approach that brings value and protection to wildlife and inextricable ecosystem services is of critical importance to working towards resiliency.

5 Examples of Disasters Involving Wildlife: Response and Management

5.1 Climate Change–Related Disasters

Global climate change and associated extreme climatic effects have the potential for significant threats to wildlife and ecosystems (see Chap. 13). Highlighted impacts associated with the direct impacts of climate change–related disasters include reduced populations, extinction, and extirpation of range-restricted species, as well as direct loss of habitat (e.g., from fire and flooding). More longer term and indirect impacts relevant to wildlife include shifts in species distributions (e.g., from repeated hurricanes), changes in phenology (timing of life-history events), introduction of invasive species, and emergence and expansion of zoonoses (Mawdsley et al. 2009). Biodiversity is expected to be enormously reduced with a sixth mass extinction event and consequent negative changes to ecosystem functioning (Pimm 2008; Bellard et al. 2012; Nunez et al. 2019) A number of Arctic ungulates, for example, have experienced large-scale population declines at least partially associated with rain-on-snow events, winter precipitation and tidal ice surges (Berger et al. 2018). South Pacific island countries are currently facing climatic changes that are threatening the existence of marine species through alterations to their habitats and food resources. Terrestrial wildlife is equally affected with similar habitat and food changes, but additionally experiencing heat stress, limitations to dry land mass, and increases in human–wildlife conflict (Bakare et al. 2020). Extreme climatic events, which have seen a rapid increase in frequency, adversely affect ecosystems by causing rapid mortality in populations and broad alterations to ecosystem structure and function (Abernathy et al. 2019). No matter the climatic modification, wildlife species will be forced to respond to these dynamic conditions by altering their geographic ranges, adapting to local conditions, or going extinct (Moritz and Agudo 2013).

5.2 Fire

Fire is a necessary natural disturbance to promote the persistence of some dependent flora and fauna worldwide, but conversely, fire events can be a major threat of species extinction. Trends indicate that many parts of the world are experiencing larger, intensified, and more frequent fire events due to climate and human influences (dos Reis et al. 2021). The Australian mega-fires of 2019–2020 were unprecedented in their spatial extent (approximately 97,000 km^2) and caused severe ecosystem damage. Approximately three billion wild animals including amphibians, reptiles, birds, and terrestrial mammals were killed or displaced due to over 15,000 fires that destroyed extensive regions of native habitat (Parrott et al. 2021). Seventy taxa had a substantial proportion (>30%) of habitat impacted, of which 21 taxa were already listed as threatened with extinction due to existing environmental conditions and invasive species (Ward et al. 2020). In the short term, these animals faced starvation, predators, and exposure to the elements in their quest to find food and shelter in the burnt and ash-laden environment. In the longer term, there could be large-scale ecological disruption due to food and habitat availability and decreased reproductive success. The fires removed access to food, water, and shelter, particularly from invasive predators, such as feral cats (*Felis catus*) and red foxes (*Vulpes vulpes*). Additional long-term negative impacts to affected species include chronic stress and exposure to ash and debris (Alexandra and Finlayson 2020).

Numerous groups were mobilized to specifically respond to the wild animal impacts of the Australian fires, particularly zoos and nongovernmental organizations. Despite recognition of massive losses, the response has been deemed a model of an effective response to a disaster focused entirely on wildlife welfare. Many organizations had been previously experienced and trained in team-based emergency response, had triage protocols, emergency response kits, and knowledge of emergency enclosures, which allowed for a rapid response albeit with a necessary scaling up to address the magnitude of impact of the fires (Parrott et al. 2021). Many strong collaborative partnerships were in place prior to the event among governments, zoos, non-governmental organizations, and communities. Ultimately the state government had jurisdiction and responsibility for the response with existing emergency governance procedures in place and an ICS structure. At the national level, a Wildlife and Threatened Species Expert Panel was mobilized that identified 119 species with high vulnerability to the fires that were prioritized as targets for government funding for recovery efforts. Weekly briefings were held amongst zoo-based conservation organizations in addition to targeted workshops held after the fires for key personnel. Individual animals were assessed in the field and either immediately euthanized if severe injuries were present, or transferred to wildlife triage units with adequate equipment and supplies for further evaluation and care by trained veterinarians and staff. Significant consideration was placed regarding species-specific appropriate housing and transport, as well as infection control measures that included quarantine and work-flow plans. After careful consideration, targeted and nutritionally appropriate food and water provisioning was provided for key species, such as the brush-tailed rock wallaby (*Petrogale peniculata*). Effective and clear communication with the media and public were highlighted as successes of the response with particular attention to how the public could assist with the response, while maintaining the ICS framework. The tremendous international and national support in terms of funding from donations to NGOs and public eyes on government agencies has been attributed to the successful communication efforts that were able to obtain resources, while still ensuring the resources were used effectively while maintaining animal welfare and human safety. Recovery efforts are ongoing and include nature-focused community programs built upon successful frameworks utilized in trauma-affected communities. After the catastrophic fires, the Australian Royal Commission into National Natural Disaster Arrangements produced a report that highlights a need to have a coordinated and collaborative approach to dealing with wildlife in future disasters, particularly including environmental assets within emergency management planning and response (Commonwealth of Australia 2020).

5.3 Floods

Intense flooding events were observed in the western Amazon basin during the high-water months of February through May in the years 2011–2015. In these hydrological events, floodwaters can cover vegetation that wildlife depends on for food and habitat, greatly impacting terrestrial species. In the Amazon basin, the increased extent and duration of flooding during the wet seasons caused a dramatic (95%) decrease in terrestrial mammal populations as they struggled with shrunken available habitat. Animals were reported to have drowned and suffered increased predation and competition for food and shelter resources as they were forced onto smaller swaths of land. Populations of arboreal species' that include birds, primates, and felids remained stable because they were presumably able to escape the physical effects of the floods. Long-term impacts of ecosystem disruption are unknown; however, there is a potential for prolonged recovery or permanent changes to terrestrial wildlife populations. Marine species populations that included fish and waterfowl increased or remained stable (Bodmer et al. 2018). Local Amazonians turn to fishing when flooding kills off terrestrial animals, which raised concerns regarding food security and sustainability of harvest of aquatic and remaining terrestrial animals, especially vulnerable species. Wildlife managers may need to adjust harvest limits to accommodate for extreme events that impact game species and include wild species in emergency response efforts (Wuczyński and Jakubiec 2013). Additional concerns with severe flooding in any region are "what the flooding leaves behind" in terms of potentially pathogenic microorganisms that have been associated with flooding events such as *Leptospira interrogans* (Smith et al. 2013) and toxins, such as mercury, that can accumulate in organisms at higher trophic levels (e.g., large predatory fish) that are consumed by humans in the region (Kasper et al. 2017). The virus that causes hepatitis A, a well-known waterborne pathogen of humans, was detected in environmental samples from the Amazon basin, presumably associated with sewage contamination of the environment (De Paula et al. 2007). Potential impacts of environmental contamination of floodplains with human-associated organisms and chemicals create an interface with wildlife that could lead to emerging infectious diseass outbreaks associated with soil disturbance, such as anthrax and botulism (Shin et al. 2010; Song et al. 2014), pathogen vectors proliferating post-flooding, such as for mosquito-borne Rift Valley fever (Fyumagwa et al. 2011), exchange of antimicrobial resistance genes among human, animal, and environmental bacteria (Pérez-Valdespino et al. 2021); morbidity and mortality in free-ranging wildlife species (Alho and Silva 2012); and other unpredictable deleterious impacts on wildlife.

5.4 Droughts and Desertification

Droughts and desertification affect a wide range of animal taxa and cause many natural resource losses based on their severity that is determined by geographical extent, duration, and intensity. Individual animal deaths of free-ranging wildlife species may be attributed to chronic dehydration and starvation with intra/interspecies competition and predation being contributing factors that could impact larger population-level effects and reproductive success. Land animals may be killed crossing roadways trying to gain access to food or water sources. Prolonged droughts may negatively impact reproduction by delaying breeding initiation, decreasing reproductive output and juvenile survival rates, and/or forcing range shifts and reducing survival and abundance (Cady et al. 2019). The drought experienced in Kruger National Park in South Africa has created conditions for the endangered southern white rhino (*Ceratotherium simum simum*) that have lead to an increased natural death rate and decreased birth rate resulting in significant declines in population size despite constant poaching rates (Ferreira et al. 2019). Within the semiarid areas of southern and eastern Africa, periodic die-offs of large herbivores are caused by droughts (Dudley et al. 2001) with a recent example in Zimbabwe (2019) that killed more than 100 African elephants (*Loxodonta africana*) (African Wildlife Foundation 2020). Prolonged changes in environmental temperature will impact available forage and potentially drive outbreaks of certain infectious diseases with an environmental component to their disease ecology, either directly, such as anthrax (Hampson et al. 2011) or associated with aggregation of species at limited watering holes, such as avian cholera (Rosen 1972) or due to changes in vector abundance, such as chikungunya emergence in East Africa (Chretien et al. 2007). Aquatic and semiaquatic species are highly sensitive to prolonged droughts due to lack of habitat and decreased opportunities for breeding (Miller et al. 2018). Creation of marine-protected areas with planning for impacts of disasters have been proposed as mechanisms to promote resiliency for marine species and ecosystems (Roberts et al. 2017). Artificial feeding or water supplementation for wildlife should be considered with a thorough risk assessment and consideration of unintentional consequences. Protection of adequate wetland habitat in land-use planning can help mitigate potential impacts of localized drought conditions for migratory species and may be more logistically feasible and economical in the long term than attempts at individual animal treatment or supplemental nutrition and water provision.

5.5 Man-Made Disasters

There are classic examples of anthropogenic disasters that have greatly impacted wildlife and their ecosystems that include periodic oil spills, and infrequent but catastrophic nuclear explosions. The largest accidental oil spill on record, the Deepwater Horizon spill, occurred in 2010. An estimated four million barrels of oil were discharged in the Gulf of Mexico (USA) (U.S. Environmental Protection Agency 2013), and nearly two million gallons of chemical dispersants were used to clean and control the spill, representing the first time such chemicals were used in a deep-sea environment (Barron 2012). Countless pelagic, tidal, and estuarine organisms, including sea turtles, marine mammals, and birds, were killed or harmed by direct toxic impacts of the spill, and their marine and coastal habitats destroyed. A decade later, this oceanic disaster is still harming wildlife (Meiners 2020).

Nuclear accidents have occurred in Chornobyl, Ukraine in 1986 and Fukushima, Japan in 2011. Both disasters killed multitudes of radiosensitive wildlife species and inflicted harmful health effects on a myriad of others. Over three decades later, wildlife and diverse flora in Chornobyl are still recovering but exhibiting increased presence and abundance (Baker and Chesser 2000). To date, there is no scientific consensus surrounding the impacts of long-term exposure to radiation for wildlife in this event (Beresford and Copplestone 2011). Blood and bone marrow abnormalities were observed in macaques (*Macaca fuscata*) that inhabited the Fukushima site, but the significance, causality, and extent of the impact to chronic exposure to ionizing radiation remain poorly understood for many species (Urushihara et al. 2018; Cannon and Kiang 2020).

5.6 COVID-19

The COVID-19 pandemic represents a stochastic shock that, despite being a disease primarily threatening human health, has caused great harm to wildlife. This has been propagated mostly through economic channels as global demand for natural resources decreased in the economic downturn and existing supply chains disrupted (Rondeau et al. 2020). Regional and international trade was further impacted through border restrictions and closures, lockdowns, and other related government and organizational policy alterations. Nature and wildlife tourism were significantly diminished or halted resulting in conservation funding losses and backsliding of sustainable nature conservation efforts. The absence of active tourism and COVID illness in conservation officers led to concerns of increased poaching events (Rondeau et al. 2020). Overall, these system-wide perturbations to existing impoverished areas may bring further habitat loss from extractive activities and harvesting of wildlife, with heightened biodiversity loss.

6 Conclusion

The preceding examples demonstrate that while some disasters may be anticipated to some extent, the true scope of disaster impact, including population size changes, species' behavior deviations, and habitat alterations can never be fully realized or predicted. Disasters affecting wildlife species and their ecosystems have devastating expansive consequences far outside our ability to quantify, that sometimes cause entire generations of species to be lost, and recovery extremely prolonged or even prevented entirely, as some fail to survive. Consideration of buffers in terms of ensuring available habitat, harvest limits, and other conservation management actions should be included to accommodate and mitigate the impacts of disasters on biodiversity. As with any disaster response, a collaborative and coordinated approach with clear roles and responsibilities in addition to effective and clear communication, as well as opportunities for local and community involvement in response, is essential to successful disaster management. Free-ranging wildlife could and should be included within disaster planning and response to best ensure sustainability of animal, human, and environmental health in a holistic systems approach.

Disasters have been a part of the world's past and present and are predicted to continue more gravely in the future given the explosive changes and complexities that exist today. Wildlife, like humans, is vulnerable to the compounding results of natural and anthropogenic disasters but remain left out of many disaster management plans and long-term responses to climate change adaptation management. We must continue to promote the value of biodiversity and work for the protection of ecosystems as a critical planetary need and immense human responsibility. Planning for and robustly responding to disasters that involve wildlife will require will and initiative, closely followed by effective communication and coordination among researchers, practitioners, policymakers, wildlife managers, nongovernmental organizations, and deeply invested communities.

References

Abernathy HN, Crawford DA, Garrison EP, Chandler RB, Conner ML, Miller KV et al (2019) Deer movement and resource selection during hurricane Irma: implications for extreme climatic events and wildlife. Proc Biol Sci 286:1916

African Wildlife Foundation (2020) Turning to conservation as severe droughts, food insecurity spike [internet]. African Wildlife Foundation [cited 2021 Jun 15]. https://www.awf.org/news/turning-conservation-severe-droughts-food-insecurity-spike

Alexandra J, Finlayson CM (2020) Floods after bushfires: rapid responses for reducing impacts of sediment, ash, and nutrient slugs. Australasian J Water Resour 24(1): 9–11

Alho CJR, Silva JSV (2012) Effects of severe floods and droughts on wildlife of the Pantanal wetland (Brazil)—a review. Animals (Basel) 2(4):591–610

Arnold JL (2002) Disaster medicine in the 21st century: future hazards, vulnerabilities, and risk. Prehosp Disaster Med 17(1):3–11

Bakare AG, Kour G, Akter M, Iji PA (2020) Impact of climate change on sustainable livestock production and existence of wildlife and marine species in the South Pacific island countries: a review. Int J Biometeorol 64(8):1409–1421

Baker RJ, Chesser RK (2000) The chornobyl nuclear disaster and subsequent creation of a wildlife preserve. Environ Toxicol Chem 19(5):1231–1232

Barron MG (2012) Ecological impacts of the Deepwater horizon oil spill: implications for Immunotoxicity. Toxicol Pathol 40(2):315–320

Bellard C, Bertelsmeier C, Leadley P, Thuiller W, Courchamp F (2012) Impacts of climate change on the future of biodiversity. Ecol Lett 15(4):365–377

Bengis RG, Leighton FA, Fischer JR, Artois M, Mörner T, Tate CM (2004) The role of wildlife in emerging and re-emerging zoonoses. Rev Sci Tech 23(2):497–511

Beresford NA, Copplestone D (2011) Effects of ionizing radiation on wildlife: what knowledge have we gained between the Chernobyl and Fukushima accidents? Integr Environ Assess Manag 7(3):371–373

Berger J, Hartway C, Gruzdev A, Johnson M (2018) Climate degradation and extreme icing events constrain life in cold-adapted mammals. Sci Rep 8(1):1156

Bodmer R, Mayor P, Antunez M, Chota K, Fang T, Puertas P et al (2018) Major shifts in Amazon wildlife populations from recent intensification of floods and drought. Conserv Biol 32(2):333–344

Burgiel SW (2020) The incident command system: a framework for rapid response to biological invasion. Biol Invasions 22(1):155–165

Buttke DE, Decker DJ, Wild MA (2015) The role of one health in wildlife conservation: a challenge and opportunity. J Wildl Dis 51(1):1–8

Cady SM, O'Connell TJ, Loss SR, Jaffe NE, Davis CA (2019) Species-specific and temporal scale-dependent responses of birds to drought. Glob Chang Biol 25(8): 2691–2702

Cannon G, Kiang JG (2020) A review of the impact on the ecosystem after ionizing irradiation: wildlife population. Int J Radiat Biol 24:1–9

Centre for Research on the Epidemiology of Disasters (2009) EM-DAT: general classification [Internet]. [cited 2 Jun 2021]. https://www.emdat.be/classification

Centre for Research on the Epidemiology of Disasters (2020) Human cost of disasters: an overview of the last 20 years 2000-2019.

Centre for Research on the Epidemiology of Disasters (CRED) (2021) Disaster year in review 2020: global trends and perspectives [Internet]. https://www.emdat.be/publications

Chardonnet PH, Des Clers B, Fisher JR, Gerhold R, Jori F, Lamarque F (2002) The value of wildlife. Rev Sci Tech OIE 2(1):15–51

Cho E (2018) Response [Internet]. Oiled Wildlife Care Network, School of Veterinary Medicine. https://owcn.vetmed.ucdavis.edu/response

Chretien J-P, Anyamba A, Bedno SA, Breiman RF, Sang R, Sergon K et al (2007) Drought-associated chikungunya emergence along coastal East Africa. Am J Trop Med Hyg 76(3):405–407

Commonwealth of Australia (2020) Royal Commission into national natural disaster arrangements report [Internet]. https://naturaldisaster.royalcommission.gov.au/system/files/2020-11/Royal%20Commission%20into%20National%20Natural%20Disaster%20Arrangements%20-%20Report%20%20%5Baccessible%5D.pdf

Cox DT, Gaston KJ (2018) Human–nature interactions and the consequences and drivers of provisioning wildlife. Philos Trans Roy Soc: Biol Sci 373(1745): 20170092

Daszak P (2005) Emerging infectious diseases and the socio-ecological dimension. Ecohealth 2(4):239–240

De Paula VS, Diniz-Mendes L, Villar LM, Luz SLB, Silva LA, Jesus MS et al (2007) Hepatitis A virus in environmental water samples from the Amazon Basin. Water Res 41(6):1169–1176

de Ruiter MC, Couasnon A, van den Homberg MJC, Daniell JE, Gill JC, Ward PJ (2020) Why we can no longer ignore consecutive disasters. Earth's Future 8(3):e2019EF001425

Decker DJ, Schuler K, Forstchen AB, Wild MA, Siemer WF (2016) Wildlife health and public trust responsibilities for wildlife resources. J Wildl Dis 52(4): 775–784

dos Reis M, de Graca PML, Yanai AM, CJP R, Fearnside PM (2021) Forest fires and deforestation in the Central Amazon: effects of landscape and climate on spatial and temporal dynamics. J Environ Manage 15(288): 112310

Dudley JP, Criag GC, Gibson DSC, Haynes G, Klimowicz J (2001) Drought mortality of bush elephants in Hwange National Park. Zimb Afr J Ecol 39(2):187–194

Federal Emergency Management Agency (2018). ICS Review Document [cited 29 Nov 2021]. https://training.fema.gov/emiweb/is/icsresource/assets/ics%20review%20document.pdf

Ferreira SM, le Roex N, Greaver C (2019) Species-specific drought impacts on black and white rhinoceroses. PLoS One 14:1

Focus Wildlife International, Ltd (2021) Focus wildlife [cited 19 Jun 2021]. https://www.focuswildlife.org

Fyumagwa RD, Ezekiel MJ, Nyaki A, Mdaki ML, Katale ZB, Moshiro C et al (2011) Response to Rift Valley fever in Tanzania: challenges and opportunities. Tanzan J Health Res 13(5)

Hampson K, Lembo T, Bessell P, Auty H, Packer C, Halliday J et al (2011) Predictability of anthrax infection in the Serengeti. Tanzania J Appl Ecol 48(6): 1333–1344

Heath SE, Linnabary RD (2015) Challenges of managing animals in Disasters in the U.S. Animals 5(2):173–192

International Federation of Red Cross and Red Crescent Societies (2020) Risk reduction: disaster preparedness training programme [Internet]. https://www.ifrc.org/Global/Risk.pdf

International Fund for Animal Welfare (2021) International Fund for Animal Welfare [internet]. IFAW. https://www.ifaw.org/eu

Iwasaki A, Noda T (2018) A framework for quantifying the relationship between intensity and severity of impact of disturbance across types of events and species. Sci Rep 8(1):795

Kasper D, Forsberg BR, Amaral JH, Py-Daniel SS, Bastos WR, Malm O (2017) Methylmercury modulation in Amazon rivers linked to basin characteristics and seasonal flood-pulse. Environ Sci Technol 51(24): 14182–14191

Kruse H, Kirkemo A-M, Handeland K (2004) Wildlife as source of zoonotic infections. Emerg Infect Dis 10(12): 2067–2072

Lindsey P, Allan J, Brehony P, Dickman A, Robson A, Begg C et al (2020) Conserving Africa's wildlife and wildlands through the COVID-19 crisis and beyond. Nat Ecol Evol 4:1300

Lunney D, Moon C (2012) Disasters for wildlife: an analysis of media attention. Australian Zool 36(1):5–19

Marjanishvili SM (2012) Mitigating disasters in the 21st century. In Ancona, Italy, p 1017–1024. http://library.witpress.com/viewpaper.asp?pcode=SC12-085-2

Mawdsley JR, O'malley R, Ojima DS. (2009) A review of climate-change adaptation strategies for wildlife management and biodiversity conservation. Conserv Biol 23(5):1080–1089

Meiners J (2020) Ten years later, BP oil spill continues to harm wildlife—especially dolphins. Animals [cited 2021 Jun 19]. https://www.nationalgeographic.com/animals/article/how-is-wildlife-doing-now%2D%2Dten-years-after-the-deepwater-horizon

Miller DAW, Grant EHC, Muths E, Amburgey SM, Adams MJ, Joseph MB et al (2018) Quantifying climate sensitivity and climate-driven change in north American amphibian communities. Nat Commun 9(1):3926

Moritz C, Agudo R (2013) The future of species under climate change: resilience or decline? Science 341(6145):504–508

Neupane D (2020) How conservation will be impacted in the COVID-19 pandemic. Wildl Biol 27:2020

Nunez S, Arets E, Alkemade R, Verwer C, Leemans R (2019) Assessing the impacts of climate change on biodiversity: is below 2 °C enough? Clim Change 154(3):351–365

One Health Commission (2021) What is one health? One Health Commission. https://www.onehealthcommission.org/en/why_one_health/what_is_one_health/

Parrott ML, Wicker LV, Lamont A, Banks C, Lang M, Lynch M et al (2021) Emergency response to Australia's black summer 2019–2020: the role of a zoo-based conservation organisation in wildlife triage, rescue, and resilience for the future. Animals 11(6):1515

Pérez-Valdespino A, Pircher R, Pérez-Domínguez CY, Mendoza-Sanchez I (2021) Impact of flooding on urban soils: changes in antibiotic resistance and bacterial community after hurricane Harvey. Sci Total Environ 766:142643

Petrosillo I, Aretano R, Zurlini G (2015) Socioecological systems. In: Reference module in Earth systems and environmental sciences. Elsevier, Amsterdam, pp 1–7

Pimm SL (2008) Biodiversity: climate change or habitat Loss — which will kill more species? Curr Biol 18(3): R117–R119

Roberts CM, O'Leary BC, McCauley DJ, Cury PM, Duarte CM, Lubchenco J et al (2017) Marine reserves can mitigate and promote adaptation to climate change. PNAS 114(24):6167–6175

Rondeau D, Perry B, Grimard F (2020) The consequences of COVID-19 and other Disasters for wildlife and biodiversity. Environ Resource Econ 76(4):945–961

Rosen MN (1972) The 1970–71 avian cholera epornitic's impact on certain species. J Wildl Dis 8(1):75–78

Shin N-R, Byun SH, Chun JH, Shin JH, Kim YJ, Kim J-H et al (2010) An outbreak of type C botulism in waterbirds: Incheon. Korea J Wildlife Dis 46(3): 912–917

Smith JK, Young MM, Wilson KL, Craig SB (2013) Leptospirosis following a major flood in Central Queensland. Australia Epidemiol Infect 141(3):585–590

Sohrabizadeh S, Yousefian S, Bahramzadeh A, Vaziri MH (2021) Health system preparedness for coincidence of natural disasters and COVID-19: a systematic literature review. BMC Public Health 21:1

Song L, Zuo T, Lu Y, He J, Yu Y, Li Y et al (2014) A post-flood co-outbreak of anthrax and enterotoxemia in beef cattle of Liaoning Province in North-Eastern China. Int J Infect Dis 19:106

Taylor-Brown A, Booth R, Gillett A, Mealy E, Ogbourne SM, Polkinghorne A et al (2019) The impact of human activities on Australian wildlife. PLoS One 14(1): e0206958

United Nations Office for Disaster Risk Reduction (2020) Disaster risk reduction [Internet]. https://www.undrr.org/terminology/disaster-risk-reduction

United States Department of Agriculture 2020. USDA APHIS, Incident Command System. https://www.aphis.usda.gov/aphis/ourfocus/planthealth/plant-pest-and-disease-programs/sa_ics/ct_incident_command_system

Urushihara Y, Suzuki T, Shimizu Y, Ohtaki M, Kuwahara Y, Suzuki M et al (2018) Haematological analysis of Japanese macaques (Macaca fuscata) in the area affected by the Fukushima Daiichi Nuclear Power Plant accident. Sci Rep 8(1):16748

US Environmental Protection Agency (2013) Deepwater horizon—BP Gulf of Mexico oil spill. US EPA, Washington

Veterinary Services, United States Department of Agriculture (2017) Final report for the APHIS veterinary services response to the 2016–2017. Outbreak of New World Screwworm (NWS) in Florida. https://www.aphis.usda.gov/animal_health/emergency_management/downloads/public-nws-usdaaphis-final-report.pdf

Ward M, Tulloch AIT, Radford JQ, Williams BA, Reside AE, Macdonald SL et al (2020) Impact of 2019-2020 mega-fires on Australian fauna habitat. Nat Ecol Evol 4(10):1321–1326

World Bank (2020) The impact of COVID-19 on global poverty under worsening growth and inequality. https://blogs.worldbank.org/opendata/impact-covid-19-global-poverty-under-worsening-growth-and-inequality

World Health Organization (2020) Weekly epidemiological update—29 December 2020 [Internet]. https://www.who.int/publications/m/item/weekly-epidemiological-update%2D%2D-29-december-2020

World Health Organization (2021) Policy [Internet]. https://www.euro.who.int/en/health-topics/health-emergencies/from-disaster-preparedness-and-response/policy

Wuczyński A, Jakubiec Z (2013) Mortality of game mammals caused by an extreme flooding event in South-Western Poland. Nat Hazards 69(1):85–97

Urbanization

Claire Jardine

Abstract

Urbanization transforms the environment and changes how humans interact with wildlife. These changes can have dramatic impacts, both positive and negative, on the health of wildlife. Predicting the effect of urbanization on wildlife health is challenging. While some wildlife species thrive in urban areas, others are less tolerant, and urbanization is often associated with a loss of native species and declines in species richness. Populations of synanthropic species that live in urban areas often reach very high densities as a result of abundant, stable anthropogenic resources. However, the health of these animals may be negatively impacted by urbanization as a result of increased exposure to pollution/ contaminants, poor quality food resources, and increased exposure to some pathogens in urban areas. Increased transmission of pathogens to vulnerable wildlife hosts or to humans in urban areas is of particular concern. Increased human population densities and changing human perceptions of wildlife in urban areas can lead to increasing human-wildlife conflicts that require careful management to protect both human and wildlife health. As urbanization increases globally, further research is required to reveal patterns and processes affecting urban wildlife health throughout the world.

Keywords

Anthropogenic · Biodiversity · Human–wildlife interactions · Synanthropic · Urban · Urbanization

1 Introduction

Urbanization transforms the environment, converting formerly rural areas into urban settlements and shifting the distribution of the human population to urban areas (United Nations 2019). Researchers use a variety of metrics for identifying urban areas, ranging from the type and degree of built areas to human population density (Moll et al. 2019). In this chapter, we use the definition of Francis and Chadwick (2013) who interpreted urban areas broadly as a combination of (1) a high proportion of built environment and (2) a relatively high human population density within a regional context.

Over the coming decades, urbanization is expected to increase. It is estimated that by 2050, 68% of the world's population will be urban, up from 55% in 2018 (United Nations 2019). The degree and rate of urbanization vary globally, with 82% of the North American population already living in urban areas while Africa

C. Jardine (✉)
Ontario Veterinary College, University of Guelph, Guelph, ON, Canada
e-mail: cjardi01@uoguelph.ca

C. Stephen (ed.), *Wildlife Population Health*,
https://doi.org/10.1007/978-3-030-90510-1_12

has 43% of its population living in urban areas (in 2018; United Nations 2019). It is projected that there will be an additional 2.5 billion people added to the world's urban population by 2050, with almost 90% of this growth happening in Asia and Africa (United Nations 2019).

Some of the most notable impacts associated with urbanization are dramatic changes in human density, land cover, climate, and resource availability. Landscape changes combined with increased human density in urban areas lead to changes in contaminant and pollution exposure, including air, noise, and light pollution. It is clear to people living in urban areas that some wildlife species thrive in urban environments, but urbanization dramatically alters the composition of wildlife communities to assemblages dominated by synanthropic species that are tolerant of, and benefit from, their close association with humans. Some synanthropic species may achieve abundances that are orders of magnitude greater than the same species in nonurban areas. For example, Smith and Engeman (2002) reported a raccoon density of 238/km^2 in an urban park, in contrast to typical rural densities of around 10/km^2 (Rosatte et al. 2010). Many species are less tolerant of urban environments and urbanization is, therefore, often associated with a decline in species richness, loss of native species, and biotic homogenization (McKinney 2002; McKinney 2006).

Changes in human/wildlife interactions in urban areas combined with the transformation of abiotic and biotic factors associated with urbanization can have dramatic impacts on wildlife population health. In this chapter, we consider multiple factors that affect health, not just disease (see Chap. 1). Our discussion will focus on the impacts of urbanization on the access to determinants of wildlife health identified by Wittrock et al. (2019) as they apply to wildlife living in urban areas. The determinants of interest here are needs for daily living (habitat, resources); abiotic environment (climate, pollution, contaminants); biologic endowment (stress, body condition, genetics, disease); social environment (competition, population age structure); direct mortality pressures (predation, hunting); and human expectations.

2 Needs for Daily Living

Access to needs for daily living, including food and suitable habitats, are all impacted by urbanization. Buildings, roads, and other infrastructure take over previous natural or agricultural habitats leading to native habitat loss and habitat fragmentation (Liu et al. 2016). These changes in habitat generally lead to a decrease in wildlife diversity in urban areas (McKinney 2002) as relatively few species can adapt to urban conditions (McKinney 2006).

Anthropogenic sources of food, including garbage, bird feed, and pet food are abundant and often available year-round, supporting high population densities of synanthropic species that can take advantage of these resources (Bradley et al. 2007). For example, garbage dumps have been associated with positive health impacts for the wildlife exploiting them for food, including increased body condition, enhanced reproduction, increased abundance, and increased survival rate (reviewed by Plaza and Lambertucci 2017). However, garbage dumps are also associated with negative consequences for wildlife health, including exposure to pathogens and poisons (Plaza and Lambertucci 2017). More generally, urban food resources may be abundant, but the value of these resources may be reduced. Thus, urban habitat use has been associated with potentially negative health metrics, including increased cholesterol (Townsend et al. 2019), increased glycated serum protein, a marker for increased blood glucose concentrations (Schulte-Hostedde et al. 2018), and changes in the microbiome (Murray et al. 2020).

Anthropogenic resources have been associated with both increased and decreased pathogen infection risk depending on the nature of provisioning and the particular host–pathogen interaction (Becker et al. 2015). Food resources in urban areas are often aggregated leading to increased inter- or intraspecific contact, including fighting. Increased or altered contact rates in urban areas can have dramatic impacts on disease ecology in urban areas. Wright and Gompper (2005) found that there was an increase in the intensity and prevalence of endoparasites in raccoons on sites with experimentally increased food resources and concluded that "anthropogenic changes which alter resource availability may have important consequences for disease transmission in wildlife".

In addition to the impacts of the built environment on resource availability, the increased intensity of human activity in urban areas may also be associated with changes in behavior and activity patterns in wildlife, which can impact their ability to access resources. For example, Gaynor et al. (2018) found that wild mammals consistently increased their nighttime activity in response to human disturbance across continents, habitats, taxa, and human activities. Shifts in activity to avoid humans may impact health if the animal's efforts lead to reduced or altered access to resources.

3 Abiotic Environment

In addition to changes in the built environment that impact habitat and resource availability (discussed previously), climate and exposure to contaminants and pollution are dramatically altered in urban ecosystems. These components of urban environments can impact wildlife health directly and indirectly. Although localized urban cool island effects can occur in some areas, urban microclimates are typically warmer than outlying areas (Oke 1982). In addition to a typically warmer climate, urban areas also tend to have reduced seasonality, with milder winters, which may increase the survival and activity period for arthropod pathogen vectors (Bradley et al. 2007) or the environmental stages of some pathogens (e.g., *Leptospira*) (Lau et al. 2010). Milder winters combined with increased resource availability in urban areas may also reduce the individual-level impacts of disease (Bradley et al. 2007) leading to increased individual survival, but also potentially increasing opportunities for pathogen transmission (Riley et al. 2014b). Climate change can be expected to exacerbate these effects (Luber and McGeehin 2008). In some parts of the world, urban areas may be increasingly vulnerable to extreme weather events such as droughts and heavy rainstorms associated with climate change (Revi et al. 2014).

Contaminant exposure represents a major threat to wildlife populations (Wilcove and Master 2005) (see Chap. 16). Urbanization is often associated with increased pollution and environmental contamination (Shifaw et al. 2018; Riley et al. 2014a, b). Murray et al. (2019) found that toxicant loads were significantly higher in urban animals than in their nonurban conspecifics across all wildlife taxa. Wildlife may be exposed to pollutants in the air via inhalation (Sanderfoot and Holloway 2017) or to contaminants, like rodenticides, directly through malicious poisonings, via inadvertent ingestion of chemicals in the environment, or through secondary exposure via consumption of other exposed animals (Rodríguez-Estival et al. 2019). Exposure to contaminants can have both direct consequences such as acute toxicity or cause sublethal effects, including reproductive impairment and decreased immune competence (reviewed by Saaristo et al. 2018).

In addition to chemical contaminants, noise (e.g., vehicular traffic) and light pollution associated with urban areas may also impact wildlife health. For example, for species that rely on vocalizations for communication, noise pollution may lead to impaired communication which could have negative impacts on reproductive success (Ditchkoff et al. 2006; Halfwerk et al. 2011). Light pollution is an important factor in collision mortality for nocturnally migrating birds that are attracted to, and disoriented by, artificial lighting (Van Doren et al. 2021). Indirect effects, such as the disruption of immune function by artificial lighting, which has been found across a range of taxa, have potentially negative synergistic effects in combination with the spread of pathogens under changed climates (Longcore and Rich et al. 2016). In addition, there may be other subtle, widespread, and cumulative effects of nighttime lighting on ecological communities that are not yet recognized (Longcore and Rich 2004).

4 Biological Endowment

The inherited or predisposed biological capacity of wildlife to cope is made up of factors including body condition, response to stress, genetics, and disease (Wittrock et al. 2019). Wildlife in urban areas is likely to encounter different stressors and resources compared to their rural counterparts. Differences in body condition and stress responses have been reported for species living in urban versus rural areas. But, in a recent meta-analysis of wildlife health and urbanization, Murray et al. (2019) found no consistent differences in body condition or stress responses associated with urbanization. The authors concluded that changes in body condition and stress associated with urbanization may be species- and location-specific. Kark et al. (2007) found that successful urban birds share a combination of life history traits, suggesting that studies of how species traits affect their ability to adapt to urban ecosystems will be important for understanding the impact of urbanization on health outcomes (Murray et al. 2019).

Losses in genetic diversity in urban areas may result from genetic drift associated with the "(i) loss of natural habitat caused by fragmentation, (ii) founder effects associated with the establishment of new urban populations, and (iii) severe bottlenecks due to direct selection pressures from humans" (Johnson and Munshi-South 2017). Reduced genetic diversity is associated with reduced immune responses and increased vulnerability to pathogens and some cancers (Ujvari et al. 2018). It is important to note that some urban features, including habitat corridors, can also increase gene flow which may increase genetic diversity within urban populations (Johnson Munshi-South 2017), so it is not surprising that the effects of urbanization on the genetic structure of populations and associated health impacts vary among species, and as was described for body condition and response to stress above, likely depends on species-specific biology (Miles et al. 2019).

The occurrence and prevalence of pathogens in, and impact of disease on, urban wildlife is also variable. The changes in landscape and resource availability associated with urbanization and the resulting loss in wildlife diversity may lead to an overall decrease in pathogen diversity (Hassell et al. 2017). However, Murray et al. (2019) found a higher likelihood of infection by parasites transmitted through close contact, along with greater parasite abundance and diversity, in urban as compared to nonurban populations. They suggested that this was perhaps because some urban-adapted hosts live at higher densities due to abundant and patchily distributed food resources. The effects of urbanization on disease ecology have implications for both wildlife and public health (Bradley et al. 2007). For example, wildlife species less adapted to urban areas, which are susceptible to infection with pathogens circulating in high-density hosts, may be at increased risk of infection and there is the potential for increased transmission of certain zoonotic pathogens in competent synanthropic wildlife hosts (Hassell et al. 2017). In addition, reverse zoonotic transmission from humans to wildlife may also occur (Britton et al. 2019) and could pose a threat to wildlife populations with increased exposure to humans (Hassell et al. 2017).

5 Social Environment

The social environment of wildlife, including competition, reflects how communities can impact an individual's capacity to cope by influencing access to various resources (Wittrock et al. 2019). Urbanization modifies species interactions through the introduction of nonnative species, changes in social behavior, and by changes in composition of ecological communities (Alberti et al. 2020) with sometimes dramatic impacts on wildlife health. For example, the introduction of non-native gray squirrels (*Sciurus carolinensis*) has been associated with dramatic declines in native red squirrel (*Sciurus vulgaris*) populations in the United Kingdom (Tompkins et al. 2003). This decline is associated with both direct competition for resources between these two species and also through indirect competition, via an introduced squirrel paramyxovirus maintained in gray squirrels, but lethal to red squirrels (Bradely et al. 2007; Tompkins et al. 2003).

Changes in population age structure may impact susceptibility to disease and disease transmission and prevalence in populations (Cross et al. 2009), but differences in age structure along urbanization gradients have only rarely been examined and there is no clear consensus on the impact of urbanization on population age structure (Kozlovsky et al. 2021). Kozlovsky et al. (2021) found a higher proportion of juvenile chickadees (*Poecile atricapillus*) in urban areas in a city in Ontario, Canada while Evans et al. (2009) found a higher proportion of adult European blackbirds (*Turdus merula*) in urban compared to nonurban environments. As Kozlovsky et al. (2021) point out, additional studies are needed to address whether the observations are related to differences between European and North American species or whether differences in age structure across urban environments are species- and location-specific.

6 Direct Mortality Pressures

Mortality pressures affecting urban wildlife differ from what is seen in more natural areas and anthropogenic-related mortality generally increases with urbanization (Rodewald and Gehrt 2014). Some synanthropic urban species, like rats or rock pigeons (*Columba livia*) may be subject to increased mortality pressure from humans who target these species as nuisance animals in urban areas. Mortality may also be higher for some bird species in urban areas related to collisions with buildings, windows, and vehicles (Rodewald and Gehrt 2014). Many mammalian species are affected by traffic-related mortality in urban areas (Ditchkoff et al. 2006). Many smaller species, including reptiles and amphibians, are not easily able to cross roads, and mortality rates can reach 100% (Riley et al. 2014a, b; Aresco 2005). Larger and more mobile urban animals, such as carnivores or ungulates, are typically able to cross roads, but the high density in urban areas still creates significant mortality risk by forcing large animals to make frequent crossings (Riley et al. 2014a, b).

Human mortality pressure for hunted species, like deer, will likely be reduced in urban areas where hunting is not allowed. In addition, the density of natural predators in urban areas may be lower compared with more rural habitats. However, actual predatory pressures may be greater in urban areas because of domestic animals, such as dogs and cats (Ditchkoff et al. 2006). Cats are estimated to kill between 1.3 and 4.0 billion birds in the United States annually (Loss et al. 2013). This mortality estimate exceeds other direct sources of anthropogenic mortality and suggests that mortality from cat predation is likely to be substantial in all parts of the world where free-ranging cats occur (Loss et al. 2013). Although it is hard to quantify, Loss et al. (2013) suggest that cats are likely causing population declines for some bird species in some regions.

7 Human Expectations

How people value and perceive wildlife can have dramatic impacts on wildlife health. In urban areas, species like song birds, that are valued and enjoyed, may receive extra food resources at backyard feeders, while other species, like rats, are targeted and killed with poisons and traps. Human perceptions of wildlife are influenced by a complex interplay of factors, including where people live (Kimmig et al. 2020). Urban residents may be more tolerant of wildlife, because they view wildlife as beings with rights rather than as a food source (Patterson et al. 2003). However, even urban residents that live in the same place tend to have a variety of perceptions about wildlife and these human dimensions tend to be the dominant drivers of wildlife management in urban areas (McCance et al. 2017) (see Chap. 19). The rapid expansion of urban areas worldwide is leading to an increase in the frequency of human–wildlife encounters (Soulsbury and White 2015), including encounters that may result in negative outcomes (Schell et al. 2021) that may require intervention. These types of urban wildlife issues in much of North America can be considered wicked problems ("problems where different values lead to different interpretations of desirable outcomes and acceptable means of achieving them" McCrance et al. 2017) and will require approaches that incorporate the variety of values that people have toward urban wildlife (see Chap. 17).

8 Future Considerations

Urbanization can generate positive and negative health effects for wildlife that may be host-, parasite-, or region-specific, and predicting the overall effect of urbanization on wildlife health is challenging (Murray et al. 2019). As urbanization increases globally, it is clear that further research is required to better understand the patterns and processes affecting urban wildlife health throughout the world.

It is often assumed that movement of humans from rural to urban areas and the resulting concentration of populations in cities would ease the pressure on natural habitats (Güneralp et al. 2017). In many parts of Africa, the migration and subsequent concentration of people in urban areas have led to reduced rural populations (Güneralp et al. 2017). However, land speculation has also driven loss and fragmentation of rural rangelands close to cities and towns in some parts of Africa (Güneralp et al. 2017). Well managed urbanization in Africa could help African nations effectively conserve biodiversity (Güneralp et al. 2017) and preserve wildlife health.

Although urbanization will continue to increase, particularly in Asia and Africa (United Nations 2019) a number of countries and cities, particularly in developed nations, are shrinking (United Nations 2017). Human population declines and urban shrinkage could lead to increased ecosystem services and positive environmental outcomes (Haase et al. 2014), but the ecological consequences of de-urbanization, including impacts on zoonotic disease and wildlife health, are not well understood (Eskew and Olival 2018).

References

Alberti M, Palkovacs EP, Roches SD, Meester LD, Brans KI, Govaert L, Grimm NB, Harris NC, Hendry AP, Schell CJ, Szulkin M (2020) The complexity of urban eco-evolutionary dynamics. Bioscience 70(9):772–793

Aresco MJ (2005) Mitigation measures to reduce highway mortality of turtles and other herpetofauna at a North Florida lake. J Wildl Manage 69(2):549–560

Becker DJ, Streicker DG, Altizer S (2015) Linking anthropogenic resources to wildlife–pathogen dynamics: a review and meta-analysis. Ecol Lett 18(5):483–495

Bradley CA, Altizer S (2007) Urbanization and the ecology of wildlife diseases. Trends Ecol Evol 22(2): 95–102

Britton AP, Trapp M, Sabaiduc S, Hsiao W, Joseph T, Schwantje H (2019) Probable reverse zoonosis of influenza a (H1N1) pdm 09 in a striped skunk (*Mephitis mephitis*). Zoonoses Public Health 66(4):422–427

Cross PC, Drewe J, Patrek V, Pearce G, Samuel MD, Delahay RJ (2009) Wildlife population structure and parasite transmission: implications for disease management. In: Management of disease in wild mammals. Springer, Tokyo, pp 9–29

Ditchkoff SS, Saalfeld ST, Gibson CJ (2006) Animal behavior in urban ecosystems: modifications due to human-induced stress. Urban Ecosyst 9(1):5–12

Eskew EA, Olival KJ (2018) De-urbanization and zoonotic disease risk. Ecohealth 15(4):707–712

Evans KL, Gaston KJ, Sharp SP, McGowan A, Simeoni M, Hatchwell BJ (2009) Effects of urbanisation on disease prevalence and age structure in blackbird *Turdus merula* populations. Oikos 118(5): 774–782

Francis RA, Chadwick MA (2013) Urban ecosystems: understanding the human environment. Routledge, New York

Gaynor KM, Hojnowski CE, Carter NH, Brashares JS (2018) The influence of human disturbance on wildlife nocturnality. Science 360(6394):1232–1235

Güneralp B, Lwasa S, Masundire H, Parnell S, Seto KC (2017) Urbanization in Africa: challenges and opportunities for conservation. Environ Res Lett 13(1):015002

Haase D, Haase A, Rink D (2014) Conceptualizing the nexus between urban shrinkage and ecosystem services. Landsc Urban Plan 132:159–169

Halfwerk W, Bot S, Buikx J, van der Velde M, Komdeur J, ten Cate C, Slabbekoorn H (2011) Low-frequency songs lose their potency in noisy urban conditions. Proc Natl Acad Sci 108(35):14549–14554

Hassell JM, Begon M, Ward MJ, Fèvre EM (2017) Urbanization and disease emergence: dynamics at the wildlife–livestock–human interface. Trends Ecol Evol 32(1):55–67

Johnson MT, Munshi-South J (2017) Evolution of life in urban environments. Science 358:6363

Kark S, Iwaniuk A, Schalimtzek A, Banker E (2007) Living in the city: can anyone become an 'urban exploiter'? J Biogeogr 34(4):638–651

Kimmig SE, Flemming D, Kimmerle J, Cress U, Brandt M (2020) Elucidating the socio-demographics of wildlife tolerance using the example of the red fox (*Vulpes vulpes*) in Germany. Conserv Sci Pract 2(7):e212

Kozlovsky DY, Jarjour CA, Morand-Ferron J (2021) Urbanization is associated with differences in age class structure in black-capped chickadees (*Poecile atricapillus*). Urban Ecosyst 24(2):405–416

Lau CL, Smythe LD, Craig SB, Weinstein P (2010) Climate change, flooding, urbanisation and leptospirosis: fueling the fire? Trans R Soc Trop Med Hyg 104(10): 631–638

Liu Z, He C, Wu J (2016) The relationship between habitat loss and fragmentation during urbanization: an empirical evaluation from 16 world cities. PLoS One 11(4): e0154613

Longcore T, Rich C (2004) Ecological light pollution. Front Ecol Environ 2(4):191–198

Longcore T, Rich C (2016) Ecological and organismic effects of light pollution. eLS 2016:1–8

Loss SR, Will T, Marra PP (2013) The impact of free-ranging domestic cats on wildlife of the United States. Nat Commun 4(1):1–8

Luber G, McGeehin M (2008) Climate change and extreme heat events. Am J Prev Med 35(5):429–435

McCance EC, Decker DJ, Colturi AM, Baydack RK, Siemer WF, Curtis PD, Eason T (2017) Importance of urban wildlife management in the United States and Canada. Mammal Study 42(1):1–6

McKinney ML (2002) Urbanization, biodiversity, and conservation. Bioscience 52(10):883–890

McKinney ML (2006) Urbanization as a major cause of biotic homogenization. Biol Conserv 127(3):247–260

Miles LS, Rivkin LR, Johnson MT, Munshi-South J, Verrelli BC (2019) Gene flow and genetic drift in urban environments. Mol Ecol 28(18):4138–4151

Moll RJ, Cepek JD, Lorch PD, Dennis PM, Tans E, Robison T, Millspaugh JJ, Montgomery RA (2019) What does urbanization actually mean? A framework for urban metrics in wildlife research. J Appl Ecol 56(5):1289–1300

Murray MH, Lankau EW, Kidd AD, Welch CN, Ellison T, Adams HC, Lipp EK, Hernandez SM (2020) Gut microbiome shifts with urbanization and potentially facilitates a zoonotic pathogen in a wading bird. PLoS One 15(3):e0220926

Murray MH, Sánchez CA, Becker DJ, Byers KA, Worsley-Tonks KE, Craft ME (2019) City sicker? A meta-analysis of wildlife health and urbanization. Front Ecol Environ 17(10):575–583

Oke TR (1982) The energetic basis of the urban heat island. Q J Roy Meteorol Soc 108(455):1–24

Patterson ME, Montag JM, Williams DR (2003) The urbanization of wildlife management: social science, conflict, and decision making. Urban For Urban Green 1(3):171–183

Plaza PI, Lambertucci SA (2017) How are garbage dumps impacting vertebrate demography, health, and conservation? Glob Ecol Conserv 12:9–20

Revi A, Satterthwaite DE, Aragón-Durand F, Corfee-Morlot J, Kiunsi RBR, Pelling M, Roberts DC, Solecki W (2014) Urban areas. In: Field CB, Barros VR, Dokken DJ, Mach KJ, Mastrandrea MD, Bilir TE, Chatterjee M, Ebi KL, Estrada YO, Genova RC, Girma B, Kissel ES, Levy AN, MacCracken S, Mastrandrea PR, White LL (eds) Climate change 2014: impacts, adaptation, and vulnerability. part a: global and sectoral aspects. Contribution of working group II to the fifth assessment report of the Intergovernmental Panel on Climate Change. Cambridge University Press, Cambridge, New York, pp 535–612

Riley SP, Brown JL, Sikich JA, Schoonmaker CM, Boydston EE (2014a) Wildlife friendly roads: the impacts of roads on wildlife in urban areas and potential remedies. In: Urban wildlife conservation. Springer, Boston, pp 323–360

Riley SP, Serieys LE, Moriarty JG (2014b) Infectious disease and contaminants in urban wildlife: unseen and often overlooked threats. In: Urban wildlife conservation. Springer, Boston, pp 175–215

Rodewald AD, Gehrt SD (2014) Wildlife population dynamics in urban landscapes. In: Urban wildlife conservation. Springer, Boston, pp 117–147

Rodríguez-Estival J, Mateo R (2019) Exposure to anthropogenic chemicals in wild carnivores: a silent conservation threat demanding long-term surveillance. Curr Opin Environ Sci Health 1(11):21–25

Rosatte R, Ryckman M, Ing K, Proceviat S, Allan M, Bruce L, Donovan D, Davies JC (2010) Density, movements, and survival of raccoons in Ontario, Canada: implications for disease spread and management. J Mammal 91(1):122–135

Saaristo M, Brodin T, Balshine S, Bertram MG, Brooks BW, Ehlman SM, McCallum ES, Sih A, Sundin J, Wong BB, Arnold KE (2018) Direct and indirect effects of chemical contaminants on the behaviour, ecology and evolution of wildlife. Proc R Soc B 285(1885):20181297

Sanderfoot OV, Holloway T (2017) Air pollution impacts on avian species via inhalation exposure and associated outcomes. Environ Res Lett 12(8):083002

Schell CJ, Stanton LA, Young JK, Angeloni LM, Lambert JE, Breck SW, Murray MH (2021) The evolutionary consequences of human–wildlife conflict in cities. Evol Appl 14(1):178–197

Schulte-Hostedde AI, Mazal Z, Jardine CM, Gagnon J (2018) Enhanced access to anthropogenic food waste is related to hyperglycemia in raccoons (Procyon lotor). Conserv Physiol 6(1):coy026

Shifaw E (2018) Review of heavy metals pollution in China in agricultural and urban soils. J Health Pollut 8:18

Smith HT, Engeman RM (2002) An extraordinary raccoon, Procyon lotor; density at an urban park. USDA National Wildlife Research Center-Staff Publications, Washington, p 487

Soulsbury CD, White PC (2015) Human–wildlife interactions in urban areas: a review of conflicts, benefits and opportunities. Wildl Res 42(7):541–553

Tompkins DM, White AR, Boots M (2003) Ecological replacement of native red squirrels by invasive greys driven by disease. Ecol Lett 6(3):189–196

Townsend AK, Staab HA, Barker CM (2019) Urbanization and elevated cholesterol in American crows. Condor 121(3):duz040

Ujvari B, Klaassen M, Raven N, Russell T, Vittecoq M, Hamede R, Thomas F, Madsen T (2018) Genetic diversity, inbreeding and cancer. Proc R Soc B Biol Sci 285(1875):20172589

United Nations, Department of Economic and Social Affairs, Population Division (2017) World population prospects: the 2017 revision, key findings and advance tables. In Working paper No. ESA/P/WP/248 2017

United Nations, Department of Economic and Social Affairs, Population Division (2019) World urbanization prospects: the 2018 revision (ST/ESA/SER.A/420). United Nations, New York

Van Doren BM, Willard DE, Hennen M, Horton KG, Stuber EF, Sheldon D, Sivakumar AH, Wang J, Farnsworth A, Winger BM (2021) Drivers of fatal bird collisions in an urban center. Proc Natl Acad Sci 15:118

Wilcove DS, Master LL (2005) How many endangered species are there in the United States? Front Ecol Environ 3(8):414–420

Wittrock J, Duncan C, Stephen C (2019) A determinants of health conceptual model for fish and wildlife health. J Wildl Dis 55(2):285–297

Wright AN, Gompper ME (2005) Altered parasite assemblages in raccoons in response to manipulated resource availability. Oecologia 144(1):148–156

Climate Change

Craig Stephen

Abstract

Climate change impacts on wildlife health are inevitable and are happening. Climate change directly and indirectly impacts wildlife determinants of health in multiple, interacting ways, across a range of scales. These impacts will amplify existing risks and create new risks. In response, wildlife health programs will need to: (1) provide services to mitigate climate change impacts; (2) reduce population vulnerability to lessen those impacts; (3) enhance population resilience to avoid or cope with the impacts, and (4) attack climate change risks at their sources. Wildlife population health programs will need to respond to amplified known problems and the impacts of surprising or emerging health impacts. Wildlife health services need to be able to respond to surges in unexpected disasters, emergencies, or disease outbreaks, including having adaptable early warning surveillance of expected health outcomes or hazards and capacity to detect circumstances that are altering population exposure, sensitivity, and/or capacity to adapt. Coping capacity will come from protecting the determinants of wildlife health so wildlife has options and capacities to cope with new or amplified stressors and changes. Wildlife population health programs cannot isolate climate change response from their responses to other threats. Wildlife is being impacted by multiple anthropogenically driven global crises simultaneously. Many of these concurrent threats are linked not only in their causes but also in their solutions.

Keywords

Climate change · Wildlife health · Vulnerability · Adaptation · Mitigation

1 Introduction

Climate change can result from human activity or natural factors but since the 1950s, these changes have been overwhelmingly driven by human activities such as alteration of the atmospheric composition and land use (IPCC 2013). The Intergovernmental Panel on Climate Change (IPCC) is the United Nations body for assessing climate change science. Its fifth assessment report in 2014 forecasted that, under all the assessed greenhouse gas emissions scenarios, it was highly likely that heatwaves and extreme weather will occur more often, last longer, and become more intense and frequent on land and in the oceans (IPPC 2014). By 2021, the IPCC described climate change as being widespread, rapid, and intensifying. The Panel proclaimed with high confidence that climate change will contribute to the extinction of terrestrial and marine species.

C. Stephen (✉)
School of Population and Public Health, University of British Columbia, Vancouver, BC, Canada

C. Stephen (ed.), *Wildlife Population Health*,
https://doi.org/10.1007/978-3-030-90510-1_13

The evidence that recent changes in climate are resulting in widespread impacts on natural and social systems is unequivocal. No longer are the risks from climate change to wildlife health in question. They are inevitable and are happening now.

Climate change is a long-term shift in global or regional climate patterns. The recent anthropogenically driven accumulation of greenhouse gases has reduced the rate of heat loss from Earth to space, with a subsequent warming effect on the global climate system (GoC 2019). The main impacts on natural systems are predominately arising from atmospheric and oceanic warming, the diminution of snow and ice, and the rising of sea levels. Worldwide, wildlife and ecosystems are experiencing increased heat, extreme weather events, declining air quality, expanded ranges of parasites and pathogens, habitat loss, wildfires, and diminished food and water security. The effects of climate change will be modified, amplified, or dampened by the availability and sustainability of ecosystem services that are supported by biodiversity and socioeconomic conditions (Hoegh-Guldberg et al. 2018).

2 Implications for Wildlife Health

Climate change will have direct and indirect impacts on wildlife determinants of health in multiple, interacting ways, across a range of scales. These impacts will amplify existing risks and create new risks. Vulnerability to the direct and indirect health impacts of climate change will vary widely by species, location, and management system, but will ultimately be determined by exposure to the effects of climate change and capacity to adapt to or cope with those effects (Yohe 2001). There are parts of the world that are seeing more rapid and impactful climate change, such as Arctic regions where polar ice and snow loss are happening faster than predicted. Other regions are suffering from more acute and dramatic impacts such as hurricanes, droughts, and fires. Spatial differences in ocean chemistry at regional and local levels are translating into different rates and magnitudes of ocean acidification in marine ecosystems. Given that climate change is happening simultaneously with other stressors such as overexploitation, habitat fragmentation, and pollution, it is hard to tease apart the effects of climate change from other hazards impacting wildlife health. This chapter can only provide an overview of some of the changes that can be expected for global wildlife populations because of the vast diversity of species–climate–environment interactions that can happen around the world.

Climate has influenced and will continue to influence the occurrence and severity of infectious and parasitic diseases. Changes in food webs, timing of lifecycles, alterations of physiological status, and weather patterns will influence infectious and parasitic disease transmission pathways and frequency (Gallana et al. 2013). Most attention has focussed on vector-borne diseases, but waterborne, windborne, and enteric infections can also be expected to increase (Forman et al. 2008). The relationship between climate change and infectious diseases will not be straightforward. Many aspects of an infectious disease's ecology can be climate sensitive including the distribution and environmental survival of pathogens and parasites and the distributions and sensitivity of their hosts. Modifying factors such as the global movement of invasive species and habitat changes altering exposure pathways and connectivity of populations will further modify infectious diseases trends and complicate predicting epidemiological changes in advance of harm. Shifting climate regimes will alter biotic communities in surprising ways, leading to new and unanticipated opportunities for pathogens and parasites to move between species (Williams and Jackson 2007).

Changes in temperature, precipitation, and weather patterns will not only alter pathogen and parasite epidemiology but also the pathways, persistence, and concentrations of pollutants in the environment (Burek et al. 2008). Climate change alterations to food webs, lipid dynamics, ice and snowmelt, and organic carbon cycling will affect pollutant levels, distributions, and toxicity in water, soil, air, plants, and animals (Noyes et al. 2009). Flooding and melting events will redistribute contaminants. Changing patterns of insect pests will affect when and how much pesticides enter the environment and the production, frequency, and distribution of mycotoxins and toxic algae will increase (Griffith and Gobler 2020; Tirado et al. 2010; Van der Fels-Klerx et al. 2016). There is a multitude of well-known health impacts of wildlife exposure to pollutants ranging from acute effects such as oil spill exposure and choking from plastic ingestion to endocrine and reproductive effects, to immunosuppression. Increasing and extreme temperatures, intense periods of drier and wetter conditions; reduced ocean pH and altered salinity dynamics have the potential to enhance organism susceptibility to chemical toxicity. In return, chemical exposures may impair the ability of organisms to cope with the shifting climate (Noyes and Lema 2015). There is compelling evidence that increasing temperatures may alter the biotransformation of contaminants to more bioactive metabolites (Noyes et al. 2009) and evidence of the synergistic effects of contaminants with parasitic or infectious diseases (e.g., Marcogliese and Pietrock 2011).

With increasing water temperature, water quality declines, harmful algae blooms become more frequent, water oxygen levels decrease, and reduced feeding and growth occur, all of which can increase the incidence of diseases in marine and aquatic systems (Handisyde et al. 2006). Heat stress will affect terrestrial species by causing metabolic alterations and suppressing the immune and endocrine system thereby enhancing disease susceptibility (Das et al. 2016) and affecting reproduction and survival. Impacts of extreme heat events are expected to become more frequent, severe, and widespread in the future as witnessed by more frequent and greater number of flying fox (*Pteropus* spp.) deaths with heatwaves in Australia (Ratnayake et al. 2019). Extreme heat has already interacted with drought and habitat alterations to kill billions of wild animals due to wildfires in South America, Europe, North America, and Australia. Typhoons, hurricanes, floods, wildfires, and drought have animal health implications that in turn will have public health and economic implications, yet many countries omit animals from their national and regional contingency planning (Garde et al. 2013).

Changes in the function and structure of ecosystems arising from snow loss, desertification, acidification, changes in precipitation and temperature patterns, and drought will have profound implications for wildlife's access to their needs for daily living and will influence other determinants of health. Changes in the annual cycles of plants and animals, including their growth and migration patterns, are affecting food webs and disease cycles. The impacts of climate change on coral reef organisms, for example, are affecting access to the needs for daily living for many species depending on reef ecosystems as well as influencing reef disease ecology (Hoegh-Guldberg et al. 2017). Anthropogenic environmental changes, including climate change, are expected to affect the reproductive success and survival of many wildlife species by multiple routes and in often unpredictable ways (Milligan et al. 2009). For example, associations have been found between temperature and growth, phenology, and survival for all the life stages of sockeye salmon (*Oncorhynchus nerka*), and the effects occurring in one life stage are likely carried over to subsequent life stages (Martins et al. 2012). Range-restricted species or species with extremely specific habitat requirements may be most vulnerable to changes in their ecosystems (Parmesan 2006).

In summary, climate change impacts on ecosystems, populations, and individuals will influence the patterns and impacts of infectious and noninfectious diseases. It will expose animals to more extreme weather in both disastrous acute situations and throughout an animal's life span. It will affect the access to the determinants of health such as food, migration fidelity, and water. Some animals will benefit from climate change, while others will suffer depending on their life histories, habitat needs, and concurrent stressors. All these effects will interact with concurrent risks and hazards in unpredictable ways. Given all the anticipated and unanticipated effects of climate change on wildlife health, the challenge of developing a useful climate change management strategy for wildlife can seem insurmountable.

3 Adapting Wildlife Population Health Programs for Climate Change

There are four general approaches to managing climate change impacts on wildlife health (Fig. 1). These four approaches need to consider the climate impacts in effect now as well as to try to forecast how to protect wildlife from what might come. Effective climate change programs, therefore, must not only have the best current intelligence but also require an indication of what the future may bring (Stephen and Duncan 2017). Little scholarly work has been dedicated to identifying evidence-based, climate change actions for animal health (Stephen and Wade 2020). Wildlife health management is constrained by a lack of systematic study of the anticipated effects of climate change on animals and how to be best prevent, recover from or cope with those effects. While there is not enough experience or published evaluations to define best practices for wildlife health climate change action, information from other sectors can inform action and planning.

Effective climate change adaptation requires an understanding of the diverse ways in which climate change impacts the populations of concern and the opportunities and obstacles to act in the population's socio-ecological system. An intersectoral approach is, therefore, fundamental to action on climate change and health. Creating teams with a common goal has been proposed as the "first and foremost" step to achieving greater collaboration between health professionals (Eussen et al. 2017). By focusing on factors that can benefit a range of species through multiple pathways, there is greater likelihood that teams will identify areas of commonality. Another component of developing strong interdisciplinary teams includes fostering both formal and informal relationships built on trust and respect (Anholt et al. 2012).

There is growing attention on climate change adaptation, in part due to the acceptance that, even with the cessation of all carbon pollution today, its impacts will last for generations. It has been estimated that anthropogenic climate change will be irreversible on centennial to millennial timescales even after a complete cessation of CO_2 emissions (Thorne et al. 2011). However, climate change mitigation cannot be forgotten. Climate change action is a generational event and not limited to funding or political cycles. Taking tangible steps to reduce further climate change by managing activities and policies that lead to global warming must still be a priority. This can range from reviewing ways to reduce the carbon footprint of a wildlife health program to lobbying for legislative changes affecting a nation's greenhouse gas pollution. Wildlife health managers must be able to investigate and communicate how the implications of climate change affect conservation, sustainable food production systems, public health, and community resilience to encourage political and multi-

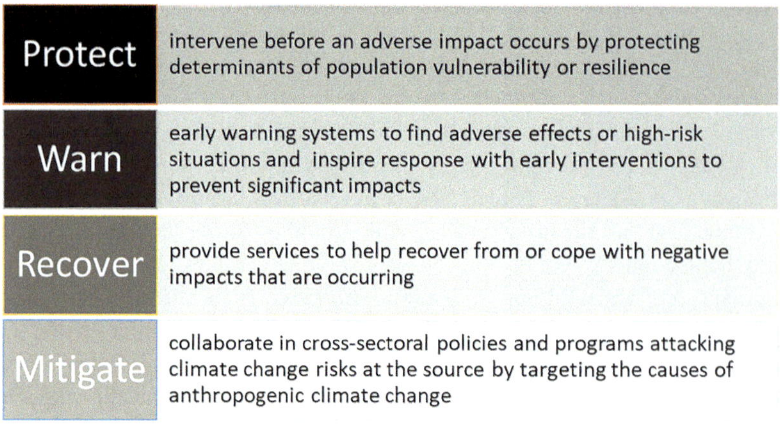

Fig. 1 Four general climate change strategies for wildlife health management

sectoral collaborations that will be essential for cross-sectoral policies and actions to attack climate change at its source.

Climate change vulnerability is determined by a population's exposure to the impacts of climate change and its adaptive capacity (Yohe 2001). Building adaptive capacity is, therefore, critical to any population's health in the face of climate change. Climate change adaptation capacity is the ability of systems, institutions, and organisms to adjust to potential damage, to take advantage of opportunities, or to respond to the consequences of climate change (IPCC n.d.). Table 1 adapts some general lessons for building adaptation capacity from public health programs (Hess et al. 2012) into a wildlife health context.

Table 1 General guidance on building climate change adaptation capacity in wildlife population health programs (adapted from Hess et al. 2012)

Action theme	General actions	Example program action
Systematically identify priority management targets for which you can feasibly and appropriately act	Identify populations, problems, and/or places with distinct climate sensitivity through vulnerability and impact assessments	Develop a wildlife health intelligence system (see Chap. 9)
	Seek synergies between existing programs and climate actions to increase program efficiency, feasibility, and acceptance	Work with public health agencies to use habitat protection for urban wildlife conservation while also reducing urban heat sinks for people
	Tailor actions to the prevailing social and ecological context	Use participatory means to understand which actions are morally and legally acceptable in the local community (see Chap. 21)
Invest in proactive efforts to cultivate and retain population and system resilience	Protect and promote populations and socio-ecological systems that can retain their essential functions and structures in the face of disturbance	Create multi-solving strategies that protect multiple determinants of health such as habitat protection which provides food and physical security for wildlife
	Integrate climate action into health promotion programs and look at opportunities for all-hazards approaches	Identify general drivers of population resilience (such as protecting genetic diversity or habitat connectivity) that will promote the capacity to cope with pandemic threats while also reducing climate change vulnerability
Build adaptive programs	Build capacity to match actions to changing spatial and temporal scales of climate impacts and have programs in place to anticipate or detect these changes	Develop horizon scanning competencies to provide insight into future training and capacity development needs
	Employ a systems perspective to program design and delivery to ensure a breadth of perspective on possible effects of climate and management actions	Apply lessons from addressing wicked problems and managing health as a cumulative effect to wildlife health programs (see Chap. 17)
	Invest in program monitoring and evaluation and feed back their findings into program delivery in a timely fashion	Develop implementation science and program evaluation capacity (see Chap. 23)
Learn from others	Build collaborative knowledge networks and social learning groups to increase access to information and innovations from within and outside of the wildlife health sector	Providc a knowledge brokering service that increases access and understanding of available research on effective climate change adaptation

Sustainable climate adaptation planning must recognize: (1) that there are multiple concurrent stressors and that adaptation against one harm can affect adaptation for another; (2) differing values and interests will affect adaptation outcomes; (3) local knowledge is needed for effective adaptation action; (4) adaptation actions should not add to climate change, limit the ability to respond or negatively impact other parts of society, the economy, or the natural environment; and (5) adaptation actions should be effective, efficient, equitable, and evidence-based (Eriksen et al. 2011; Prutsch et al. 2010). There is currently no certainty that climate change adaptation recommendations will be socially or environmentally sustainable or how they are likely to contribute to preserving wildlife health. A 2019 review of animal health literature revealed a critical gap in research intended to share experiences or provide the evidence needed to confidently select health adaptation actions (Stephen and Wade 2020).

Climate change will impact wildlife health by both amplifying existing health problems and creating unanticipated threats. For amplified problems, climate change adaptation is essentially a matter of ensuring accessible health services that can be deployed and/or enhanced in response to locally changing epidemiological situations. Responding to amplified problems will be achieved, in part, by increasing investment to cope with the increased frequency, distribution, or impacts of existing climate-sensitive problems and ensuring that resources can be adaptably deployed to changing local conditions (Table 2).

The drivers of climate change maladaptation will not simply be mediated by changing hazard–climate interactions but will also be modified by other global trends such as landscape change, wildlife exploitation, globalization, agricultural intensification, and urbanization. The interaction of these concurrent forces will lead to unanticipated or surprising effects. Surprise preparation is a product of finding vulnerable populations or settings and building capacity to cope with harms that cannot be prevented (see Chap. 24). An all-hazards approach focuses on integrating

Table 2 Key elements of climate change management for wildlife health

For amplified preexisting health problems	For unanticipated threats
Adaptable early warning surveillance and monitoring of expected health outcomes or hazards	Health intelligence to identify and address preexisting socioeconomic and ecological circumstances that exacerbate climate change vulnerability
Capacity to detect circumstances that are altering population exposure, sensitivity, and/or capacity to adapt to known climate change hazards and risk (i.e., tracking population vulnerability)	Proactively build capacity to cope with multiple interacting threats and stressors before an impact occurs
Strengthen core wildlife health program capacities, and build sustainable policies and infrastructures to manage increased effects and changing distributions of existing or reasonably expected threats	Adaptable wildlife health services that can respond to surges in unexpected disasters, emergencies, or disease outbreaks
Be attentive to how changes in known problems change the species affected and the secondary ecological, economic, or social impacts	Integrate wildlife health climate change adaptation into public health, agriculture planning, and ecosystem management
Adaptable wildlife health services that can respond to surges in disasters, emergencies, or disease outbreaks from known agents or hazards	Incorporate climate change literacy in wildlife health program personnel to encourage and enable proactive adaptive and mitigative behaviors within and outside the wildlife health sector
Innovative leadership, partnerships, and governance that support cross-sectoral, collaborative actions	Investigate and communicate the implications of climate change for wildlife health affecting conservation, sustainable food production systems, public health, and community resilience to encourage political and multi-sectoral collaboration
	Innovative leadership, partnerships, and governance that support cross-sectoral, collaborative actions

capacities and capabilities to reduce vulnerability for a full spectrum of impacts rather than focusing on hazards in a piecemeal fashion. Coping capacity comes from attention to protecting the determinants of animal health so they have options (e.g., habitat connectivity, genetic diversity) and capacities (e.g., food security, reduced immunocompromising stressors) to cope with new stressors and changes. Coping capacity will be limited by the state and availability of critical ecosystem services and/or financial resources, policies, and collaborations available to manage their determinants of health. Table 2 proposes some key elements of climate change strategy concerned with unanticipated wildlife health impacts.

4 Summary

The essential elements of climate change management for effective adaptation to the future will be unpredictable and emergent rather than predictable and planned (Hanlon and Carlisle 2008) because of the unprecedent rate of social and environmental change and the complexity of interactions between co-occurring global threats. Wildlife population health programs cannot isolate their climate change response from their responses to other threats. Wildlife is being impacted by multiple anthropogenically driven global crises simultaneously. Many of these concurrent threats are linked not only in their causes but also in their solutions. Growing experience is pointing to the need for action on the "causes-of-the-causes" that are shared between climate change, the extinction crisis, habitat degradation, and other mega-threats. If wildlife health programs are to protect, maintain, and improve wildlife health, they are obliged to provide a continuum of climate change actions. This continuum goes from: (1) providing services to mitigate climate change impacts; (2) reducing population vulnerability to lessen those impacts; (3) enhancing population resilience to avoid or cope with the impacts; and (4) attacking climate change risks at their sources.

References

Anholt RM, Stephen C, Copes R (2012) Strategies for collaboration in the interdisciplinary field of emerging zoonotic diseases. Zoonoses Public Health 59(4): 229–240

Burek KA, Gulland FM, O'Hara TM (2008) Effects of climate change on Arctic marine mammal health. Ecol Appl 18(sp2):S126–S134

Das R, Sailo L, Verma N, Bharti P, Saikia J (2016) Impact of heat stress on health and performance of dairy animals: a review. Vet World 9(3):260

Eriksen S, Aldunce P, Bahinipati CS, Martins RD, Molefe JI, Nhemachena C, O'brien K, Olorunfemi F, Park J, Sygna L, Ulsrud K (2011) When not every response to climate change is a good one: identifying principles for sustainable adaptation. Clim Dev 3(1):7–20

Eussen BG, Schaveling J, Dragt MJ, Blomme RJ (2017) Stimulating collaboration between human and veterinary health care professionals. BMC Vet Res 13(1):1–1

Forman S, Hungerford N, Yamakawa M, Yanase T, Tsai HJ, Joo YS, Yang DK, Nha JJ (2008) Climate change impacts and risks for animal health in Asia. Rev Sci Tech Off Int Epiz 1(27):581–597

Gallana M, Ryser-Degiorgis MP, Wahli T, Segner H (2013) Climate change and infectious diseases of wildlife: altered interactions between pathogens, vectors and hosts. Curr Zool 59(3):427–437

Garde E, Pérez GE, Acosta-Jamett G, Bronsvoort BM (2013) Challenges encountered during the veterinary disaster response: an example from Chile. Animals 3(4):1073–1085

GoC. Government of Canada (2019) Canada's changing climate report. https://www.nrcan.gc.ca/sites/www.nrcan.gc.ca/files/energy/Climate-change/pdf/CCCR_FULLREPORT-EN-FINAL.pdf. Accessed 3 Aug 2020

Griffith AW, Gobler CJ (2020) Harmful algal blooms: a climate change co-stressor in marine and freshwater ecosystems. Harmful Algae 91:101590

Handisyde NT, Ross LG, Badjeck MC, Allison EH. The effects of climate change on world aquaculture: a global perspective. Aquaculture and fish genetics research programme, Stirling Institute of Aquaculture. Final technical report, DFID, Stirling. p 151 2006

Hanlon P, Carlisle S (2008) Do we face a third revolution in human history? If so, how will public health respond? J Public Health 30(4):355–361

Hess JJ, McDowell JZ, Luber G (2012) Integrating climate change adaptation into public health practice: using adaptive management to increase adaptive capacity and build resilience. Environ Health Perspect 120(2): 171–179

Hoegh-Guldberg O, Jacob D, Taylor M, Bindi M, Brown S, Camilloni I et al (2018) Impacts of 1.5°C global warming on natural and human systems. In: Masson-Delmotte V, Zhai P, Pörtner H-O, Roberts D, Skea J, Shukla PR et al (eds) Global warming of 1.5°C. Cambridge University Press, Cambridge, New York. https://www.ipcc.ch/sr15/. Accessed 3 Aug 2020

Hoegh-Guldberg O, Poloczanska ES, Skirving W, Dove S (2017) Coral reef ecosystems under climate change and ocean acidification. Front Mar Sci 29(4):158

IPCC (2013) Summary for policymakers. In: Stocker TF, Qin D, Plattner G-K, Tignor M, Allen SK, Boschung J, Nauels A, Xia Y, Bex V, Midgley PM (eds) Climate change 2013: the physical science basis. Contribution of working group I to the fifth assessment report of the intergovernmental panel on climate change. Cambridge University Press, Cambridge, New York

IPCC (2014) In: Pachauri RK, Meyer LA (eds) Climate change 2014: synthesis report. Contribution of working groups I, II and III to the fifth assessment report of the intergovernmental panel on climate change. IPCC, Geneva, p 151

IPPC (n.d.) Definition of terms used within the DDC pages. https://www.ipcc-data.org/guidelines/pages/glossary/glossary_a.html

Marcogliese DJ, Pietrock M (2011) Combined effects of parasites and contaminants on animal health: parasites do matter. Trends Parasitol 27(3):123–130

Martins EG, Hinch SG, Cooke SJ, Patterson DA (2012) Climate effects on growth, phenology, and survival of sockeye salmon (Oncorhynchus nerka): a synthesis of the current state of knowledge and future research directions. Rev Fish Biol Fish 22(4):887–914

Milligan SR, Holt WV, Lloyd R (2009) Impacts of climate change and environmental factors on reproduction and development in wildlife. Philos Trans R Soc B 364: 3313–3319

Noyes PD, Lema SC (2015) Forecasting the impacts of chemical pollution and climate change interactions on the health of wildlife. Curr Zool 61(4):669–689

Noyes PD, McElwee MK, Miller HD, Clark BW, Van Tiem LA, Walcott KC, Erwin KN, Levin ED (2009) The toxicology of climate change: environmental contaminants in a warming world. Environ Int 35(6): 971–986

Parmesan C (2006) Ecological and evolutionary responses to recent climate change. Annu Rev Ecol Evol Syst 1(37):637–669

Prutsch A, Grothmann T, Schauser I, Otto S, McCallum S (2010) Guiding principles for adaptation to climate change in Europe. ETC/ACC technical paper, pp 1–32

Ratnayake HU, Kearney MR, Govekar P, Karoly D, Welbergen JA (2019) Forecasting wildlife die-offs from extreme heat events. Anim Conserv 22(4): 386–395

Stephen C, Duncan C (2017) Can wildlife surveillance contribute to public health preparedness for climate change? A Canadian perspective. Clim Change 141(2):259–271

Stephen C, Wade J (2020) Missing in action: sustainable climate change adaptation evidence for animal health. Can Vet J 61(9):966

Thorne PW, Willett KM, Allan RJ et al (2011) Guiding the creation of a comprehensive surface temperature resource for twenty-first-century climate science. Bull Amer Meteor 92(11):40

Tirado MC, Clarke R, Jaykus LA, McQuatters-Gollop A, Frank JM (2010) Climate change and food safety: a review. Food Res Int 43(7):1745–1765

Van der Fels-Klerx HJ, Liu C, Battilani P (2016) Modelling climate change impacts on mycotoxin contamination. World Mycotoxin J 9(5):717–726

Williams JW, Jackson ST (2007) Novel climates, no-analog communities, and ecological surprises. Front Ecol Environ 5(9):475–482

Yohe GW (2001) Mitigative capacity—the mirror image of adaptive capacity on the emissions side. Clim Change 49(3):247

Landscapes Supporting Wildlife Health

Colin Robertson

Abstract

Wildlife health is frequently considered through bifurcated foci of either disease (through spillover, risk factors, transmission, and mortality) or a species-level conservation and recovery lens (emphasizing parameters such as critical habitat, abundance, and fecundity). The intersection of these two perspectives remains rare largely due to different disciplinary and professional traditions. This chapter proposes that the concept of the landscape can serve as a unifying scale for understanding and managing wildlife health. We present how processes of landscape change can influence wildlife determinants of health and overview some common approaches for measuring and monitoring landscapes as they pertain to wildlife health. Inherent to a landscape perspective is the role of human processes and values in deciding what locations and contingent biotic and abiotic components receive disproportionate impacts from disturbance or conservation. Through a whole-of-landscape approach, similar to neighborhood-level analysis and planning in human health, a greater synthesis of the shared dimensions of health can provide concrete tools to aid in understanding, maintaining, and promoting wildlife health.

Keywords

Landscape · Wildlife health · Space · Place · Fragmentation · Wildlife health geography

1 Landscapes and Wildlife Health

Connections between environment and health are fundamental to understand health at the population level. As anthropogenic transformations to the earth continue at an accelerating global rate (Lewis and Maslin 2015), wildlife is vulnerable to a wide variety of impacts due to habitat loss and ecosystem change (Butchart et al. 2010). Some primary wildlife responses observed globally include range shifts northward and upward (Buntgen et al. 2017; Cristofari et al. 2018), adaptations to hunting pressure (Fariss et al. 2015) and migratory behaviors (Dollman and Sutherland 1995), and in some cases population decline (Flockhart et al. 2015) and/or collapse (Powers and Jetz 2019). How do we gauge how wildlife are responding to landscape change? What key variables should be monitored? What are the appropriate units for management and mitigation? This chapter uses a focus on "landscapes" as a central unit of analysis and requirement for healthy wildlife, examining how interacting landscapes can amplify, dampen, and

C. Robertson (✉)
Department of Geography and Environmental Studies, Wilfrid Laurier University, Waterloo, ON, Canada
e-mail: crobertson@wlu.ca

C. Stephen (ed.), *Wildlife Population Health*,
https://doi.org/10.1007/978-3-030-90510-1_14

transmit positive and negative impacts of landscape change through a broad suite of wildlife determinants of health (see Chap. 1).

1.1 What Are Landscapes?

Landscape-level approaches to understand habitat quality, threats, and conservation priorities require a clear understanding of what we mean by the word "landscape." In short, a landscape defines a region of spatial heterogeneity in terms of habitat quality or characteristics (Turner 1989; Picket and Cadenasso 1995). Spatial differentiation provides variation in habitat and how this variation is arranged and configured can have a significant impact on population-level ecological processes. These ideas are the cornerstones of landscape ecology research (Turner 1989).

Landscapes vary considerably in size depending on species, season, and other geographic and ecological factors. Landscapes are dynamic. They fluctuate and reconfigure as human, environmental, and ecological processes interact. Global assessments of biodiversity decline have found landscape change and related habitat loss due to agriculture expansion, urbanization/sprawl, deforestation, and resource extraction to be critical drivers of declines.

Landscapes are not only the geographic expression of ecosystems but envelopes within which the full scope of natural and human activities take place. In a geographic sense, this includes the processes and outcomes of human activity, which reflects the values and cultures of societies. At the individual level, deeply held personal connections to landscapes play an important role in how humans value locations differently, which in turn shapes management priorities, disturbance regimes, and impacts on wildlife. These internal dimensions of human connection to location are described in health geography as 'places' as opposed to the physical features of spaces (Kearns 1993). Places are widely known to influence human health but only recently have ideas of place and health been turned to wildlife (e.g., Robertson 2020). The way animals relate to the land they inhabit may provide important insights into how to preserve and promote wildlife health.

1.2 How to Measure and Understand Landscapes

Specific characteristics of landscapes are sometimes called "landscape features" (often describing aspects of habitat fragmentation, land use/landcover heterogeneity, disturbance regime impacts). Landscape features are typically evaluated in a species-specific study, for example, comparing disease emergence and spread (e.g., Frank et al. 1998; Gardner et al. 2020), abundance (e.g., Lewis et al. 2011), or resource selection (e.g., Stewart et al. 2013) using location and environmental contextual data. These models can provide detailed insights into how individuals move through space and interact with different aspects of their environment. Recent tools such as CircuitScape (Hall et al. 2021) and Marxan Connect (Daigle et al. 2020) allow such insights to be used to identify and preserve landscape connectivity. Multi-species or ecosystem-wide study of landscape features and their impacts are needed to meet the demand for conservation strategies with the maximum ecosystem and wildlife health benefits. But both disease-centric and species-centric approaches alone will fail to capture the complexity of processes shaping wildlife health. Considering the wider set of wildlife determinants of health through a landscape approach may provide mechanisms for intervention and mitigation that support wildlife health across diverse taxa.

Landscape conservation planning tools can integrate data on multiple species, their habitat requirements, human activity, and climate change into a comprehensive spatial plan. The Landscape Species Approach developed by the Wildlife Conservation Society for example provides a ten-step process for conservation planning centered around the designation of a suite of selected representative Landscape Species (Sanderson et al. 2002). Such approaches require a priori understanding of how landscape relates to occupancy and/or health. However, developing an understanding of how multi-species assemblages interact with landscapes is extremely complex and difficult to carry out. Even primary species interactions such as predator/prey or host/parasite dynamics are typically excluded from assessments of landscape-level habitat quality and fragmentation. These interactions themselves are structured by and influence landscape patterns within which they occur. As populations respond to changes in land configuration due to human activity and/or climate change, the impacts of these interactions multiply and have the potential to cascade through and across entire ecosystems (Mahon et al. 2019). Thus, moving to landscape conservation planning on landscape species necessarily oversimplifies processes critical for sustaining and promoting wildlife health.

Cumulative effects assessment is another approach focused on regional disturbance impacts on wildlife (see Chap. 17). In this approach, multiple stressors are enumerated with an aim to provide an overall synthesis of the potential impacts to identified valued ecosystems components (e.g., moose)—often as part of a regulatory review process for resource extraction. Human values are explicitly considered in terms of what the key management values are. However, cumulative effects assessments tend to be focused on a single target species, and often fail to capture basic species interactions that are vital to understand wildlife health (Mahon et al. 2019). Thus, two of the more commonly used approaches either consider species in isolation, assuming their landscape health requirements are simply additive and independent or consider a suite of potential threats to specific species. A wildlife determinant of health model at the landscape scale asks the question; how do the determinants of health impact collective wildlife health across the landscape?

2 Linking Determinants of Health and Landscapes

2.1 Wildlife Determinants of Health and the Landscape Scale

Inherent to a landscape approach to wildlife health are the interdependencies between the various processes which either give rise to or detract from conditions supporting healthy wildlife. Thinking about the landscape setting as a dynamic outcome of processes that shape the health of wildlife provides a lens through which to conceptualize alternate ways to understand, manage, and protect wildlife and ecosystems (Robertson 2020). At a practical level, we can use determinants of health model to specify the spatial and temporal boundaries of a landscape of wildlife health. The geographic and temporal expression provides an analytical frame within which to operate.

Central to the concept of landscape is the idea that elements within that landscape interact and are interdependent. The effect of cascading ecological changes in ecosystems, such as changes to a top predator shape populations of prey and related species at lower trophic levels, is well-established in the ecological literature (Ripple et al. 2016; Daskalov et al. 2007; Paine 1980). However, the idea of cascading change can be applied more broadly to interactions between determinants of health. For example, wind erosion is driven by the interaction of atmospheric processes with terrain shape and influence that terrain, and in turn, its hydrology and plant communities, which in turn affect the wildlife assemblage. Similarly, road access follows valleys carved out by rivers which in part determine hunting areas and wildfire risk. Wildfire risk increases with increasing temperature, extreme weather, and previously applied fire suppression and forest management practices, which create a patchwork forest pattern allowing new species to move in and interact in novel ways. This can shape and can give rise to disease transmission and/or pest movement which impact energy expenditure and fecundity (Gardner et al. 2020). That such interdependencies are played out across a variety of spatial and temporal scales, add significantly to the complexity of incorporating them into wildlife health planning and assessment (Robertson 2017).

Due to the complexities inherent in social–ecological systems, attempts to do integrated ecological modeling that incorporates detailed data on landscape variables, biotic interactions, abiotic physical processes, biological mechanisms, and human values and policies have been limited to simulation-based approaches rather than empirical modeling. The majority of ecosystem modeling approaches operate on multi-decadal time scales and focus on possible outcomes under a variety of climate change and adaptation scenarios (Pereira et al. 2010). Ecological simulation tools such as grid-based landscape simulators (e.g., Scheller et al. 2007) or individual-level tools such as agent-based models (e.g., Railsback and Grimm 2011) have been used for decades to explore the ecosystem and landscape-level questions and scenario modeling. These tools require either highly detailed data and parameterizations or highly generalized simplifying assumptions; neither of which are conducive to informing policy (Cartwright et al. 2016). Recent trends in data availability, statistical modeling, and supporting computational tools and cyberinfrastructure may offer a new era of iterative ecological forecasting which is adaptive and tailored to decision-making questions and scales (Dietze et al. 2018). However, even with the latest computational modeling tools, capturing the complexity of processes in enough detail to accurately represent and predict interactions over varied scales remains unlikely. Data for wildlife health remains biased and unrepresentative, and reliance on such sources with computer-based modeling tools risks reinforcing data biases via policy and management actions (Robertson et al. 2016).

An alternate approach to explicitly modeling individual processes that interact to create landscapes that support healthy wildlife is to instead consider their vulnerability to common drivers of change. For example, despite the complexities and interdependencies discussed above, a key and major threat to wildlife health and global biodiversity is habitat loss. Land-use changes such as converting natural forest to agriculture or mining have widespread impacts on the wildlife determinants of health which in turn influence landscape-wide processes.

Figure 1 presents four common landscape drivers of change and their linkages with wildlife determinants of health and wildlife responses.

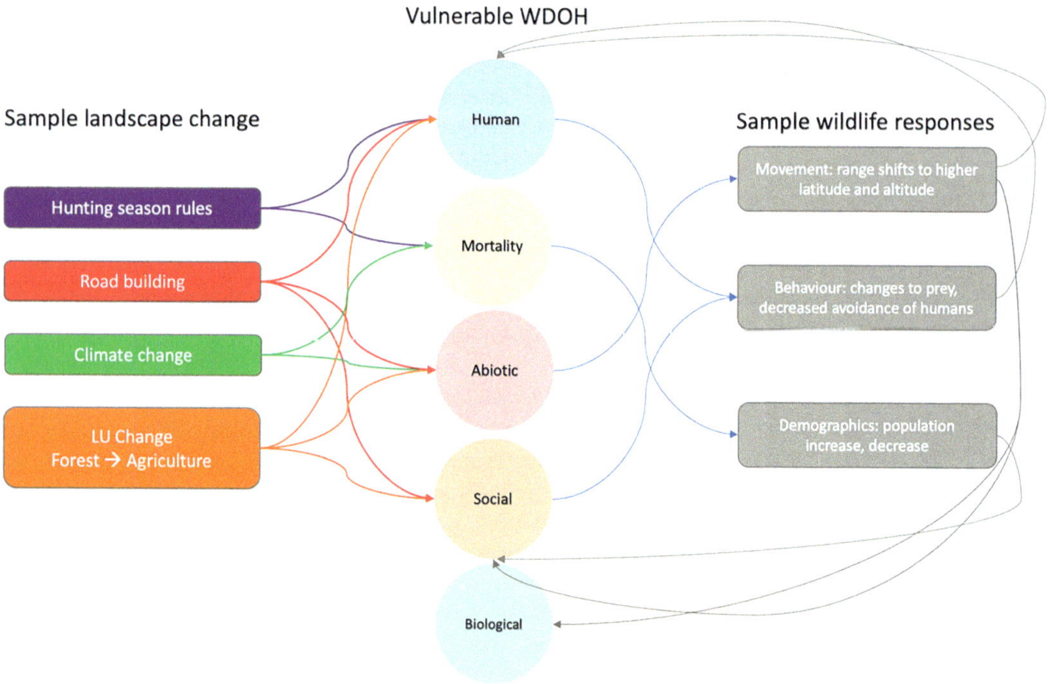

Fig. 1 Examples of landscape changes and their interactions with wildlife determinants of health and wildlife responses. (*LU* land use, *WDOH* wildlife determinants of health categories)

Hunting season changes, such as extending the season, can significantly increase overall harvest in some cases (Heuseman and McDonald 2002) and have very little impact in other cases (George et al. 1980). In addition to direct mortality from hunting, regulatory changes also change hunter perceptions of access and rights to wildlife. Wildlife responses to hunting may include adapting behavioral responses to avoid human hunters (Thurfjell et al. 2017). Road building, whether for resource extraction or more general transportation, can remove habitat, increase access to humans, and therefore the likelihood of hunting and/or other human/wildlife interactions. For example, Langevelde et al. (2009) found that traffic-related mortality of badgers was significantly higher on minor roads compared to major roads. Roads also act as linear disturbance features which increase edge habitat, fragmentation, and alter solar radiation regimes (Reed et al. 1996) as well as perturb movement patterns and predator–prey dynamics (Dickie et al. 2017). The temperature increase resulting from global climate change can cause direct mortality and heat stress, but more commonly results in shifts in species distributions to novel areas. Activities associated with resource extraction are often associated with habitat loss and changes in habitat quality. In barren-ground caribou habitat, for example, surface mining increased dust deposition in nearby areas, with increased metals concentration in lichen, a major component of caribou diet, sampled up to 40 km away (Watkinson et al. 2021), with potential for long-term physiological impacts. Such changes have transformative ecological impacts that cascade and circulate through ecosystems and landscapes (Cristofari et al. 2018).

3 Maintaining and Monitoring Healthy Wildlife Landscapes

To envisage what research and practice in support of healthy wildlife landscapes might look like we need only consider how social determinants of health have transformed public health. Some of the key insights into social determinants of health in human populations were found from the influence of contextual effects on health outcomes. Latent variables operating at the aggregate scale (e.g., neighborhood) such as social capital, neighborhood cohesion, access to education, income inequality, have been consistently found to have significant predictive power in explaining health disparities (Viner et al. 2012). The practical outcome of this large body of research has been an expansion of programming at the community level to a much wider set of activities and interventions. Such programs based on evidence at the neighborhood scale need not even target specific health outcomes, but rather focus on supporting the conditions that lead to health, or healthy settings. The same approach can be taken for creating landscapes that support wildlife health through the creation of healthy settings that provide and protect wildlife determinants of health.

Despite the complexity and interdependence of landscape-level threats, we now have unprecedented sources of data to monitor changes over vast spatial and temporal scales. Developing landscape monitoring tools in support of conservation priority, regulatory enforcement, and wildlife management can also act as indicators and monitoring tools for determinants of wildlife health. Jackson et al. (2006), for example, demonstrated how herbaceous forest edge in the northeast US was significantly linked to Lyme disease risk, suggesting that landscapes could be actively managed to produce patterns to reduce risk (Jackson et al. 2006). In the Amazon of South America, deforestation and agricultural expansion are associated with increased contact rates between humans and wildlife and linked to cases of human rabies (Ellwanger et al. 2020). Landscape changes such as snowmelt, shrub expansion, conversion of forest to agriculture, road building into remote areas, etc. can all be detected using earth observation data in an ongoing matter.

The target of conservation action in support of wildlife is often species-based, largely through the endangered species framework and legal mechanisms (e.g., Bird and Hodges 2017). This framework reinforces a species-specific approach to research and habitat conservation that often pits proximal environmental values against each other. A landscape-centric model could supplement this view by focusing on landscape units as decision-making entities, similar to how neighborhood units have been the central organizing tool for research and practice around community-level social determinants of health in human populations.

In addition to moving towards landscape rather than species-specific research and programming in support of wildlife health, research funding and targets need to be broadened to examine wildlife determinants of heath more concretely. In the context of global biodiversity decline and significant impacts of climate change, there is an urgent need for greater focus on antecedent drivers of landscape changes that are ripe for intervention, adaptation, and mitigation efforts. This comprises a compendium of drivers, from combatting global climate change to resisting local proposals for land-use change from forests to agricultural to urban. The compendium of drivers can be tailored to specific and localized wildlife health concerns. New tools are needed to visualize and communicate linkages across landscapes and landscape features driving wildlife health; where awareness of linkages can more effectively compel action by communities. Actors working at national and global scales can focus efforts on drivers at the largest scales, informed and illustrated by efforts on local scales. By focusing on the understanding of the disease, health, and population status at the landscape scale, it is possible to influence wildlife health and disease dynamics while also considering other valued components of the landscape. This can address feelings of helplessness which can fatigue action given the scale of global change.

4 Conclusion

The deeply interdependent nature of ecosystems is at odds with the highly specific and independent approach to the characterization of threats and protection of wildlife. Given decades of research and the massive scale of current global environmental change, there is a need for a renewed landscape-based approach to wildlife health. New sources of data can track and monitor ecosystems in unprecedented detail over large areas. As well, communities are increasingly aware of the threats posed by biodiversity loss, climate change, and the critical importance of habitat protection. The recent attention brought to wildlife and the intricate balance of humans and wildlife brought on by the COVID-19 pandemic may compel action to protect and safeguard habitat where has been lacking in the past. There is an immediate need to make clear the connections between human health, animal health, and environmental health and provide tools, programs, and evidence in support of this shared vision of health.

References

Bird SC, Hodges KE (2017) Critical habitat designation for Canadian listed species: slow, biased, and incomplete. Environ Sci Policy 71(May):1–8. https://doi.org/10.1016/j.envsci.2017.01.007

Büntgen U, Greuter L, Bollmann K, Jenny H, Andrew Liebhold J, Galván D, Stenseth NC, Andrew C, Mysterud A (2017) Elevational range shifts in four mountain ungulate species from the Swiss Alps. Ecosphere 8(4):e01761. https://doi.org/10.1002/ecs2.1761

Butchart SHM, Walpole M, Collen B, van Strien A, Scharlemann JPW, Almond REA, Baillie JEM et al (2010) Global biodiversity: indicators of recent declines. Science 328(5982):1164–1168. https://doi.org/10.1126/science.1187512

Cartwright SJ, Bowgen KM, Collop C, Hyder K, Nabe-Nielsen J, Stafford R, Stillman RA, Thorpe RB, Sibly RM (2016) Communicating complex ecological models to non-scientist end users. Ecol Model 338 (October):51–59. https://doi.org/10.1016/j.ecolmodel.2016.07.012

Cristofari R, Liu X, Bonadonna F, Cherel Y, Pistorius P, Le Maho Y, Raybaud V, Stenseth NC, Le Bohec C, Trucchi E (2018) Climate-driven range shifts of the king penguin in a fragmented ecosystem. Nat Clim Chang 8(3):245–251. https://doi.org/10.1038/s41558-018-0084-2

Daigle RM, Metaxas A, Balbar AC, McGowan J, Treml EA, Kuempel CD, Possingham HP, Beger M (2020) Operationalizing ecological connectivity in spatial conservation planning with Marxan connect. Methods Ecol Evol 11(4):570–579. https://doi.org/10.1111/2041-210X.13349

Daskalov GM, Grishin AN, Rodionov S, Mihneva V (2007) Trophic cascades triggered by overfishing reveal possible mechanisms of ecosystem regime shifts. Proc Natl Acad Sci 104(25):10518–10523. https://doi.org/10.1073/pnas.0701100104

Dickie M, Robert Serrouya R, McNay S, Boutin S (2017) Faster and farther: Wolf movement on linear features and implications for hunting behaviour. J Appl Ecol 54(1):253–263. https://doi.org/10.1111/1365-2664.12732

Dietze MC, Fox A, Beck-Johnson LM, Betancourt JL, Hooten MB, Jarnevich CS, Keitt TH et al (2018) Iterative near-term ecological forecasting: needs, opportunities, and challenges. Proc Natl Acad Sci 115(7):1424–1432. https://doi.org/10.1073/pnas.1710231115

Dolman PM, Sutherland WJ (1995) The response of Bird populations to habitat loss. Ibis 137(s1):S38–S46. https://doi.org/10.1111/j.1474-919X.1995.tb08456.x

Ellwanger JH, Kulmann-Leal B, Kaminski VL, Valverde-Villegas JM, Ana BG, Veiga D, Spilki FR, Fearnside PM et al (2020) Beyond diversity loss and climate change: impacts of Amazon deforestation on infectious diseases and public health. An Acad Bras Cienc 92:1. https://doi.org/10.1590/0001-3765202020191375

Farris ZJ, Golden CD, Karpanty S, Murphy A, Stauffer D, Ratelolahy F, Andrianjakarivelo V, Holmes CM, Kelly MJ (2015) Hunting, exotic carnivores, and habitat loss: anthropogenic effects on a native carnivore community, Madagascar. PLoS One 10(9):e0136456. https://doi.org/10.1371/journal.pone.0136456

Flockhart DT, Tyler J-BP, Ryan Norris D, Martin TG (2015) Unravelling the annual cycle in a migratory animal: breeding-season habitat loss drives population declines of monarch butterflies. J Anim Ecol 84(1):155–165. https://doi.org/10.1111/1365-2656.12253

Frank DH, Fish D, Moy FH (1998) Landscape features associated with Lyme disease risk in a suburban residential environment. Landsc Ecol 13(1):27–36. https://doi.org/10.1023/A:1007965600166

Gardner AM, Pawlikowski NC, Hamer SA, Hickling GJ, Miller JR, Schotthoefer AM, Tsao JI, Allan BF (2020) Landscape features predict the current and forecast the future geographic spread of Lyme disease. Proc R Soc B Biol Sci 287(1941):20202278. https://doi.org/10.1098/rspb.2020.2278

George RR, Wooley JB, Kienzler JM, Farris AL, Berner AH (1980) Effect of hunting season length on ring-necked pheasant populations. Wildlife Soc Bull 8(4):279–283

Hall KR, Anantharaman R, Landau VA, Clark M, Dickson BG, Jones A, Platt J, Edelman A, Shah VB (2021) Circuitscape in Julia: empowering dynamic approaches to connectivity assessment. Landarzt 10(3):301. https://doi.org/10.3390/land10030301

Heusmann HW, McDonald JE (2002) Distribution of wood duck harvest in the Atlantic and Mississippi flyways in relation to hunting season length. Wildlife Soc Bull 30(3):666–674

Jackson LE, Hilborn ED, Thomas JC (2006) Towards landscape design guidelines for reducing lyme disease risk. Int J Epidemiol 35(2):315–322. https://doi.org/10.1093/ije/dyi284

Kearns RA (1993) Place and health: towards a reformed medical geography. Prof Geogr 45(2):139–147. https://doi.org/10.1111/j.0033-0124.1993.00139.x

Lewis CW, Hodges KE, Koehler GM, Mills LS (2011) Influence of stand and landscape features on snowshoe hare abundance in fragmented forests. J Mammal 92(3):561–567. https://doi.org/10.1644/10-MAMM-A-095.1

Lewis SL, Maslin MA (2015) Defining the anthropocene. Nature 519(7542):171–180. https://doi.org/10.1038/nature14258

Mahon CL, Holloway GL, Bayne EM, Toms JD (2019) Additive and interactive cumulative effects on boreal Landbirds: winners and losers in a multi-stressor landscape. Ecol Appl 29(5):e01895. https://doi.org/10.1002/eap.1895

Paine RT (1980) Food webs: linkage, interaction strength and community infrastructure. J Anim Ecol 49(3):667–685. https://doi.org/10.2307/4220

Pereira HM, Leadley PW, Proença V, Alkemade R, Scharlemann JPW, Fernandez-Manjarrés JF, Araújo MB et al (2010) Scenarios for global biodiversity in the 21st century. Science 330(6010):1496–1501. https://doi.org/10.1126/science.1196624

Pickett STA, Cadenasso ML (1995) Landscape ecology: spatial heterogeneity in ecological systems. Science 269(5222):331–334. https://doi.org/10.1126/science.269.5222.331

Powers RP, Jetz W (2019) Global habitat loss and extinction risk of terrestrial vertebrates under future land-use-change scenarios. Nat Clim Chang 9(4):323–329. https://doi.org/10.1038/s41558-019-0406-z

Railsback, Steven F, Grimm V (2011) Agent-based and individual-based modeling: a practical introduction. Princeton University Press, New Jersey

Reed RA, Johnson-Barnard J, Baker WL (1996) Contribution of roads to forest fragmentation in the rocky mountains. Conserv Biol 10(4):1098–1106. https://doi.org/10.1046/j.1523-1739.1996.10041098.x

Ripple WJ, Estes JA, Schmitz OJ, Constant V, Kaylor MJ, Lenz A, Motley JL, Self KE, Taylor DS, Wolf C (2016) What is a trophic cascade? Trends Ecol Evol 31(11):842–849. https://doi.org/10.1016/j.tree.2016.08.010

Robertson C (2017) Towards a Geocomputational landscape epidemiology: surveillance, modelling, and interventions. GeoJ 82(2):397–414. https://doi.org/10.1007/s10708-015-9688-5

Robertson C (2020) Traversing the eco-healthscape: the final frontier in understanding shared determinants of health at the animal-society interface. In: Animals, health, and society. CRC Press

Robertson C, Yee L, Metelka J, Stephen C (2016) Spatial data issues in geographical zoonoses research. Canadian Geogr/Le Géographe Canadien 60(3):300–319. https://doi.org/10.1111/cag.12280

Sanderson EW, Redford KH, Vedder A, Coppolillo PB, Ward SE (2002) A conceptual model for conservation planning based on landscape species requirements. Landsc Urban Plan 58(1):41–56. https://doi.org/10.1016/S0169-2046(01)00231-6
Scheller RM, Domingo JB, Sturtevant BR, Williams JS, Rudy A, Gustafson EJ, Mladenoff DJ (2007) Design, development, and application of LANDIS-II, a spatial landscape simulation model with flexible temporal and spatial resolution. Ecol Model 201(3):409–419. https://doi.org/10.1016/j.ecolmodel.2006.10.009
Stewart BP, Nelson TA, Laberee K, Nielsen SE, Wulder MA, Stenhouse G (2013) Quantifying grizzly bear selection of natural and anthropogenic edges. J Wildl Manage 77(5):957–964. https://doi.org/10.1002/jwmg.535
Thurfjell H, Ciuti S, Boyce MS (2017) Learning from the mistakes of others: how female elk (Cervus Elaphus) adjust behaviour with age to avoid hunters. PLoS One 12(6):e0178082. https://doi.org/10.1371/journal.pone.0178082
Turner MG (1989) Landscape ecology: the effect of pattern on process. Annu Rev Ecol Syst 20:171–197
van Langevelde F, van Dooremalen C, Jaarsma CF (2009) Traffic mortality and the role of minor roads. J Environ Manage 90(1):660–667. https://doi.org/10.1016/j.jenvman.2007.09.003
Viner RM, Ozer EM, Denny S, Marmot M, Resnick M, Fatusi A, Currie C (2012) Adolescence and the social determinants of health. Lancet 379(9826):1641–1652. https://doi.org/10.1016/S0140-6736(12)60149-4
Watkinson AD, Virgl J, Miller VS, Anne Naeth M, Kim J, Serben K, Shapka C, Sinclair S (2021) Effects of dust deposition from diamond mining on subarctic plant communities and barren-ground caribou forage. J Environ Qual 50(4):990–1003. https://doi.org/10.1002/jeq2.20251

An Emerging Disease Agenda for Wildlife Health Management

Craig Stephen

Abstract

Wildlife has both been the victim and source of emerging infectious diseases throughout history. But the nature of future disease emergence under global change, and how that might be influenced by our decisions and actions, is uncertain. The pragmatic aim of a wildlife emerging disease program is not to predict the next emergence but to identify the approaches and tools that will effectively reduce the likelihood and impact of the next inevitable emerging disease using the given resources, in a given region at a given time. A hazard-by-hazard approach will be insufficient to prepare for the next emergence when the next hazard cannot be reliably predicted. Maintaining the status quo or reverting to historical rates of emerging diseases, rather than promoting resilience against the unexpected will need to be re-evaluated as a management strategy considering climate disruption and other environmental trends are driving emerging disease risks in surprising ways. Despite general agreement that the interplay of individual, environmental, ecological, and social factors influence emerging diseases, many wildlife health programs remain bound to disciplinary conventions. There is a need for a more unified set of guidelines and actions to promote emerging disease management by concurrently and collaboratively tending to the determinants of health while also developing adaptive and collaborative response programs to minimize harms during and after a disease emerges.

Keywords

Emerging disease · Wildlife · Pragmatic · A cycle of action · Adaptive · Drivers

1 Introduction

> Things that have never happened before, happen all the time. (Sagan 1993).

This chapter was written during the surge of cases in the third European and North American waves of the COVID-19 pandemic. The predominance of that pandemic in global political, media, and community conversations ensured that at no other times in recent history had wildlife been so widely associated with massive numbers of sick and dead people along with global social disruption. That wildlife can be the source of zoonotic infections is not new. People and domestic animals have been afflicted by endemic wildlife diseases such as rabies and foodborne infections and suffered from epidemics linked to wildlife such as the plague and tuberculosis for centuries. Wildlife has been the etiologic source of global pandemics such as HIV and influenzas. Wildlife

C. Stephen (✉)
School of Population and Public Health, University of British Columbia, Vancouver, BC, Canada

C. Stephen (ed.), *Wildlife Population Health*,
https://doi.org/10.1007/978-3-030-90510-1_15

too has suffered from epidemics and pandemics with serious conservation consequences such as the plague of chytridiomycosis afflicting global frog populations, white-nose syndrome in bats in North America, and infectious facial tumors in Tasmanian devils in Australia. The literature on the occurrences, impacts, and causes of emerging diseases is massive and growing. The interested reader will have no problem is assembling an extensive reading list on a wide variety of emerging diseases originating from and affecting wildlife. This chapter does not summarize that literature. Instead, it introduces some perspectives and approaches that can help shape an emerging disease agenda for wildlife health management in a post COVID era.

2 The Emerging Disease Forecast

Before forecasting what might be, it is important to define what is an emerging disease. The answer to that question can vary by discipline, agency, and experience. For this book, an emerging disease does not have to be an infectious disease. It can be toxic, metabolic, neoplastic, infectious, or other. An emerging disease will be considered a new occurrence of a disease, infection, infestation, or intoxication, causing significant health, conservation, or social impact resulting from: (1) a change of a known etiologic agent or its spread to a new geographic area or species; (2) a previously unrecognized agent or disease diagnosed for the first time; or (3) the creation of new social or ecological circumstances leading to a significant increase in a health impact(s) in a given population, in a given area over a given period. An emerging disease may be a case of real disease emergence, emergence of knowledge, the emergence of a disease already recognized in some places that appears in a new area of the world, emergence in a species formerly not considered susceptible, or an unexpected increase of disease in a known area and a known species (Moutou and Pastoret 2015).

One common feature of emerging diseases is that they are hard to predict and often come to us as a surprise. There is presently not enough predictive power to accurately forecast disease outcomes resulting from the environmental and social changes we anticipate in the twenty-first century (Whitmee et al. 2015a, b). This makes forecasting the emerging disease future with sufficient precision and reliability to plan programs exceedingly difficult. It may be feasible to determine the general conditions that are favorable to the emergence and spread of a disease, but it is still not yet possible to anticipate the date, place, population, and etiology that will combine to start the next outbreak. The number of times the conditions conducive to epidemics and pandemics exist without an outbreak occurring far outnumber the few occasions when a disease does emerge (Stephen et al. 2004). We are, therefore, left with general predictions such as, the rate of emergence of new infectious diseases is expected to increase in unexpected ways as climate change accelerates (Zell 2004) or epidemics will become more frequent, more complex, and harder to prevent and contain due to rapidly changing ecology, urbanization, climate change, and increased travel (Bedford et al. 2019).

Emerging disease research has made great strides in discovering threats, mapping their consequences, and mounting responses when the threat is known. It has been much less successful in "getting ahead of the curve" to inspire action in advance of impacts or to prepare for multiple, uncertain, or unknown threats. Despite increasing numbers and sophistication of risk factor analyses, risk modeling, and dynamic modeling, predicting emerging diseases is still the "art of the possible" (Woolhouse 2011). These possibilities can help direct research efforts towards the more promising question and circumstances to identify drivers of disease emergence. For example, combining an understanding of the pathogenicity of the micro-organism and the potential for the pathogen to spread and become established, with knowledge of existing capacity to control it can help direct surveillance or intervention resources to particular places for particular threats (Ogden et al. 2017). A variety of publications have implicated a diversity of possible predictors of emerging zoonotic diseases such as the amount of forest being converted into other land uses, mammalian species richness, the species present and their interactions, or a population's contaminant burden (ex. Fuller et al. 2013; Hodges and Tomcej 2016; Allen et al. 2017; Olival et al. 2017). While such information can give insights into areas that may be more vulnerable to a disease emerging or to the effects of an emerged disease, they are yet to be able to identify what will emerge when and where and what can be done to manage the threat. Management actions, therefore, need to maintain adaptable functionality in natural systems and management systems so we can cope with expected, unexpected, and interrelated threats by preserving the social and environmental assets that build health and reduce vulnerability.

3 Emerging Disease Management Principles

3.1 Be Pragmatic

In the ideal situation, a wildlife population health program would have the capacity to deal with proximate and distal (or downstream and upstream) causes and risks (see Chap. 6 on causation). This ideal is rarely experienced. The historic approach to an emerging wildlife disease that lacked a public health or agriculture implication has been to launch surveillance, monitoring, and research on species susceptibility and disease pathology rather than identifying management actions that can be implemented prior to the arrival of disease or realization of its harms (Bernard and Grant 2019; Stephen et al. 2018). Three reasons could explain this situation. First, there are comparatively few tools and resources available to detect, respond to, and contain new diseases in wildlife and to sustain the necessary actions to contain the spatial and temporal impacts of an emerged disease. Second, there are generally large uncertainties about the effectiveness and unintended consequences of intervening in a wildlife health event. This is accompanied by a desire to delay action until the uncertainties can be resolved. Third, regulatory and expertise limitations can often place management of the social and ecological drivers of emerging diseases outside of the scope of a wildlife health program.

Population health practitioners will be less frustrated in their efforts if they match the reality of their management context and capacity with their emerging disease management needs. Table 1 offers a set of questions that can define a pragmatic agenda for emerging disease response and preparedness. A pragmatic approach is not intended to find a way to compromise on aspirations or avoid unpleasant actions, but rather to help program planners see what can be done "today" with the existing resources, partnerships, and capacities. The aim is to identify the approaches and tools that will be effective for a disease with the given resources, in a given region at a given time (Langwig et al. 2015). Finding pragmatic, feasible, and acceptable actions to address some harms and impacts quickly is a better approach than delaying until an ideal program to contain all harms can be launched or a new reality can be created to allow planners to have their ideal program.

Effective emerging disease responses require timely action. Preemptive investment in wildlife health capacity to adaptively respond to surges in unexpected diseases or disease outbreaks is needed to support timely responses. Wildlife health managers will need to ensure there are policies, agreements, education, and flexible fiscal arrangements to ensure they can mobilize extra and adjustable human and material resources in a fast and productive way (Khorram-Manesh 2020). This may necessitate recruiting, training, and coordinating people outside of a wildlife health program as too often wildlife health resources and personnel are limited.

3.2 Evolve Away from Crisis Management

Too often a real or perceived crisis is the signal to launch an emerging disease management action. Systems are rarely in place to find subclinical early warning signals, identify circumstances

Table 1 Guiding question to help formulate a pragmatic and realistic plan for emerging disease prevention, response, and recovery that account for the prevailing circumstances for wildlife population health practice

Question	Details
In what phase of emergence do you find yourself?	The needs for action depend on if you need to prevent emergence, stop an initial outbreak, prevent spread, or aid in recovery. A comprehensive program considers actions across all phases, but often, you will be called to initially target one
What is the objective?	What is the specific measurable outcome you are aiming for and is that objective feasible, allowable, and acceptable in your management context? What would success look like in the short, medium, and long term?
What do you know?	What actions can be taken to make incremental improvements to prevent or reduce harms with the existing knowledge?
Can you recognize when impacts and changes happen?	Do you have surveillance, intelligence, or networks to detect harms and measure the responses to your actions to assess if the actions are having the desired effects?
Who is allowed to act?	What agencies, organizations, and/or individuals have the legal and regulatory capacity and authority to act on priority issues?
Who can act?	What agencies, organizations, and/or individuals have the resources to launch and sustain the necessary actions and interventions to meet the goals and objectives? Can they act in the timely fashion needed to meet the goals?
With whom can I work?	What agencies, organizations, and/or individuals can bring expertise, capacities, and resources needed to meet the goals? Do you have a history or agreements in place to facilitate collaboration and a co-management approach to the problem? Can the affected communities be engaged in planning and response? How could collaboration help others achieve their goals?
What are the consequences of acting or intervening?	What are the implications for social welfare, wildlife population conservation, ecosystem function, or economics from the proposed actions? Will these implications be socially acceptable, equitable, legal, and ethical? What can be done to reduce undesired consequences while still achieving the emerging disease program goals?

conducive to a disease emergence or detect an outbreak before significant numbers of animals are sick or dead (Grogan et al. 2014). Effective epidemic responses require programs to evolve from crisis response during discrete outbreaks to an integrated cycle of actions deployed before, during, and after a disease emerges (Bedford et al. 2019). There are four components to this cycle of action: (1) prevent or mitigate the impact of a hazard; (2) ensure that a population, agency, or community is prepared to act if an impact or risk emerges; (3) provide an effective response immediately after a hazard's impact is detected; and (4) help the effected populations, communities, and systems recover from the impacts. The importance of each component and its implementation sequence will vary with the situation, priorities, and resources. Actions taken in each component might be unique, overlap, or be the same. For example, effective surveillance will be important for all components but physical activities to block the spread of a hazardous agent may only be essential for two.

Salter's disaster management recommendations (Salter 1998 as cited in Crondstedt 2002) to shift away from delivering a limited range of response services to risk management and outcome-based approach provides some useful perspectives from which to formulate wildlife emerging disease management principles (Table 2). To "get ahead of the curve" emerging disease management must be able to identify socio-ecological changes in vulnerability of the wildlife to the diseases and vulnerability of societies, populations, and/or ecosystems to cope with or manage the effects of that disease. This necessitates a connected system of human intelligence distributed throughout the causal chain of emergence that can combine hazard

Table 2 Shifts in perspectives that can be used for designing an integrated cycle of actions to be deployed before, during, and after a disease emerges (adapted Salter 1998 as cited by Crondstedt 2002)

Perspective shifts	
From	To
Manage the hazardous agent	Manage population vulnerability
Rapid reaction to emergencies to constrain impacts	Proactive plan for an integrated program of prevention, response, mitigation, and recovery to manage risks and harms
Manage the response to the disease	Manage the social, population, and ecological harms
A single agency is responsible	A cross-sectoral, collaborative team is needed
Science-driven	Use a multi- and interdisciplinary approach
Plan for others	Plan with others
Communicate to others	Communicate with others

surveillance with population and environmental reconnaissance.

Combining the necessary knowledge, expertise, and resources to implement a comprehensive emerging disease system is often beyond the resources or capacity of a wildlife health agency because much of the needed information and resources reside outside of their delegated authority. Partnerships are needed. Despite an urgency for interdisciplinary work to respond to the Anthropocene, many wildlife health programs remain bound to disciplinary conventions. Preemptive and strategic risk management is unlikely unless someone is assigned to integrate diverse warning signals and bring them back to a collaboration of organizations able to deploy their shared capacities towards shared goals. Program impact and sustainability will decline quickly when program activities fall victim to departmentalization and specialization (Stephen 2020).

3.3 Work Upstream

A hazard-by-hazard approach to risk management is insufficient to ensure emerging disease preparedness when the next hazard cannot be reliably predicted. Management objectives that rely heavily on maintaining the status quo or reverting to historical rates of emerging diseases, rather than promoting resilience against the unexpected, will need to be reevaluated considering climate disruption and other environmental trends are driving emerging disease risks in surprising ways. A twenty-first-century strategy for emerging diseases requires flexible organizations with learning networks that cut across and move beyond traditional sectoral boundaries.

Although emerging diseases may be surprising, they do not arise due to chance. Changes in how people interact with animals and the natural world affect the frequency and severity of emerging diseases (Davidson 2016). The drivers of emerging diseases lie in sociology and ecology. Landscape change, wildlife exploitation, globalization, agricultural intensification, and urbanization are influencing wildlife exposure to infectious and noninfectious hazards. Mounting evidence recognizes the role of healthy ecosystems in risk reduction, recovery, and resilience, including for emerging diseases. Degradation of nature's life support systems arising from globalization, climate change, and other megatrends challenges population coping capacity which in turn reduces resilience to emerging diseases and pandemics (Whitmee et al. 2015a, b). As Hanlon and Carlisle (2008) characterized public health, so too is wildlife health often "prisoners of the proximate" in the sense that it prioritizes studies of relatively proximate causes of disease, at the expense of synthesis from a spectrum of fields to help see the breadth of the emerging challenge and be exposed to other ways of conceiving solutions.

Much work has gone into reducing emerging diseases to predictable and preventable events. A recognized challenge of the Anthropocene is that effective adaptation to the future will be unpredictable and emergent rather than predictable and planned (Hanlon and Carlisle 2008). Emerging disease planners could learn from and adapt work being undertaken to build resilience as part of adaptation to surprises (see Chap. 24). Emerging disease programs usually focus on reducing the total amount of harm by reducing exposure and sensitivity to hazardous agents. A surprise-oriented approach would also address the total impact of harm by promoting the populations' capacity to cope with an emerging disease and by managing how other social and ecological outcomes influence vulnerability (Stephen et al. 2018).

4 Learning from Failure

Galaz et al.'s (2011) examination of institutional and political leadership dimensions in ecological crisis was adapted by Stephen (2020) as a framework to categorize failures in past approaches to emerging diseases (Table 3).

The need for adaptive policies and flexible institutions cannot be over-emphasized, but the challenges of incorporating them into existing discipline-based programs should not be underestimated. Rarely is it anyone's job to "pull it all together" and form the connections and teams to acquire, assess, and act on changing signals coming from different sectors. Overconfidence in big science and artificial intelligence to reduce risk by forecasting the next emerging disease creates vulnerabilities when trying to prepare wildlife health programs in a surprising and rapidly changing world. Preparedness decision-making will benefit from scoping the problem broadly to generate a richer understanding of complex system dynamics, a more accurate and comprehensive assessment of uncertainties, and deeper insights into potential health threats (Polaski et al. 2011).

5 Summary

This chapter began by stating that its purpose is to help set an emerging disease agenda for wildlife health. Sudden, rare, and harmful events, such as an emerging disease, can focus groups to address the policy and program failures that may be revealed by such events and open the "window of opportunity" for change (Birkland and Schwaeble 2019). The COVID-19 pandemic, as well as wildlife emerging diseases such as chronic wasting disease, amphibian chytridiomycosis, and bat white-nose syndrome, all have been sudden harmful events that we failed to foresee and struggled to manage. Wildlife health professionals must continuously address emerging issues using the latest strategies and research. However, the large gaps in implementation science and program evaluation leave little basis from which to create an evidence-based agenda (see Chap. 23). The ideas presented in this chapter are largely drawn from outside of

Table 3 Why emerging disease programs fail (Stephen 2020)

Theme	Why does this occur?
Individuals or organizations are unreceptive to warning signals outside of their usual scope of practice	Cross-sectoral communication breaks down, and bureaucratic conflicts prevent information flow. Inadequate protocols exist to collectively assess signals external to an organization
Collaboration is discouraged	Organizations have their own priorities and overcrowded agendas that prevent them from working on issues beyond their immediate interest. This is compounded when there is no investment in developing partnerships to understand or influence socio-ecological causes of emerging diseases
Failure to act on early warning signs	Power dynamics can make people unwilling to accept that others can provide action signals for them. Education and training may limit the cognitive capacity of the individual to see the value of a signal outside of their domain. Wishful thinking may lead people to devalue early signals as a warning

the wildlife health literature. The nature of future disease emergence under global change, how that might be influenced by our decisions and actions, and the effectiveness of the proposed actions presented within this chapter are uncertain. It will be up to the current and future wildlife health professionals to apply and assess this or other proposed agendas to remedy the lack of validated examples of more effective ways to prevent, contain and recover from emerging diseases.

Although there is general agreement that the interplay among individual, environmental, ecological, and social factors influence emerging diseases, there is a disconnection between the multiple policies, practices, and perspectives influencing each aspect of these threats. There is a need for a more unified set of guidelines and actions to promote emerging disease management by concurrently and collaboratively tending to the determinants of health while also developing adaptive and collaborative response programs to minimize harms during and after a disease emerges. Wildlife health leaders have an important role to play in responding to environmental degradation and loss of biodiversity that will inevitably change the social structures and systems that influence the emergence of new diseases. The growing literature on the overlaps between social and biophysical determinants of disease outbreaks makes ecosystem stewardship increasingly relevant to wildlife health in an era of emerging diseases (Parkes et al. 2003).

References

Allen T, Murray KA, Zambrana-Torrelio C, Morse SS, Rondinini C, Di Marco M, Breit N, Olival KJ, Daszak P (2017) Global hotspots and correlates of emerging zoonotic diseases. Nat Commun 8(1):1

Bedford J, Farrar J, Ihekweazu C, Kang G, Koopmans M, Nkengasong J (2019 Nov) A new twenty-first century science for effective epidemic response. Nature 575(7781):130–136

Bernard RF, Grant EH (2019) Identifying common decision problem elements for the management of emerging fungal diseases of wildlife. Soc Nat Resour 32(9):1040–1055

Birkland TA, Schwaeble KL (2019) Agenda setting and the policy process: focusing events. In: Oxford research encyclopedia of politics. Oxford University Press, New York

Cronstedt M (2002) Prevention, preparedness, response, recovery-an outdated concept? Australian J Emerg Manage 17(2):10

Davidson P (2016) Rockefeller Foundation-lancet commission report: a call to action for human health. Int J Nurs Stud 53:2

Fuller TL, Gilbert M, Martin V, Cappelle J, Hosseini P, Njabo KY, Aziz SA, Xiao X, Daszak P, Smith TB (2013) Predicting hotspots for influenza virus reassortment. Emerg Infect Dis 19(4):581

Galaz V, Moberg F, Olsson E, Paglia E, Parker C (2011) Institutional and political leadership dimensions of cascading ecological crises. Pub Admin 89(2):361–380

Grogan LF, Berger L, Rose K, Grillo V, Cashins SD, Skerratt LF (2014) Surveillance for emerging biodiversity diseases of wildlife. PLoS Pathog 10(5):e1004015

Hanlon P, Carlisle S (2008) Do we face a third revolution in human history? If so, how will public health respond? J Public Health 30(4):355–361

Hodges E, Tomcej V (2016) Is there a link between pollutant exposure and emerging infectious disease? Can Vet J 57(5):535

Khorram-Manesh A (2020) Flexible surge capacity–public health, public education, and disaster management. Health Promot Perspect 10(3):175

Langwig KE, Voyles J, Wilber MQ, Frick WF, Murray KA, Bolker BM, Collins JP, Cheng TL, Fisher MC, Hoyt JR, Lindner DL (2015) Context-dependent conservation responses to emerging wildlife diseases. Front Ecol Environ 13(4):195–202

Moutou F, Pastoret PP (2015 Apr) Defining an emerging disease. Revue scientifique et technique (Int Officef Epizootics) 34(1):41–52

Ogden NH, AbdelMalik P, Pulliam JR (2017) Emerging infections: emerging infectious diseases: prediction and detection. Can Commun Dis Rep 43(10):206

Olival KJ, Hosseini PR, Zambrana-Torrelio C, Ross N, Bogich TL, Daszak P (2017) Host and viral traits predict zoonotic spillover from mammals. Nature 546(7660):646–650

Parkes M, Panelli R, Weinstein P (2003) Converging paradigms for environmental health theory and practice. Environ Health Perspect 111(5):669–675

Polaski S, Carpenter S, Folke C, Keeler B (2011) Decision-making under great uncertainty: environmental management in an era of global change. TREE 26(8):398–404

Sagan SD (1993) The limits of safety. Princeton University Press, New Jersey. Cited in Stikeleather J, Masys AJ (2020) Global health security innovation. In Global health security (pp. 387-425). Springer, Cham

Salter J (1998) In: Smith DJ (ed) Risk management in the emergency management context. Risk Management for Safer Communities, Melbourne

Stephen C (2020) Rethinking pandemic preparedness in the Anthropocene. Healthc Manage Forum 33(4): 153–157

Stephen C, Artsob H, Bowie WR, Drebot M, Fraser E, Leighton T et al (2004) Perspectives on emerging zoonotic disease research and capacity building in Canada. Can J Infect Dis Med Micro 15(6):339–344

Stephen C, Wittrock J, Wade J (2018) Using a harm reduction approach in an environmental case study of fish and wildlife health. Ecohealth 2018:1–4

Whitmee S, Haines A, Beyrer C, Boltz F, Capon AG, de Souza Dias BF, Ezeh A, Frumkin H, Gong P, Head P, Horton R (2015a) The Rockefeller Foundation–Lancet Commission on planetary health. Safeguarding human health in the Anthropocene epoch: report of The Rockefeller Foundation–Lancet Commission on planetary health. Lancet 386(10007):1973–2028

Whitmee S, Haines A, Beyrer C, Boltz F, Capon AG, de Souza Dias BF et al (2015b) Safeguarding human health in the Anthropocene epoch: report of The Rockefeller Foundation–Lancet Commission on planetary health. Lancet 386(10007):1973–2028

Woolhouse M (2011) How to make predictions about future infectious disease risks. Philos Trans Royal Soc 366(1573):2045–2054

Zell R (2004) Global climate change and the emergence/re-emergence of infectious diseases. Int J Med Microbiol 293:16–26

Pollution and Wildlife Health

Thierry M. Work

Abstract

Pollution is a pervasive and growing threat to wildlife health. This chapter discusses two broad groups of pollution, those whose abatement could have immediate beneficial effects including light, air, and noise pollution, and those that will take relatively longer to address due to their environmental persistence or their continuing discharge. Whilst we are very good at detecting the presence of pollutants in tissues or the environment, making a convincing link between the presence of these compounds and mortality events in the field or population effects will remain a challenge for the foreseeable future. Creative new approaches are also being considered to mitigate the effects of pollution on wildlife and ecosystems. Depending on the source of pollution, the beneficial outcomes of mitigation measures, if properly implemented, could have immediate effects. Given the plethora of potential adverse pollution effects, frameworks to prioritize which threats are most likely to cause adverse effects and develop means to address or manage them are an imperative. In the interim, focusing on preserving existing habitats and reducing our footprint by adjusting human activities to minimize the release of pollutants into the environment will go a long way toward promoting healthy wildlife and ecosystems.

Keywords

Pollution · Contaminant · Wildlife · Impact · Disease · Light · Noise · Oil · Air · Pesticides · Drugs · Plastic · Metals · Nitrogen

1 Introduction

Pollution as defined in this chapter is "an activity of humans, which directly or indirectly results in the addition to water, air or soil, of matter or energy which has a deleterious effect on living organisms or structures it is desirable to preserve, or which reduces the quality of water, air, or soil for any subsequent use" (Russell 1974). More pithily, pollution is the introduction of contaminants into the natural environment that cause adverse change (Merriam-Webster 2020). Throughout their presence on earth, humans, as the ultimate ecosystem engineers, have succeeded in significantly altering their environment, a side effect of which has been pollution. Some historical examples of adverse effects of pollution to humans go as far back as the era of the Roman Empire where awareness existed that exposure to lead used for ship construction resulted in illnesses in shipbuilders (Hernberg 2000). Excess mortalities in England due to poor air quality

T. M. Work (✉)
U.S. Geological Survey's National Wildlife Health Center
Honolulu Field Station, Hawaii, USA
e-mail: thierry_work@usgs.gov

C. Stephen (ed.), *Wildlife Population Health*,
https://doi.org/10.1007/978-3-030-90510-1_16

from factories during the industrial revolution in the early nineteenth century is another example (Anderson 2009). At the same time, it became evident that pollutants could affect wildlife as evidenced by lead poisoning in birds from ingestion of lead shot, and poisoning of wildlife from mine tailings (Rattner 2018). However, it was not until the 1960s when there was an increase in broader public awareness that pollutants could have adverse effects to human health and wildlife. Massive mercury poisoning originating from effluents of industrial waste in Minimata, Japan illustrated the problem in people (Harada 1995). Contemporaneously, Rachel Carson's book Silent Spring seeded the popular imagination with the concepts of environmental contamination of persistent organochlorine pesticides that bioaccumulated in the food chain and led to reproductive failure in a variety of wild birds. These events led to creations of agencies and international accords such as the Environmental Protection Agency in the United States (Rattner 2009), and the 1972 United Nations Conference on the Human Environment in Stockholm that formed the basis for international community environmental law to protect humans and ecosystems (Sands 1991). Since that time, in certain regions like North America and Europe, the use of organochlorines has stopped thereby leading to declines in environmental concentrations (van den Berg et al. 2017). However, these compounds continue to be used in many countries to control insects that spread various infectious diseases such as malaria that impose severe human health burdens (van den Berg et al. 2017), and societies are faced with new threats and challenges such as plastic (Worm et al. 2017) and noise (Duarte et al. 2021) pollution.

Given the continued increase in human populations, which is expected to exceed 9.5 billion by the mid-2050s (Bongaarts 2009), human pressures on wildlife and the tolls taken by pollution are only bound to increase concomitantly. Moreover, whilst organochlorines, acid rain, and heavy metals historically dominated ecotoxicologists' attention (Rattner 2009), newer aspects of human activity are being recognized as having potential negative effects on wildlife. Arguably, a chief source of pollution now affecting humans and wildlife is carbon dioxide leading to climate change, but this particular pollutant is beyond the scope of this chapter (Crowley and Berner 2001) (see Chap. 13). At the same time, creative new approaches are also being considered to mitigate the effects of pollution on wildlife and ecosystems. Depending on the source of pollution, the beneficial outcomes of these mitigation measures, if properly implemented, could have immediate effects. Accordingly, this chapter divides the source of pollution into two groups: Those whose abatement would have more immediate beneficial effects including light, air, and noise pollution, and those that will take relatively longer to address due to their nature of persisting in the environment either due to physical properties or their continuing discharge. The latter would include chemicals used in or waste products from agriculture, mining, pharmaceuticals, plastics, and other industries.

2 Sources of Pollution for Which Mitigation Measures Would Have Immediate Effects

2.1 Light

The term "light pollution" was originally coined by Longcore and Rich (2004) to describe how light might affect behaviors of wildlife that are predominantly active at night, expected to be about 30% of vertebrates and 60% of invertebrates (Gaston 2019). Artificial light at night has particularly pernicious effects on insects. For instance, artificial lights can interfere with mating signals for fireflies (Owens et al. 2018). Light fixtures attract various insects making them more susceptible to predation (Owens et al. 2020), and polarized light pollution, where light reflects from flat shiny surfaces mimicking water, leads to large numbers of aquatic insects ovipositing in unsuitable areas like parking lots (Horváth et al. 2009). Artificial light can also adversely affect vertebrates. Examples include disorientation of sea turtle nestlings that emerge from nests and are unable to migrate towards the ocean (Longcore and Rich 2004), and seabirds confused by urban lighting and stranding when returning to their nests from foraging at sea (Rodríguez et al. 2014). Indeed, artificial light at night will likely have some of the greatest impacts on birds and their nocturnal migrations (La Sorte et al. 2017). Given that going forward, light pollution will be most prominent near coastal areas where most humans live (Gallaway et al. 2010), the effects on coastal marine nocturnal wildlife will probably be most severe.

Most studies of the effects of light on wildlife are correlative, and clear population effects have yet to be identified (Schowalter et al. 2019). However, it is possible that the effects of light on wildlife are underestimated, in part due to "diurnal bias" where most investigations by humans of biological phenomena are done during the day (Gaston 2019). There appears, however, to be increasing recognition that light pollution can also adversely affect humans leading to obesity, cancer, and immune dysfunction (Navara and Nelson 2007). Moreover, light pollution is globally estimated to increase 6% per annum, so there will be a need to limit this, in part because of lighting's energy demands and contribution to global warming (Gallaway et al. 2010). Increasing awareness of this issue is reflected in various international "dark sky" initiatives to reclaim the night from urban light pollution (Wartmann et al. 2019). Abating light pollution would have immediate benefits to wildlife, and measures to implement this include eliminating unnecessary lighting, limiting the duration of lighting, limiting light to where it is needed, reducing intensity, and adjusting wavelengths to minimize effects of artificial light at night on sensitive species (Gaston et al. 2012).

2.2 Air

Air pollution defined here includes the discharge of man-made substances into the air that have potential adverse environmental effects. The classic effect of air pollution on wildlife was the melanization of peppered moths in nineteenth-century England pursuant to severe air pollution that led to sooty staining of surfaces. This made melanized moths less visible to predators thereby giving them a selective advantage; as air pollution abated, the relative frequency of melanized moths declined (Cook 2003). Aerial pollutants are particularly problematic because this mode of transport has the potential to carry contaminants long distances relatively rapidly compared to other modes of dissemination such as water or biota (Newman 1979). Mining operations are important sources of long-distance transport of metals and metalloids, and adverse health effects of lead and arsenic poisoning in humans from exposure to metal-contaminated dust from mines have been documented (Csavina et al. 2012). Also, veterinary pharmaceuticals could be spread by airborne dust from feedlots (Sandoz et al. 2018). Long-range transport of organochlorines by air to the arctic from industrialized areas is thought to be partially responsible for the exposure of arctic wildlife to these compounds with attendant adverse health effects (Norstrom and Muir 1994). Finally, acidification of watersheds from acid rain, mainly through sulfate pollution has led to forest diebacks and declines of aquatic invertebrates. Acidification of soil can also leach heavy metals into the environment and increase their uptake in fish (Newman et al. 1992, Greaver et al. 2012). Air pollution, in the form of particulate matter, imposes significant health burdens on humans leading to cardiovascular diseases and lung cancer, particularly in low- to middle-income countries where air pollution is increasing due to economic development (Cohen et al. 2017). Presumably, similar effects would be seen in urban wildlife; however, to date studies examining this possibility are lacking (see Chap. 12). A review on measures to mitigate air pollution originating from transportation, agricultural, and industrial sectors was recently done by Sofia et al. (2020). A vivid example of the relatively rapid response to mitigation actions of air pollution was on display during global lockdowns imposed due to COVID19. Within 3 months of lockdowns, Venter et al. (2020) documented global mean declines of atmospheric nitrogen by 60% and particulate matter by 31%.

2.3 Noise

Public awareness of noise as an irritant has been with humanity historically (Coates 2005). However, the soundscape surrounding humanity has undergone profound changes over time with a general increase in exposure to noise with increased populations and industrialization (Maheshwari et al. 2012). Wildlife depends on sound for a variety of life-sustaining activities. For instance, birds vocalize to attract mates and other species listen to calls of different species to make choices about habitat, a phenomenon known as acoustical eavesdropping. Sound is also used by birds, mammals, and invertebrates to locate prey. It is anticipated that increased human generated sounds from transportation in terrestrial ecosystems will impede animals from effective communications, known as masking (Barber et al. 2010). Although experimentation with sound has shown that human noise alters habitat use by wildlife (Shannon et al. 2016), some animals are also able to adapt to noisier ecosystems by vocalizing louder or at different frequencies, for example (Lowry et al. 2013). The issue of noise pollution in aquatic ecosystems may be more serious, however, because noise travels farther in water than in the air. It is becoming apparent that noise may have detrimental effects on marine wildlife. Many marine vertebrates and invertebrates depend on sound to locate settlement habitats, communicate, and avoid predation. Shipping-induced noise has increased 32 fold over the past 50 years and along with other human activities (sonar, seismic surveys, mining exploration) contributes significantly to sound pollution in oceans (Duarte et al. 2021). The problem is likely to worsen with global warming as warmer water carries sounds further. To date, most evidence of the effects of noise involve behavior and physiologic changes mainly in fish and marine mammals; however there is less evidence that noise affects fitness on population levels, and more research is needed in this arena (Duarte et al. 2021). Eliminating or reducing noise that is produced as a byproduct of activities (e.g., shipping) is likely to be a more tractable solution for the near term, particularly if measures align with economic incentives. For instance, efforts to mitigate sounds from shipping by designing quieter more efficient propellers, dampening vibrations, and use of electric motors could benefit shippers as a result of reduced fuel costs (Duarte et al. 2021).

3 Sources of Pollution for Which Mitigation Measures Would Have Medium to Long Term Effects

3.1 Nitrogen-Containing Compounds

Biologically available nitrogen (e.g., NH4 and NO3 as opposed to gaseous N2) is produced naturally through various natural processes, such as nitrogen fixation by plants and bacteria. Since the invention of the Haber process to make synthetic nitrogen-containing compounds, human activity has increased biologically available nitrogen in the environment by 30–50% (Hernández et al. 2016). Sources of nitrogen originating from human activities include the use of synthetic fertilizers, nitrogen fixation through agriculture, fossil fuel combustion, and food/animal waste. Half of the synthetic nitrogens used on earth have been produced since 1985 (Howarth 2008). In terrestrial and freshwater ecosystems, nitrogen-containing compounds are known to be toxic to fish, invertebrates, and amphibians, promote eutrophication and algal blooms and promote the growth of invasive plant species that could displace native plants or alter prey base for native animals (Hernández et al. 2016). In marine ecosystems, nitrogen is a limiting element, and given that 65% of large cities are near the coasts, globally, eutrophication of coastal ecosystems is only likely to increase leading to increased algal blooms and potential wildlife mortalities (Glibert and Burkholder 2018). A good example is the large areas of the Gulf of Mexico known as the dead zone that has undergone eutrophication and hypoxia events leading to ablation of most biota over a large geographic scale (Rabalais et al.

2002). There is also a strong link between coastal eutrophication and diseases of sessile corals. Sedimentation and light attenuation from algal blooms leads to increased mucus production and reduced photosynthesis of symbiotic algae (zooxanthellae) in corals (Cooper et al. 2009). Elevations in nitrogen-containing compounds also can directly compromise photosynthesis in corals and interfere with calcification (Zhao et al. 2021). In addition to the direct effects of nitrogen-containing compounds on marine biota, there is an emerging consensus that coastal eutrophication will lead to increased frequencies of blooms of toxic algae which could have important adverse health ramifications to humans (Heisler et al. 2008), marine mammals (Broadwater et al. 2018), marine birds (Gibble and Hoover 2018), shellfish (Basti et al. 2018), and food webs (Burkholder et al. 2018).

Methods exist to reduce or eliminate the influx of nitrogen-containing compounds into ecosystems. These include the construction of wetlands to filter out nitrogen-containing compounds from agricultural runoff, injecting animal waste into soil, applying fertilizers near roots, rotating crops to help lock nitrogen-containing compounds into the soil through fixation, and amending soil with wood pulp or biochar to aid denitrification of soil (Bednarek et al. 2014; Ghaly and Ramakrishnan 2015; Shaaban et al. 2018). Improving wastewater treatment infrastructure globally might aid in reducing discharges of nitrogen-containing compounds in the environment (Van Drecht et al. 2009).

3.2 Pesticides

Since the 1990s, pesticide use globally has been increasing steadily but has plateaued since about 2012; however, the global cost/benefits of pesticide use have declined since 2007 (Zhang 2018). Wildlife gets exposed to pesticides mainly through ingestion, and for birds, organochlorines, organophosphates, and carbamates continue to pose the greatest threats, mainly as endocrine and reproductive disrupters for the first and neurotoxins for the latter two (Mitra et al. 2011). Because of the nontarget effects of older pesticides such as organochlorines, efforts have been made to develop classes of pesticides that are more selective to their intended target. Pyrethroids, a derivative of chrysanthemums, are minimally toxic to mammals but very toxic to fish and nontarget invertebrates (Li et al. 2017). Neonicotinoids such as fipronil, are highly toxic to fish whereas others such as imidacloprid are moderately toxic to birds and mammals but not to fish and amphibians (Gibbons et al. 2015). Neonicotinoids are also thought to interfere with the migration patterns of songbirds (Eng et al. 2019). Herbicides are shown to be toxic to seagrass beds and mangroves in marine ecosystems (Peters et al. 1997). It is also increasingly recognized that in addition to direct effects, pesticides can have indirect effects such as decreasing food/prey availability and quantity and loss of plant habitat for invertebrates (Gibbons et al. 2015).

Evaluating the direct and indirect effects of pesticides on wildlife continues to be a daunting task complicated by the fact that there are likely synergistic or antagonistic interactions with pesticide mixtures (Köhler and Triebskorn 2013). One proposed way to address the complexities of the interaction between pesticides and the physiology of organisms is the concept of adverse outcome pathways that take into account knowledge of the interaction of molecules with bioreceptors (Spurgeon et al. 2020). For instance, knowing the location of a receptor to a molecule, its effects, and its conservation across taxa may aid in assessing how a particular chemical might affect communities of animals. Given the continued threats that pesticides will pose to ecosystems, there will need to be ongoing "pestidovigilance" (Spurgeon et al. 2020) and continued use of ecological risks assessments (Norton et al. 1992).

3.3 Pharmaceuticals

Although pharmaceuticals are not known to persist in the environment, they are included here because it is likely that pharmaceuticals will increase in use as human populations increase, age, and have greater access to medical care. There is also now better capacity to monitor the presence of pharmaceuticals because of better analytical techniques. Wastewater is the chief source of pharmaceuticals that are dominated by antimicrobials and painkillers (Hughes et al. 2013). Pharmaceuticals are considered pseudo persistent in the environment due to their continuous discharge from wastewater treatment plants. Unlike compounds like organochlorines, pharmaceuticals do not bioaccumulate, but residues are found in wildlife. For example, antibiotics and antidepressants are the dominant residues in freshwater fish and invertebrates (Miller et al. 2018) reflecting global discharge statistics outlined earlier. Pharmaceuticals are also being detected in marine ecosystems, and their presence in those ecosystems is influenced by the size of adjoining urban areas, type of wastewater treatment, hydrodynamics, and proximity to terrestrial animal husbandry and aquaculture operations (Gaw et al. 2014). To date, the effect of pharmaceutical exposure to wildlife is unclear, in part because most efforts to sample for residues have focused on water and not biota (Miller et al. 2018). However, there are two instances where pharmaceuticals directly harmed wildlife. Massive declines of vultures in India as a result of consumption of dead livestock containing residues of the anti-inflammatory drug Diclofenac (Oaks et al. 2004) and feminization of fish exposed to oral contraceptives through exposure to the water column (Jobling et al. 2006). The vulture incident led to regulatory changes in the use and disposition of Diclofenac to minimize hazards to birds of prey (Arnold et al. 2013). Challenges going forward will be to find ways to assess the effects of other pharmaceuticals in a manner that provides the necessary data to assess whether additional management and reductions of discharges are justified. Using current practices, extrapolating from laboratory studies to effects in the field are difficult because laboratory exposures often do not reflect levels seen in the field. Recommendations to rectify this include standardizing sampling and analytical techniques and more comprehensive sampling spatially and temporally (Hughes et al. 2013; Miller et al. 2018).

3.4 Metals and Metalloids

Heavy metal poisoning has been documented in wildlife since the nineteenth century with examples of lead poisoning in birds due to lead shot ingestion or exposure to mine tailings and arsenic poisoning in deer near silver foundries. Because lead poisoning from lead shot ingestion was threatening populations of bald eagles in the United States, regulations were enacted limiting the use of lead shot for hunting in public lands (Rattner 2009) with additional international agreements aimed to safeguard wild birds from lead poisoning such as the African Eurasian Migratory Waterbird Agreement (AEWA) (Lewis 2016). Sources of metals in the environment include natural weathering, industrial activities (textiles, electroplating), phosphate fertilizers, and fossil fuel combustion. Heavy metals can also accumulate in crops such as rice. Exposure in wildlife is via ingestion. In addition to the lead poisoning example with eagles, other metals or metalloids have had important adverse effects on wildlife populations. For instance, tributyltin found in ship paint led to a decline in marine molluscs that have since recovered since these compounds were eliminated from use (Wells and Gagnon 2020), and selenium poisoning from agricultural wastewater that led to mass mortalities of waterbirds (Ohlendorf 1989). With increasing economic development and industrialization, it is likely that heavy metal exposure, and in some instances, poisoning of wildlife, will continue; however, methods to monitor and assess this are well established (Blus et al. 1987). More intriguing are ways to remediate soils to remove heavy metals or limit their bioavailability from the environment. Phytoremediation, the use of plants to absorb and remove metals from the soil, is a very dynamic field of research (Ali et al. 2013). There is also now a greater understanding of soil characteristics that promote plants, fungi, and microbes to remove heavy metals from soil

(Gall et al. 2015). Merging these methods along with the better understanding of genetic and molecular mechanisms that promote metal accumulation in plants might be a way forward in minimizing environmental exposure of wildlife to heavy metals.

3.5 Plastic

Since the 1950s, the use of plastics has increased to the point that global plastic production annually equals the weight of all humans on earth, and we are dumping the equivalent of one garbage truckload of plastic into the ocean every minute (Worm et al. 2017). Most studies looking at the effects of plastic on wildlife have focused on marine ecosystems with relatively fewer efforts in freshwater systems (Blettler and Wantzen 2019) even though freshwater ecosystems (rivers and streams) are important conduits of plastics to the marine environment. Aside from their unsightliness, plastics are thought to affect wildlife through entanglement, ingestion, transfer of contaminants, and rafting of invasive marine species (Gall and Thompson 2015). Studies of microplastics that seem to be pervasive in sediments and biota are also gaining attention in the scientific community (Matsuguma et al. 2017; Hale et al. 2020). However, linking exposure to plastics to measurable adverse effects of wildlife populations remains elusive. For instance, although a consensus is emerging that visible plastic entanglement adversely affects marine mammals, it is difficult to assess trends in entanglement due to skewed reporting, variable research effort, and uncertainty of the severity of plastic pollution (Jepsen and de Bruyn 2019). The effects of plastic ingestions on wildlife are even harder to assess. For instance, it is impossible based on necropsy to assess whether plastic ingestion leads to emaciation in birds or whether underlying disease leads to the inability to excrete plastic (Roman et al. 2020). Indeed, a recent meta-analysis of research on the effects of plastics on wildlife concluded that there was little evidence of harm from exposure to microplastics and that more effects were detected with macroplastics (Bucci et al. 2020). Even hypotheses about ingested plastics playing a role in leaching organic contaminants into wildlife are open to question and it is likely that ingested plastics play a minor role if any at all in the translocation of organic compounds (Koelmans et al. 2016). Guidelines are available on how to standardize studies on the effects of plastics ingestion to gain more robust conclusions as to their effects on wildlife (Provencher et al. 2017). Salient questions to prioritize research in plastics were outlined by Vegter et al. (2014). In the longer term, addressing plastic pollution will require reducing production, use, and waste generation (Worm et al. 2017). There are also new research avenues looking into microbial (Shen et al. 2019) and chemical (Gewert et al. 2015) degradation of plastics as well as development of biodegradable plastics (Narancic et al. 2018) which might contribute to longer term abatement of plastic pollution.

4 Conclusion

Diagnosis of contaminants exposure and adverse effects that result in wildlife mortality events is a process of systematically ruling out more obvious causes of death using pathology and laboratory diagnostics. A final diagnosis of contaminants is made based on the presence of the compound, history of plausible exposure, and absence of other evidence causes of death. For chronic contaminant effects or contaminant interactions, the process is more uncertain because of confounding factors that become introduced through time. Whilst we are very good at detecting the presence of pollutants in tissues or the environment, making a convincing link between the presence of these compounds and mortality events in the field or population effects will remain a challenge for the foreseeable future. Additionally, given the plethora of potential adverse effects pollutants play in wildlife, an imperative will be to develop frameworks to prioritize which threats are most likely to cause adverse effects in wildlife such as embryotoxicity, cancer, impaired growth, ideally before population declines are documented.

Developing the means to address or manage these threats will also be important. In the interim, focusing on preserving existing habitats and reducing our footprint by adjusting human activities to minimize the release of pollutants into the environment will go a long way toward promoting healthy wildlife and ecosystems.

Acknowledgments Barnett Rattner and Jennifer Lynch provided constructive comments in earlier versions of this MS. Any use of trade, firm, or product names is for descriptive purposes only and does not imply endorsement by the U.S. Government.

References

Ali H, Khan E, Sajad MA (2013) Phytoremediation of heavy metals—concepts and applications. Chemosphere 91(7):869–881

Anderson HR (2009) Air pollution and mortality: a history. Atmos Environ 43(1):142–152

Arnold KE, Boxall AB, Brown AR, Cuthbert RJ, Gaw S, Hutchinson TH, Jobling S, Madden JC, Metcalfe CD, Naidoo V, Shore RF (2013) Assessing the exposure risk and impacts of pharmaceuticals in the environment on individuals and ecosystems. Biol Lett 9:20130492

Barber JR, Crooks KR, Fristrup KM (2010) The costs of chronic noise exposure for terrestrial organisms. Trends Ecol Evol 25(3):180–189

Basti L, Hégaret H, Shumway SE (2018) Harmful algal blooms and shellfish. In: Harmful algal blooms: a compendium desk reference, vol 2. John Wiley & Sons, Hoboken, pp 135–190

Bednarek A, Szklarek S, Zalewski M (2014) Nitrogen pollution removal from areas of intensive farming—comparison of various denitrification biotechnologies. Ecohydrology & Hydrobiology 14(2):132–141

Blettler MC, Wantzen KM (2019) Threats underestimated in freshwater plastic pollution: mini-review. Water Air Soil Pollut 230(7):1–1

Blus LJ, Henny CJ, Mulhern BM (1987) Concentrations of metals in mink and other mammals from Washington and Idaho. Environ Pollut 44(4):307–318

Bongaarts J (2009) Human population growth and the demographic transition. Philos Trans Royal Soc 364(1532):2985–2990

Broadwater MH, Van Dolah FM, Fire SE (2018) Vulnerabilities of marine mammals to harmful algal blooms. Harmful Algal Blooms 2(8):191–222

Bucci K, Tulio M, Rochman CM (2020) What is known and unknown about the effects of plastic pollution: a meta-analysis and systematic review. Ecol Appl 30(2): e02044

Burkholder, J.M., Shumway, S.E., and Glibert, P.M. 2018. Food web and ecosystem impacts of harmful algae, S.E. Shumway, J.M. Burkholder, S.L. Morton, Harmful algal blooms: a compendium desk reference, John Wiley, Hoboken pp. 243–301

Coates PA (2005) The strange stillness of the past: toward an environmental history of sound and noise. Environ Hist 10(4):636–665

Cohen AJ, Brauer M, Burnett R, Anderson HR, Frostad J, Estep K, Balakrishnan K, Brunekreef B, Dandona L, Dandona R, Feigin V (2017) Estimates and 25-year trends of the global burden of disease attributable to ambient air pollution: an analysis of data from the global burden of diseases study 2015. Lancet 389(10082):1907–1918

Cook LM (2003) The rise and fall of the carbonaria form of the peppered moth. Q Rev Biol 78(4):399–417

Cooper TF, Gilmour JP, Fabricius KE (2009) Bioindicators of changes in water quality on coral reefs: review and recommendations for monitoring programmes. Coral Reefs 28(3):589–606

Crowley TJ, Berner RA (2001) CO2 and climate change. Science 292(5518):870–872

Csavina J, Field J, Taylor MP, Gao S, Landázuri A, Betterton EA, Sáez AE (2012) A review on the importance of metals and metalloids in atmospheric dust and aerosol from mining operations. Sci Total Environ 433: 58–73

Duarte CM, Chapuis L, Collin SP, Costa DP, Devassy RP, Eguiluz VM, Erbe C, Gordon TA, Halpern BS, Harding HR, Havlik MN (2021) The soundscape of the Anthropocene Ocean. Science 371:6529

Eng ML, Stutchbury BJ, Morrissey CA (2019) A neonicotinoid insecticide reduces fueling and delays migration in songbirds. Science 365(6458):1177–1180

Gall JE, Boyd RS, Rajakaruna N (2015) Transfer of heavy metals through terrestrial food webs: a review. Environ Monit Assess 187(4):1–21

Gall SC, Thompson RC (2015) The impact of debris on marine life. Mar Pollut Bull 92(1–2):170–179

Gallaway T, Olsen RN, Mitchell DM (2010) The economics of global light pollution. Ecol Econ 69(3):658–665

Gaston KJ (2019) Nighttime ecology: the "nocturnal problem" revisited. Am Nat 193(4):481–502

Gaston KJ, Davies TW, Bennie J, Hopkins J (2012) Reducing the ecological consequences of night-time light pollution: options and developments. J Appl Ecol 49(6):1256–1266

Gaw S, Thomas KV, Hutchinson TH (2014) Sources, impacts and trends of pharmaceuticals in the marine and coastal environment. Philos Trans Royal Soc 369(1656):20130572

Gewert B, Plassmann MM, MacLeod M (2015) Pathways for degradation of plastic polymers floating in the marine environment. Environ Sci: Processes Impacts 17(9):1513–1521

Ghaly AE, Ramakrishnan VV (2015) Nitrogen sources and cycling in the ecosystem and its role in air, water and soil pollution: a critical review. J Pollut Effects Contr 27:1–26

Gibble CM, Hoover BA (2018) Interactions between seabirds and harmful algal blooms. Harmful Algal Blooms 2:223–242

Gibbons D, Morrissey C, Mineau P (2015) A review of the direct and indirect effects of neonicotinoids and fipronil on vertebrate wildlife. Environ Sci Pollut Res 22(1):103–118
Glibert PM, Burkholder JM (2018) Causes of harmful algal blooms. Harmful Algal Blooms 2:1–38
Greaver TL, Sullivan TJ, Herrick JD, Barber MC, Baron JS, Cosby BJ, Deerhake ME, Dennis RL, Dubois JJ, Goodale CL, Herlihy AT (2012) Ecological effects of nitrogen and sulfur air pollution in the US: what do we know? Front Ecol Environ 10(7):365–372
Hale RC, Seeley ME, La Guardia MJ, Mai L, Zeng EY (2020) A global perspective on microplastics. J Geophys Res Oceans 125(1):e2018JC014719
Harada M (1995) Minamata disease: methylmercury poisoning in Japan caused by environmental pollution. Crit Rev Toxicol 25(1):1–24
Heisler J, Glibert PM, Burkholder JM, Anderson DM, Cochlan W, Dennison WC, Dortch Q, Gobler CJ, Heil CA, Humphries E, Lewitus A (2008) Eutrophication and harmful algal blooms: a scientific consensus. Harmful Algae 8(1):3–13
Hernández DL, Vallano DM, Zavaleta ES, Tzankova Z, Pasari JR, Weiss S, Selmants PC, Morozumi C (2016) Nitrogen pollution is linked to US listed species declines. Bioscience 66(3):213–222
Hernberg S (2000) Lead poisoning in a historical perspective. Am J Ind Med 38(3):244–254
Horváth G, Kriska G, Malik P, Robertson B (2009) Polarized light pollution: a new kind of ecological photopollution. Front Ecol Environ 7(6):317–325
Howarth RW (2008) Coastal nitrogen pollution: a review of sources and trends globally and regionally. Harmful Algae 8(1):14–20
Hughes SR, Kay P, Brown LE (2013) Global synthesis and critical evaluation of pharmaceutical data sets collected from river systems. Environ Sci Technol 47(2): 661–677
Jepsen EM, de Bruyn PN (2019) Pinniped entanglement in oceanic plastic pollution: a global review. Mar Pollut Bull 145:295–305
Jobling S, Williams R, Johnson A, Taylor A, Gross-Sorokin M, Nolan M, Tyler CR, van Aerle R, Santos E, Brighty G (2006) Predicted exposures to steroid estrogens in UK rivers correlate with widespread sexual disruption in wild fish populations. Environ Health Perspect 114(Suppl 1):32–39
Koelmans AA, Bakir A, Burton GA, Janssen CR (2016) Microplastic as a vector for chemicals in the aquatic environment: critical review and model-supported reinterpretation of empirical studies. Environ Sci Technol 50(7):3315–3326
Köhler HR, Triebskorn R (2013) Wildlife ecotoxicology of pesticides: can we track effects to the population level and beyond? Science 341(6147):759–765
La Sorte FA, Fink D, Buler JJ, Farnsworth A, Cabrera-Cruz SA (2017) Seasonal associations with urban light pollution for nocturnally migrating bird populations. Glob Chang Biol 23(11):4609–4619
Lewis M (2016) AEWA at twenty: an appraisal of the African-Eurasian Waterbird agreement and its unique place in international environmental law. J Int Wildl Law Pol 19(1):22–61
Li H, Cheng F, Wei Y, Lydy MJ, You J (2017) Global occurrence of pyrethroid insecticides in sediment and the associated toxicological effects on benthic invertebrates: an overview. J Hazard Mater 324:258–271
Longcore T, Rich C (2004) Ecological light pollution. Front Ecol Environ 2(4):191–198
Lowry H, Lill A, Wong BB (2013) Behavioural responses of wildlife to urban environments. Biol Rev 88(3): 537–549
Maheshwari R, Prasad M, Yadav RK, Sharma S, Chauhan V, Rani B (2012) Noise pollution: An annoyance to endurance. Bull Environ Pharmacol Life Sci 11:50–52
Matsuguma Y, Takada H, Kumata H, Kanke H, Sakurai S, Suzuki T, Itoh M, Okazaki Y, Boonyatumanond R, Zakaria MP, Weerts S (2017) Microplastics in sediment cores from Asia and Africa as indicators of temporal trends in plastic pollution. Arch Environ Contam Toxicol 73(2):230–239
Merriam-Webster (2020) Dictionary. Merriam-Webster.com
Miller TH, Bury NR, Owen SF, MacRae JI, Barron LP (2018) A review of the pharmaceutical exposome in aquatic fauna. Environ Pollut 239:129–146
Mitra A, Chatterjee C, Mandal FB (2011) Synthetic chemical pesticides and their effects on birds. Res J Environ Toxicol 5(2):81–96
Narancic T, Verstichel S, Reddy Chaganti S, Morales-Gamez L, Kenny ST, De Wilde B, Babu Padamati R, O'Connor KE (2018) Biodegradable plastic blends create new possibilities for end-of-life management of plastics but they are not a panacea for plastic pollution. Environ Sci Technol 52(18):10441–10452
Navara KJ, Nelson RJ (2007) The dark side of light at night: physiological, epidemiological, and ecological consequences. J Pineal Res 43(3):215–224
Newman JR (1979) Effects of industrial air pollution on wildlife. Biol Conserv 15(3):181–190
Newman JR, Schreiber RK, Novakova E (1992) Air pollution effects on terrestrial and aquatic animals. In: Barker JR, Tingey DT (eds) Air pollution effects on biodiversity. Springer, Boston
Norstrom RJ, Muir DC (1994) Chlorinated hydrocarbon contaminants in arctic marine mammals. Sci Total Environ 154(2–3):107–128
Norton SB, Rodier DJ, van der Schalie WH, Wood WP, Slimak MW, Gentile JH (1992) A framework for ecological risk assessment at the EPA. Environ Toxicol Chem 11:1663–1672
Oaks JL, Gilbert M, Virani MZ, Watson RT, Meteyer CU, Rideout BA, Shivaprasad HL, Ahmed S, Chaudhry MJ, Arshad M, Mahmood S (2004) Diclofenac residues as the cause of vulture population decline in Pakistan. Nature 427(6975):630–633

Ohlendorf HM (1989) Bioaccumulation and effects of selenium in wildlife. Selenium Agric Environ 23: 133–177
Owens AC, Cochard P, Durrant J, Farnworth B, Perkin EK, Seymoure B (2020 Jan) Light pollution is a driver of insect declines. Biol Conserv 1(241):108259
Owens AC, Meyer-Rochow VB, Yang EC (2018) Short- and mid-wavelength artificial light influences the flash signals of Aquatica ficta fireflies (Coleoptera: Lampyridae). PLoS One 13(2):e0191576
Peters EC, Gassman NJ, Firman JC, Richmond RH, Power EA (1997) Ecotoxicology of tropical marine ecosystems. Environ Toxicol Chem Int J 16(1):12–40
Provencher JF, Bond AL, Avery-Gomm S, Borrelle SB, Rebolledo EL, Hammer S, Kühn S, Lavers JL, Mallory ML, Trevail A, Van Franeker JA (2017) Quantifying ingested debris in marine megafauna: a review and recommendations for standardization. Anal Methods 9(9):1454–1469
Rabalais NN, Turner RE, Wiseman WJ Jr (2002) Gulf of Mexico hypoxia, aka "the dead zone". Annu Rev Ecol Syst 33(1):235–263
Rattner BA (2009) History of wildlife toxicology. Ecotoxicology 18(7):773–783
Rattner BA (2018) Wildlife and environmental pollution. In: Scanes CG, Toukhsati SR (eds) Animals and human society. Academic Press, San Diego, pp 472–478
Rodríguez A, Burgan G, Dann P, Jessop R, Negro JJ, Chiaradia A (2014) Fatal attraction of short-tailed shearwaters to artificial lights. PLoS One 9(10):e110114
Roman L, Gilardi K, Lowenstine L, Hardesty BD, Wilcox C (2020) The need for attention to confirmation bias and confounding in the Field of plastic pollution and wildlife impacts: comment on "clinical pathology of plastic ingestion in marine birds and relationships with blood chemistry". Environ Sci Technol 55(1):801–804
Russell VS (1974) Pollution: concept and definition. Biol Conserv 6:157–161
Sandoz MA, Wooten KJ, Clendening SL, Hensley LL, Smith LR, Smith PN (2018) Transport mechanisms for veterinary pharmaceuticals from beef cattle feedyards to wetlands: is aerial deposition a contributing source? Agric Ecosyst Environ 252:14–21
Sands P (1991) European community environmental law: the evolution of a regional regime of international environmental protection. Yale Law J 100:2511–2523
Schowalter T, Willig MR, Presley SJ, Pandey M (2019) Warnings of an "insect apocalypse" are premature. Front Ecol Environ 17:547–547
Shaaban M, Van Zwieten L, Bashir S, Younas A, Núñez-Delgado A, Chhajro MA, Kubar KA, Ali U, Rana MS, Mehmood MA, Hu R (2018) A concise review of biochar application to agricultural soils to improve soil conditions and fight pollution. J Environ Manage 15(228):429–440
Shannon G, McKenna MF, Angeloni LM, Crooks KR, Fristrup KM, Brown E, Warner KA, Nelson MD, White C, Briggs J, McFarland S (2016) A synthesis of two decades of research documenting the effects of noise on wildlife. Biol Rev 91(4):982–1005
Shen M, Zeng G, Zhang Y, Wen X, Song B, Tang W (2019) Can biotechnology strategies effectively manage environmental (micro) plastics? Sci Total Environ 20(697):134200
Sofia D, Gioiella F, Lotrecchiano N, Giuliano A (2020) Mitigation strategies for reducing air pollution. Environ Sci Pollut Res 27(16):19226–19235
Spurgeon D, Lahive E, Robinson A, Short S, Kille P (2020) Species sensitivity to toxic substances: evolution, ecology and applications. Front Environ Sci 1:8
Van Den Berg H, Manuweera G, Konradsen F (2017) Global trends in the production and use of DDT for control of malaria and other vector-borne diseases. Malar J 16(1):1–8
Van Drecht G, Bouwman AF, Harrison J, Knoop JM (2009) Global nitrogen and phosphate in urban wastewater for the period 1970 to 2050. Global Biogeochem Cycles 23:4
Vegter AC, Barletta M, Beck CA, Borrero JC, Burton H, Campbell ML, Costa MF, Eriksen M, Eriksson C, Estrades A, Gilardi KV, Hardesty BD, do Sul JAI, Lavers JL, Lazar B, Lebreton L, Nichols WJ, Ribic CA, Ryan PG, Schuyler QA, SDA S, Takada H, Townsend KA, Wabnitz CCC, Wilcox C, Young LC, Hamann M (2014) Global research priorities to mitigate plastic pollution impacts on marine wildlife. Endanger Species Res 25:225–247
Venter ZS, Aunan K, Chowdhury S, Lelieveld J (2020) COVID-19 lockdowns cause global air pollution declines. Proc Natl Acad Sci 117(32):18984–18990
Wartmann FM, Mackaness WA, Bauer N, Bolliger J, Kienast F (2019) Towards an interdisciplinary understanding of landscape qualities: wilderness, tranquillity and dark skies. In: Current trends in landscape research. Springer, Cham, pp 191–220
Wells FE, Gagnon MM (2020) A quarter century of recovery of the whelk Thais orbita from tributyltin pollution off Perth, Western Australia. Mar Pollut Bull 1(158): 111408
Worm B, Lotze HK, Jubinville I, Wilcox C, Jambeck J (2017) Plastic as a persistent marine pollutant. Annu Rev Env Resour 17(42):1–26
Zhang W (2018 Mar) Global pesticide use: profile, trend, cost/benefit and more. Proc Int Acad Ecol Environ Sci 8(1):1
Zhao H, Yuan M, Strokal M, Wu HC, Liu X, Murk A, Kroeze C, Osinga R (2021) Impacts of nitrogen pollution on corals in the context of global climate change and potential strategies to conserve coral reefs. Sci Total Environ 1:145017

Working in a Complex, Wicked, and Messy World of Wildlife Health

Craig Stephen

Abstract

Wildlife health management is part of a complex system. The decision about the best approach to manage a wildlife health problem in a socially acceptable manner that considers the impacts of actions on intended and unintended targets over time is influenced by multiple interacting socioecological systems, complicated trade-offs, and limited knowledge. Looking at wildlife health issues as complex and cumulative outcomes can help shift risk assessment and management toward assessing the total effects of interacting stressors over several dimensions and into the future, thus encouraging an assessment focusing on impacts to resilience rather than on specific causes of wildlife death, disease, or disappearance. While it may be hard to find the right answer in such situations, better answers can be encouraged by adapting recent thinking and approaches to looking at complex, wicked, and cumulative problems. Recognizing the wickedness of a problem opens you up to looking at the multiple factors and forces that comprise the problem and seek out stakeholders and partners willing to engage in the problem-solving process. You cannot have a resolution of wicked problems by treating them as scientific or technical puzzles. They need integrated, adaptable, and consultative approaches from the start.

Keywords

Complexity · Health · Wildlife · Wicked problem · Intractable problem · Cumulative effects

1 Wildlife Health Is Messy

Wildlife population health practitioners cannot escape complex, complicated, and wicked systems. There are physiological systems affecting individual animal diseases, socioecological systems affecting population health, governance systems that affect how we make decisions and work together, surveillance systems, regulatory system, social systems affecting how we value and treat wildlife, and ecosystems on which wildlife depend. Each of these systems has fuzzy boundaries that spill over and affect each other. These messy interactions can limit our ability to discover how to best manage a problem and avoid unintended side effects of our actions. For example, multiple studies have found several factors linked to global declines in amphibian populations, but it is increasingly clear that interactions among local (e.g., habitat modifications), regional (e.g., contaminants), and global scales (e.g., climate change) with

C. Stephen (✉)
School of Population and Public Health, University of British Columbia, Vancouver, BC, Canada

C. Stephen (ed.), *Wildlife Population Health*,
https://doi.org/10.1007/978-3-030-90510-1_17

variations in local biotic interactions make it exceedingly difficult to anticipate how populations will respond to interventions or environmental changes (Blaustein et al. 2002). Those who grapple with managing wildlife health are often confronted with varying interpretations of evidence and by conflicting political, cultural, and economic interests. Wildlife health management is subject to the inherent variability of socio-ecological systems and is further complicated by trade-offs between multiple objectives, values, and interests and limited by uncertainties and ambiguities about what causes the problem and the effects of interventions.

In the face of human inducted climate change, the extinction crisis, global pandemics, and more, people are asking if we can continue acting as if tomorrow will be just like yesterday (Ison and Shelley 2016). By using the same approaches that created our twenty-first-century challenges, are we growing less prepared for the next threat? Are new ways of thinking and working needed to accommodate the complexity of wildlife health challenges and the growing urgency for better responses? As seen throughout this book, ideas of complexity, dynamic systems, and wicked problems are being evoked as new frameworks for health management. This chapter asks what makes a problem complex or wicked and how we should act in response to them?

2 When Does a Wildlife Health Problem Become Complex and Wicked?

Wildlife health problems can be simple, complicated, complex, or wicked (Table 1). Wildlife health issues and global biodiversity conservation, in general, have been called wicked or even super wicked problems (Walzner 2017). Wildlife health interventions are often complex to manage because they are embedded in intricate networks of physical, biological, ecological, technical, economic, social, political, and other relationships. Complex problems become more wicked when a wider range of interests make decision-making more contentious, when the information and concepts become more numerous and interrelated, and when there is a lack of clarity or overlaps in processes to manage the problem (Peters and Tarpley 2019). Problems become "super-wicked" when time is running out, those who cause the problem also seek to provide a solution, the central authority needed to address it is weak or nonexistent, and policy responses irrationally discount the future (Levin et al. 2012).

3 Working with Complex and Wicked Problems

By their varied nature, complex and wicked problems have no "recipe book" of predefined management strategies. As there is no root cause of "wickedness," there can be no single best approach to tackling such problems. However, wildlife population health experts can expect to face at least three challenges (adapted from Levin et al. 2012):

1. How to develop an initial agreement to do something despite different goals and understanding of the problem?
2. How to make early interventions or actions "stick" by changing people's behaviors over long enough time and covering sufficient places and populations to result in a positive change?
3. How can they show that the small changes being made are leading to transformative changes that address the evolving problem in an acceptable, effective, and sustainable way?

Many One Health and population health leadership skills are needed to confront these challenges including, conflict management, leading change, building coalitions, being able to compromise on some less critical areas to gain what is essential, being less focused on personally being right than focusing on finding the right people to get the process right, thinking in systems, and remembering that the essential obligation is to not make things worse (Butler-Jones, 2020; Stephen and Stemshorn 2016).

Table 1 A taxonomy of problems (based on Pearce and Merletti 2006; Glouberman and Zimmerman 2016; Anon 2018)

Problem type	Problem description	Management implication
Simple	There is a clear cause-and-effect relationship that is easily identifiable and fixable using standard processes and procedures (e.g., stop hawk predation on backyard chicken by keeping the chickens in an enclosure)	No special expertise is needed. If you can follow the well-known process, you can fix the problem
Complicated	There are multiple causes affecting the outcome. Some may be easy to identify, other causes may be hidden, and some may interact. These problems tend to be technical in nature and benefit from analysis, past knowledge, and dealing with similar problems in the past. There are several steps that you can go through to potentially solve the problem (e.g., designing a fox rabies vaccination campaign in a peri-urban forest)	Specialized and coordinated expertise is often needed to see cause–effect relationships and their interactions. Set processes and procedures are helpful but the scale of the problem will need them to be adapted to the unique situation. People with the right expertise can usually design solutions that can be implemented. More than one right answer may be found, and trade-offs may need to be made to select the best answer for the situation
Complex	Often only the symptoms, not the causes, are visible. There are large elements of ambiguity and uncertainty due to interacting simple and complicated problems that make up the complex problem. They are a collection of individual problems and influences that act in ways that are not always predictable and whose interconnections change the context for the subparts. There are multiple and possibly conflicting goals (e.g., managing wildlife-farmed animal disease transmission on pastureland used by wild sheep and farmed sheep).	There is no standard procedure for resolving these problems. They can be approached from multiple, sometimes competing, perspectives and there may be multiple possible solutions. Multiple perspectives and types of knowledge are needed. Innovative approaches tailored to the situation and systems thinking are required. Solutions may not be transferrable between places and times because of the changing and unique nature of subsystems of interactions causing the problem. There will be uncertainty about the effects of interactions
Wicked	Complex problem for which there is no agreement on the definition of the problem because of multiple values, perceptions, and perspectives, a wide array of possible solutions with trade-offs associated with each, and multiple potential causes, jurisdictions, stakeholders, regulators, or implications. Subproblems within the wicked problem are linked and changing one can affect another. There may be no final resolution, but there can be incremental gains in reducing the problem (e.g., eliminating emerging disease risks associated with the wildlife trade)	There is no single correct solution. These problems demand a participatory approach, adaptive management, and collaborative interorganizational work. Managing wicked problems requires holistic/systems thinking rather than linear thinking. Managers need to think and work in new, flexible, creative, and innovative ways. At their core, wicked problems are often wicked because their social dimensions lead to conflicts. Box 1 provides an example

At the outset, it is important to recognize and acknowledge the extent of a problem's wickedness. This opens you up to looking at the multiple factors and forces that comprise the problem and seek out stakeholders and partners willing to engage in the problem-solving process. In doing so, not only might you be able to find components of the complex or wicked problem that might have a simple or complicated part that can be solved by your partners but also you will be open to new ideas and avenues to conceive and approach a better solution (Kreuter et al. 2004).

Another step is to develop a shared conception of the problem. The goal is not to establish strict analytic boundaries but instead to instigate reflection on the nature of the problems and how standard approaches may fail (Termeer et al. 2019). Problem "wickedness" can begin when the scope of the problem is misinterpreted or, worse, underestimated (Woodford et al. 2016). The wickedness of the problem is influenced by a host of factors that need to be mapped, discussed, and understood.

Box 1 Controlling Emerging Zoonotic Disease Risk in the Wildlife Trade—A Super Wicked Problem (Based on Stephen et al. 2021)

Legislative, public, scientific, and political interests in the wildlife trade overlap at least 4 domains: **Conservation,** which strives to protect species distribution and abundance and protect ecosystem function. **Public health**, which aims to prevent human exposure to zoonotic agents from wildlife. **Human development agenda** that wishes to ensure the nutrition and the income derived from wildlife is sustainably available to vulnerable communities. **Law agencies** wanting to control the growing and significant illegal trade of wildlife and the associated benefits it provides to criminals and harms to communities.

Some features of the super wicked problem	Wildlife trade management attributes
No stopping rules	No safety thresholds for types, prevalence of levels of risk factors, or pathogens deemed to be acceptable in the trade.
Solutions are not true and false, but better or worse due to trade-offs	Banning the trade could reduce access to nutritious food and critical income for millions of vulnerable people and drive the demand for illegal harvest but also can help protect wild species and reduce emerging disease risks.
Every wicked problem is unique.	The trade is highly heterogenous, involving a wide variety of species, products, and places. Risk management is highly context specific.
Solutions depend on how the problem is framed	The priority of different problems and preferences for solution linked to this trade varies with the perspective from which the trade is viewed as a conservation, public health, human development, or crime problem or as a scientific, policy, or human behavior issue.
Stakeholders have radically different frames for understanding the problem.	Some citizens in the global north view this as a high-risk trade selling unnecessary products while some impoverished communities in the global south see this as a culturally essential contribution to community well-being.
Data may be uncertain, missing, or ambiguous	Much of the trade has not been monitored. There is a massive gap in implementation science to reveal effective, acceptable, and sustainable interventions.
Time is running out	Overexploitation is a major driver of species extinction and the wildlife trade is a significant contributor.
There is no central authority	No "all-of-government" approaches have been developed. The different aspects of this problem are regulated by different agencies, often separately.

Generating ownership of the problem through stakeholder participation and transparency is an important step toward collaborative solutions. Wicked wildlife health issues often involve wider society and the media, both of which influence the problem perception. This moves problem definitions away from a purely scientific understanding to a social construct interpreted in political and moral terms (Osmundsen 2017). Wicked problems are best approached through a planned process with input from multiple sources in an atmosphere that puts scientific inquiry on par with the community and stakeholder's knowledge (Kreuter et al. 2004). Relationships will need to be built to give voice to those who are less powerful. Wildlife networks that allow for information sharing and mutual adjustments in understanding can help shape processes and actions that fit with the lived-experiences of those affecting and being affected by the problem.

The outcome may not be consensus but hopefully, there will be a coherent and transparent set of opinions on the nature of the problem, why it matters and to whom.

"Wicked problems cannot be "solved," as such, but they can be much better managed, and significant improvements can be achieved" (Head 2018). Given that there are no completely right solutions to the wicked problem and that each problem is different, forward momentum is supported when there is a shared idea of where we want to go; in other words what makes a solution better or worse. You cannot have a resolution of wicked problems by treating them as scientific or technical puzzles. Integrated, adaptable approaches need people engaged in a consultative way from the start. This will be greatly assisted by developing a shared vision of how the problem is understood (Mertens 2015). Those managing wicked problems must be adept at understanding the context and experiences of diverse communities involved in the problem in a culturally appropriate way.

A key obstacle to making progress with intractable problems has been the tendency to act as if wicked problems were all the same in terms of their "wickedness," and therefore a "one-size-fits-all" approach will work (Alford and Head 2017). A problem is not either wicked or not. There will be a spectrum of problem types with wicked features and intensities of problems (Alford and Head 2017). Take for example the problems of invasive species versus the wildlife trade. The former is wicked because there are no clear-cut solutions, but the problem is clearly definable. The wildlife trade suffers from being both conceptually and technically ill-defined. Cognitive and social knowledge about the wildlife trade is unconsolidated (Arroyave et al. 2021). Actions for the wildlife trade may therefore be more contentious and easily neglected than for invasive species due to the combination of this lack of knowledge consolidation with high uncertainty and a small amount of research on viable solutions in the trade.

In the absence of a preexisting roadmap, demonstration and pilot projects can help discover what works for whom, why, and how by exploring what is technically and ethically feasible, working out regulatory and collaboration issues, and addressing perceived risks. These experiments must not be limited to tackling pieces of the problems with single objectives in simplified environments but instead must try to explore and balance competing objectives in complex, whole environments. Value differences cannot be effectively sidelined or depoliticized using evidence and data. This means being ready to explicitly explore trade-offs and their implications, use multiple methods to work on the problem, and being open to sharing failures openly and transparently (Mason et al. 2018).

The precautionary principle has been characterized as a wicked problem tool as it calls for all stakeholders to act in a way that protects population health against a backdrop of scientific uncertainty (Kreuter et al. 2004). This is consistent with a harm reduction approach that asks us to seek incremental gains in well-being by acting on shared goals despite uncertainties and conflicts. The precautionary approach asks us to take preventive action in the face of uncertainty, shift the burden of proof to the proponents of a potentially harmful activity, explore a wide range of alternatives to avoid harms, and increase public participation in decision-making.

4 Cumulative Health Effects

Because of the increasing significance of systemwide risk factors such as climate change, degradation of ecosystem services, and modification and loss of habitats, there is a growing need for cumulative impact assessment in wildlife management. The need is to shift assessments toward assessing the total effects of an environmental, policy, or other change over several dimensions and into the future, thus encouraging an assessment focusing on impacts to resilience rather than on specific causes of wildlife death, disease, or disappearance (Krausman and Harris 2011). However, the lack of monitoring data, lack of research on wildlife health as a cumulative effect, and significant uncertainties and data gaps on the status and interactions of these determinants can present great challenges to anyone wanting to assess impacts from a cumulative rather than unidimensional perspective.

Cumulative impact assessment is an evolving concept for which there is no single accepted state of global practice. Cumulative impacts assessments are typically used to assess the social and ecological effects of a project, such as building a dam, on valued ecosystem components, such as a fishery. Wildlife population health experts may be called on to assess the impacts of a project on wildlife, the results of which will be used by others to understand cumulative impacts. The pressing question is often about how the health trajectory for an animal will change upon exposure to a new or changed set of interacting stressors. The question becomes more complex when trying to determine the combined effect of past, present, and potential future human activities and natural processes over long time periods and large areas.

A cumulative effects assessment can be overwhelming due to the potential scope of effects and range of spaces that must be considered. Take for example the definition of "effects" used in the Canadian Environmental Assessment Act as any change that a project may cause in the environment, including any change on health and socioeconomic conditions, on physical and cultural heritage, on the current use of lands and resources for traditional purposes by aboriginal persons, or on any structure, site, or thing that is of historical, archaeological, paleontological, or architectural significance and any change to the project that may be caused by the environment, whether any such change occurs within or outside of Canada (Hegmann et al. 1999). The Act further requires assessments to study an area that is large enough to assess effects on valued ecosystem components, an area that is considerably larger than a project's "footprint". Generic frameworks for cumulative impact assessments have been developed (Fig. 1) but their application and utility will be determined by the partnerships, knowledge, and resources available. A critical early step will be to develop agreement on how diverse data will be integrated, how threats will be characterized and estimated over time, how uncertainty will be handled, and which thresholds will be used to declare the cumulative effects harmful or undesirable.

Cumulative effects impact assessments want to know about the incremental impact of the action when added to other past, present, and reasonably foreseeable future actions. Humans impact wildlife systems in a multitude of ways, yet the cumulative effect of multiple stressors on wildlife population health remains grossly understudied. While there are articles that explore the interacting effects of multiple stressors, such as the compounding effects of contaminants on the effects of parasitism or infectious disease susceptibility, the wildlife health literature rarely looks at health as a cumulative effect. Cumulative Health Risk Assessment methods exist but are most often used to look at the interactions of a small number of discrete hazards (such as the effects of being exposed to two different pesticides at once) but rarely look at the implications of a change of action on the suite of determinants of health as a whole. A scoping review of 458 wildlife health articles published between June 1994 and June 2014, for example, found only 7% looked at health as a multifactorial entity (most still looking at health as the absence of specific diseases), and none looked at health as a cumulative effect that develops and is modified over time (Sinclair 2020). The concept of life-course analysis is missing in wildlife health. Life-course health analysis is a public health framework used to organize research from several fields to explain how individual and population health develops and how developmental trajectories are determined by interactions between biological and environmental factors during the lifetime (Halfon and Hochstein 2002). While it well known that early exposure to some hazards (such as endocrine disruptors) can affect a wild animal later in its life (such as transgenerational reproductive or development effects), most wildlife health research looks at proximal risks from wildlife diseases (Sinclair 2020). There is an opportunity space for wildlife health researchers to adopt human health life-course epidemiology methods that aim to elucidate biological, behavioral, and environmental

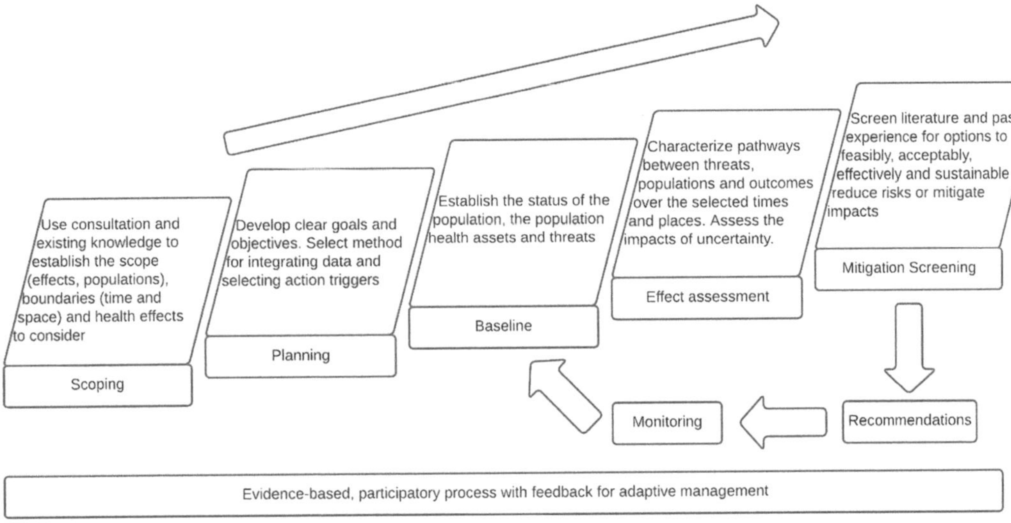

Fig. 1 A schematic of a generic cumulative effects assessment process

processes that operate across an individual's life course, or across generations, to influence the development of disease risk and health outcomes.

5 Summary

Health is a cumulative effect of interacting determinants of health that exist in an everchanging landscape of assets, threats, and resources. Health managers need to be able to appreciate the changing health landscape while at the same time deal with multiple, sometimes conflicting, expectations and, in the case of wildlife health, large uncertainties about the root causes of the social problems people want managed, and the most effective and acceptable way to meet management goals. This chapter only introduces some of the ideas and approaches to thinking about wildlife health as a complex and often wicked problem. Those interested in pursuing these ideas further will need to turn outside the wildlife health literature and experience and seek insights and methods from natural resource management, conservation, and human health spheres.

The need to look at wildlife health as a complex or wicked problem is not an academic issue. In a world of concurrent problems, seeking unique solutions for individual problems is neither effective and feasible nor efficient (Fried et al. 2012). We will miss opportunities to find the "levers" to pull to keep populations healthy if we remain focused on interdisciplinary approaches to problems in isolation rather than thinking about the interacting hazards, populations, and determinants of health existing in shared places and spaces. The goal is not to get a more and more refined understanding of all the components parts of the complexity of wildlife health and the relationships between those parts but rather to look at health problems to find ways we can intervene in socially acceptable and scientific justifiable ways to make populations better able to deal with today's threats and be more resilient to tomorrows.

References

Alford J, Head BW (2017) Wicked and less wicked problems: a typology and a contingency framework. Polic Soc 36(3):397–413

Anon. (2018) Tackling wicked problem: a public policy perspective. Australian Public Services Commission, Sydney. https://legacy.apsc.gov.au/tackling-wicked-problems-public-policy-perspective

Arroyave F, Goyeneche OY, Gore M, Heimeriks G, Jenkins J, Petersen A (2021) On the social and cognitive dimensions of wicked environmental problems

characterized by conceptual and solution uncertainty. arXiv preprint arXiv:2104.10279
Blaustein AR, Kiesecker JM (2002) Complexity in conservation: lessons from the global decline of amphibian populations. Ecol Lett 5(4):597–608
Butler-Jones D (2020) Practical approaches to leadership and one health. In: Stephen C (ed) Animals, health, and society: health promotion, harm reduction, and health equity in a one health world. CRC Press, New York
Fried LP, Piot P, Frenk JJ, Flahault A, Parker R (2012) Global public health leadership for the twenty-first century: towards improved health of all populations. Glob Public Health 7(sup1):5–15
Glouberman S, Zimmerman B (2016) 1 complicated and complex systems: what would successful reform of medicare look like? University of Toronto Press, Toronto
Halfon N, Hochstein M (2002) Life course health development: an integrated framework for developing health, policy, and research. Milbank Q 80(3):433–479
Head B (2018) Understanding "wicked" policy problems. Policy Options. https://policyoptions.irpp.org/magazines/january-2018/understanding-wicked-policy-problems/. Accessed 9 Jan 2021
Hegmann G, Cocklin C, Creasey R, Dupuis S, Kennedy A, Kingsley L, Ross W, Spaling H, Stalker D (1999) Cumulative effects assessment Practitioners' guide. Prepared for the Government of Canada. https://www.canada.ca/en/impact-assessment-agency/services/policy-guidance/cumulative-effects-assessment-practitioners-guide.html#s1-1
Ison R, Shelley M (2016) Governing in the Anthropocene: contributions from systems thinking in practice? Syst Res Behav Sci 33(5):589–594
Krausman PR, Harris LK (2011) Cumulative effects in wildlife management: impact mitigation. CRC Press, New York
Kreuter MW, De Rosa C, Howze EH, Baldwin GT (2004) Understanding wicked problems: a key to advancing environmental health promotion. Health Educ Behav 31(4):441–454
Levin K, Cashore B, Bernstein S, Auld G (2012) Overcoming the tragedy of super wicked problems: constraining our future selves to ameliorate global climate change. Pol Sci 45(2):123–152
Mason TH, Pollard CR, Chimalakonda D, Guerrero AM, Kerr-Smith C, Milheiras SA, Roberts M, R. Ngafack P, Bunnefeld N. (2018) Wicked conflict: using wicked problem thinking for holistic management of conservation conflict. Conserv Lett 6:e12460
Mertens DM (2015) Mixed methods and wicked problems. J Mixed Methods Res 9(1):3–6
Osmundsen TC, Almklov P, Tveterås R (2017) Fish farmers and regulators coping with the wickedness of aquaculture. Aquacult Econ Manage 21(1):163–183
Pearce N, Merletti F (2006) Complexity, simplicity, and epidemiology. Int J Epidemiol 35(3):515–519
Peters BG, Tarpey M (2019) Are wicked problems really so wicked? Perceptions of policy problems. Polic Soc 38(2):218–236
Sinclair D (2020) Exploring opportunities to modernize Ontario's approach to wildlife health through understanding literature, stakeholders, networks, legislation, and policy. Doctoral dissertation. . University of Guelph
Stephen C, Carmo LP, de las Valle DNM, Friker B, Sousa FM, Vidondo B, Berezowski J (2021) The implementation gap in emerging disease risk management in the wildlife trade
Stephen C, Stemshorn B (2016) Leadership, governance and partnerships are essential one health competencies. One Health 2:161–163
Termeer CJ, Dewulf A, Biesbroek R (2019) A critical assessment of the wicked problem concept: relevance and usefulness for policy science and practice. Polic Soc 38:167–179
Walzer C (2017) Beyond one health—zoological medicine in the anthropocene. Front Veterin Sci 4:102
Woodford DJ, Richardson DM, MacIsaac HJ, Mandrak NE, Van Wilgen BW, Weyl OL (2016) Confronting the wicked problem of managing biological invasions. NeoBiota 31:63–86

Applying a Health Lens in Managing Species at Risk Under Threat of Alien Invasive Species

Joy Wade and Paul Grant

Abstract

Wildlife health typically focuses on the elimination of pathogens and parasites and controlling the effects of pollutants on morbidity and mortality. Alien invasive species (AIS) also contribute to the health of wildlife populations, both negatively and positively. Once introduced, species have the potential to rapidly increase in abundance and spread. In ecosystems where there are also species at risk, controlling AIS is further complicated. This chapter explores the importance of managing the health of an ecosystem under the threat of AIS and how this challenge is complicated when there are species at risk. Using a case study and the application of the determinants of health model, we provide an example of how managing such a complex health system can be beneficial to many species, including humans.

Keywords

Alien invasive species · Mussels · Wildlife health · Wildlife management · Species at risk

1 Introduction

Wildlife health typically focuses on the elimination of pathogens and parasites and controlling the effects of pollutants on morbidity and mortality. Alien invasive species (AIS) which includes any non-native plants, animals, pathogens, or other organisms in an ecosystem also influence wildlife population health, both negatively and positively. Climate change and increased global movement of people, products, wastes, and animals have made it easier for species to move out of their native habitats into new ecosystems (Seebacher and Post 2015; Seebens et al. 2018; Smith et al. 2018). Once introduced, species have the potential to rapidly increase in abundance and spread, as AIS are often not restricted by natural limiting factors such as predators, parasites, disease, or competition in new environments. In most cases, the effects of AIS introductions are negative on the receiving environment.

1.1 Health Impacts of Alien Invasive Species

Depending on their proliferation and interactions, AIS can have many different direct and indirect impacts on their new environment or the wildlife species within, including reducing diversity, health, and resilience in species and ecosystems.

J. Wade (✉)
Fundy Aqua Services, Nanoose Bay, BC, Canada

P. Grant
Fisheries and Oceans Canada, Sidney, BC, Canada
e-mail: paul.grant@dfo-mpo.gc.ca

C. Stephen (ed.), *Wildlife Population Health*,
https://doi.org/10.1007/978-3-030-90510-1_18

These negative impacts can occur through altering food webs by predation or competing with native species for food resources or space. An example of this type of impact includes the introduction of stoats (*Mustela ermine*), a mustelid native to Eurasia and North America, to New Zealand in the nineteenth century to control invasive rabbits. Unintentionally, this AIS became a significant threat to native bird populations, including the kakapo (*Strigops habroptilus*), kiwi (*Apteryx* spp.), and rock wren (*Salpinctes obsoletus*), by predation on nests and young (Hanley and Roberts 2019).

AIS can also be the source of pathogens or parasitize native species, impacting their health and resilience. The West Nile virus, for example, is a mosquito-borne flavivirus native to Africa, Europe, and Western Asia. This AIS was introduced to North America in 1999 where it has continued to expand its range across the United States and into Canada, Mexico, and Central and South America impacting wildlife health and causing significant and sometimes severe human diseases including West Nile fever and encephalitis (Colpitts et al. 2012). The impacts of AIS-related diseases on the health of wildlife species can be highly variable due to variations in environmental factors, resilience, and transmission. Within avian populations, the West Nile virus resulted in declines in some species, but overall impacts were less severe than initially anticipated due to spatial and temporal variability in transmission intensity, providing refuges to species with a range of habitats (Kilpatrick and Wheeler 2019).

There are also numerous examples of AIS out-competing native species for food resources or space, resulting in the decline of native species. European starling (*Sturnus vulgaris*), for example, which is native to Europe, Asia, and Northern Africa, was introduced to North America, Southern Africa, Australia, and New Zealand, where it is highly aggressive and outcompetes native species for food, territory, and nest sites.

AIS can also physically alter habitats making it unsuitable for native species, disrupt essential ecosystem functions, and impact water quality and quantity. The European green crab (*Carcinus maenas*) is described as an ecosystem engineer (Jones et al. 1994). Native to Europe, this AIS has invaded new environments in North America, South America, South Africa, Australia, and New Zealand. In these new environments, it directly alters habitat by removing eelgrass (*Zostera marina*) or other foundation plant species, resulting in impacted ecosystems and food webs (Howard et al. 2019). It also has reduced biodiversity through predation and competition with native species for food resources and space. Eurasian watermilfoil (*Myriophyllum spicatum*) native to Europe, Asia, and North Africa is considered a highly invasive species that has been introduced to every continent (Eiswerth et al. 2000). It outcompetes and crowds out native plants and creates dense mats that alter habitat and create an ecosystem with less food sources and space for native species (Smith and Barko 1990). Eurasian watermilfoil (milfoil) can also reduce the amount of oxygen creating further stressors for the survival of native species (GISD 2017).

The introduction and spread of AIS can carry economic implications through effects which impact the environment and human and wildlife health (Hanley and Roberts 2019). In addition to some of the abovementioned cases, notable examples of these economic impacts include the sea lamprey (*Petromyzon marinus*), which was introduced to the Great Lakes in Canada in 1955, prior to a general awareness that AIS could pose both environmental and economic problems. In this case, sea lamprey decimated native fish species including lake trout (*Salvelinus namaycush*), whitefish (Coregoninae spp.), and lake chub (*Couesius plumbeus*) populations resulting in the collapse of lake trout and severe losses to recreation and commercial fisheries.

Through these various pathways, AIS can negatively impact and threaten many habitats and species, including those that are already rare and at some risk of extinction (McCune et al. 2013; Swan et al. 2018). The current exponential loss in biodiversity, estimated to be 1000 times higher than the normal background extinction rate (De Vos et al. 2015), is due in large part to widespread habitat loss/degradation, and the negative impact of invasive species (McCune et al. 2013; Swan et al. 2018). Often these impacts are intensified by climate change and irreversible as AIS are difficult to eradicate once established. The resulting loss of biodiversity has broader implications for ecosystem functioning and the determinants of health (see Chap. 1).

Box 1 Case Study: Rocky Mountain Ridged Mussel

Species status: The Rocky Mountain ridged mussel (*Gonidea angulata*) (Fig. 1) is a freshwater bivalve mussel species whose Canadian range includes only in the Okanagan Valley in British Columbia, the northernmost limit of its patchy distribution (Wade et al. 2020). Freshwater mussels are one of the most endangered groups of animals in North America (Williams et al. 1993). The Committee on the Status of Endangered Wildlife in Canada (COSEWIC) assessed the species as Special Concern in 2005 and reassessed it as Endangered in 2010, concurrently identifying AIS as the most serious potential threat. Rocky Mountain ridged mussel is currently listed under Canada's *Species at Risk Act* (SARA) as Special Concern, as a decision to list it as Endangered under SARA is still pending. Species that are listed as Threatened or Endangered under SARA are provided specific protections. The decision to list as Endangered is pending as there is a perceived conflict between protection of the species and habitat and the control of invasive milfoil for the benefits of tourism (Wade et al. 2020).

Species biology: Rocky Mountain ridged mussel is a burrowing species normally found in water depths less than 3 m (COSEWIC 2003; Stanton et al. 2012). It requires a fish to aid in reproduction. After fertilization, the female mussel incubates the eggs within her shell. The eggs hatch into glochidia which are then released into the water where they attach to the gills of a suitable host fish (Fisheries and Oceans Canada 2011), typically a sculpin. The glochidia develop on the gills before they drop off and settle into the substrate. Recent age validation of the species indicates that individuals are long-lived with some surviving up to 50 years.

Habitat change: This case study focuses on the Rocky Mountain ridged mussel population in Okanagan Lake, the largest part of the species' distribution in Canada. Okanagan Lake is 135 km long, 4–5 km wide, and has a maximum depth of 232 m. Since the early 1900s, a series of dams and channelized rivers have been built to control the water levels in the area including in Okanagan Lake (Symonds 2000). A control dam was built in 1953 at the terminus of Okanagan Lake before Okanagan River empties into Skaha Lake to the south. Because of a series of dams throughout this water system, sockeye salmon (*Onchorhynchus nerka*) and chinook salmon (*O. tshawytscha*), which once had plentiful returns to this area, were unable to access upstream habitats. These salmon populations were in decline from other threats such as pollution and overfishing, but access to upstream habitats was a direct impact from the construction of control structures. With remediation of water control structures in recent years to include features such as fish ladders and the input of juvenile fish through enhancement, salmon have been able to migrate further and further upstream. In 2019, the dam on Okanagan River was remediated to include a fish ladder and water can now freely flow between Okanagan and Skaha lakes for the first time in almost 70 years.

AIS threats: Established AIS in Okanagan Lake include: Eurasian watermilfoil, tench (*Tinca tinca*), mysis shrimp, brook trout (*Salvelinus fontinalis*), lake trout, lake whitefish, smallmouth bass (*Microperis dolomieui*), pumpkinseed

(continued)

Fig. 1 A Rocky Mountain ridge mussel (*Gonidea angulata*) nestled in the sediment of Okanagan Lake, British Columbia

(*Lepomis gibbosus*), yellow perch (*Perca flavescens*), black crappie (*Pomoxis nigromaculatus*) (Naito, 2000). Species of concern, but not yet established include Quagga mussel (*Dreissena bugensis*) and zebra mussel. Naito (2000) also included largemouth bass (*M. salmoides*), walleye (*Sander vitreus*) and bullhead (*Ameiurus nebulosis*) of concern as they are present directly downstream of Okanagan Lake but not present in the lake.

2 Managing Alien Invasive Species

Around the globe, governments and other international, national, and regional organizations are going to great lengths to both stop new AIS introductions and try to deal with the impacts of current AIS. Controlling AIS is becoming increasingly important for society. As the number of AIS and the complexities of control increase, so does the importance of management principles and prioritization of control actions.

Managing AIS can be tricky, and approaches vary widely. Most management options can be summarized as either pre- or post-establishment of the AIS. The most effective principle for AIS management is to take a precautionary approach which assumes new species introductions pose a risk and aims to minimize or ideally prevent new invasions. Other effective strategies include early detection of new invaders, and rapid response to new invaders, as well as management of established and spreading invaders (containment, eradication, and control). Multijurisdictional approaches are required for effective prevention, early detection, response, and management to protect the health and resilience of native species and ecosystems.

Some of the most effective approaches for prevention involve managing the pathways by which AIS could enter new environments and become established. For aquatic species, these pathways can include shipping (e.g., ballast water discharge has introduced zebra mussel, green crab, and spiny water flea), recreational and commercial boating (e.g., these activities can inadvertently help spread AIS such as milfoil or zebra mussel), the use of live bait (e.g., this activity has been responsible for the spread of many species including the rusty crayfish), the aquarium trade (e.g., aquatic species such as yellow floating heart was introduced as a water garden species and spread to natural environments), live food fish, unauthorized introductions and transfers (e.g., purposefully introduction of AIS into new environments), and canals and water diversions (e.g., the sea lamprey spread into the upper Great Lakes after the Welland Canal was opened). For terrestrial species, these pathways can include cargo transport (e.g., increase in global trade has resulted in an increased risk of transporting living plants and animals that could become established in new environments), horticulture (e.g., plant varieties imported that become established and spread into natural environments), accidental release, pet trade, and transport of wood products.

For circumstances where AIS are already established, Buckley and Han (2014) proposed three options; (1) manage the impacts of the invader and accept collateral damage; (2) abandon management of the invader and accept its impacts or; (3) seek a compromise that allows both goals to be attained (Buckley and Han 2014). In the case study described in this chapter, it is option 3 that is the most likely to be implemented to address most AIS issues.

2.1 AIS and Rocky Mountain Ridged Mussel Management

Managing the health of Rocky Mountain ridged mussels is a complex and even wicked problem (see Chap. 17) (Box 1). To exacerbate the issue, the mussels have no social importance in the region but are biological and scientifically valued and contribute to ecosystem complexity. The lake itself is a highly important tourist and leisure destination. According to the most recent census (2016), the region is home to more than 360,000 residents and is a very popular tourist destination both in summer and in winter. The tourism industry in the Thompson Okanagan region is worth more than a billion dollars per year with the top five reasons for visiting relating to outdoor activities (beach, hiking, boating etc.; Destination BC News 2017). Some other species in the lake have high social value including sockeye and chinook salmon, making them a more frequent management priority. Consequently, the management of this species and its habitat is contentious and has resulted in different opinions of whether it should be listed under Schedule 1 of the SARA.

We propose that by examining the interaction between AIS and Rocky Mountain ridged mussels through determinants of health model (see Chap. 1), it may be possible to prioritize and undertake actions before the habitat is no longer able to support the species, regardless of political pressures (Table 1). This model can help elucidate ways to reduce harm to the species while at the same time supporting the continued use of the lake in support of tourism and economic development. It can provide an entry point to see the issue from not just a single species or issue point of view, but as a system.

Rocky Mountain ridged mussels are facing direct and indirect threats from many different AIS, both plant and animal. Although these effects are largely biological (e.g., direct mortality, decreased fecundity, increased predation) or physical (e.g. reduction in habitat quality and quantity); there are, fortunately, some critical actions that can be taken to minimize these harms for the direct benefit of the species as a part of the ecosystem.

For example, milfoil is a direct threat to mussel health (Table 1), it also negatively impacts tourism creating beaches with long "weeds" that are undesirable for swimmers and boaters. Eradication of milfoil from the lake is not possible as it is currently well established. Its spread can however be controlled through mechanical removal and weed mats. Mechanical removal can harm burrowed mussels, but if beds are protected before milfoil can establish, this can be mitigated. Weed mats can also be deployed over the substrate to inhibit milfoil growth. Inhibiting the spread of milfoil is good for mussel beds as well as tourism. Although removal of milfoil from mussel beds can be deleterious to individual mussels, mussel beds can be protected before milfoil establishes. There has been concern that the removal of these plants would not be possible if the mussel were protected under SARA legislation, but these activities are not necessarily contradictory. Methods of control can be modified depending on proximity of the milfoil to mussel beds. Direct losses of mussels occurring through appropriate milfoil control would be offset by long-term habitat quality benefits because the damage caused to the biotic and abiotic environment from milfoil is far more deleterious to the mussel long-term. A balance can be negotiated between local, provincial, and federal governments to the benefit of all.

Table 1 Examples of alien invasive species (AIS) impacts on the determinants of health of the Rocky Mountain ridged mussel (RMRM)

Determinant of health	Examples of the impacts of AIS on RMRM determinants of health
Direct mortality	Native RMRM is under direct threat of mortality from the reduction of habitat and water quality from encroaching Eurasian watermilfoil (AIS), predation by AIS of obligate hosts for successful reproduction, and habitat destruction by humans. Zebra and quagga mussels have also been identified as the most serious threat to the species but are not currently in the lake
Human expectations	RMRM are important culturally to local First Nations but have no commercial or recreational value. Policy conflicts exist on habitat protection through AIS control and the implications for more valued species and more valued recreational use of the lake
Abiotic environment	Changes to water regulation through the opening of the dam on Okanagan River has increased access for potential predators (sockeye salmon) and opened the waterway to AIS known to be present immediately downstream of the lake (largemouth bass, walleye, etc.) Increased recreational water traffic increases the potential to introduce invasive mussel species to the lake including zebra and quagga mussels
Social environment	Much of the social requirements of the species are unknown including population densities, demographics, host densities for successful reproduction. It is not known what effects AIS will have on these features
Needs for daily living	Habitat availability and quality of food web productivity, water temperature, ecologic resilience, and access to hosts for successful reproduction are all being impacted by invasive aquatic plants and animals
Biological endowment	With the introduction of new fish species and the continued proliferation of those already introduced there are unknown effects on population age structure, potential parasites and diseases, growth rates, and reproductive success. This has implications for the obligate fish host involved in RMRM reproduction

Another example is that of invasive piscivorous fish species which can prey on the intermediate hosts necessary for glochidia survival. There are several alien invasive fish species established in the lake already which are unlikely to be eradicated (e.g., smallmouth bass, pumpkinseed, etc.); however, the introduction of known AIS from Skaha Lake (largemouth bass, walleye, and bullhead) can be prevented through water management. The issue is further complicated as a dam was built between the two lakes and has been opened to provide angling opportunities for sockeye salmon in Okanagan Lake. The desire to have the return of recreational fishing opportunities was stronger than protecting the ecology of Okanagan Lake. Now that it is done, however, it is possible to put in exclusion devices at dams to prevent AIS from entering the lake. Their exclusion will aid in the survival of the mussel as well as many other native fish and invertebrate species in the lake as they are aggressive predators.

The determinants of the health model (Chap. 1) provided a framework to examine this AIS from a systems perspective and identify critical management entry points. Most notably, human values could be shown to work across many determinants of health and thus were essential to consider when trying to find win-win solutions. The complexities of managing Rocky Mountain ridged mussels in concert with economic development and milfoil control alone illustrate the need to address human values (see Chap. 19) in conservation management. Although the removal of milfoil is beneficial to both the mussel and the economic development, conservationists and local governments have not been able to come to a mutually beneficial solution due largely to the lack of a process to negotiate shared solutions that protect and preserve both conservation and economic values. Without a resolution, the protection of the mussel species will remain tenuous.

3 Summary

Although there has been a big push in society to work together in a cross-disciplinary fashion to address complex conservation and wildlife health issues, the degree of success is often based on individuals not organizational mandates. This case study demonstrated how a health model can be used to provide an integrated perspective to complex wildlife management issues. It is particularly useful to remind ourselves that addressing issues from a single hazard point of view is less likely to have positive outcomes.

This case study has illustrated that the legal protection of Rocky Mountain ridged mussels and their habitat can be beneficial to both the persistence of the species and control of AIS. The hurdle rests with humans. Are we willing to work together to reduce harm to the watershed or will we remain entrenched in our different mandates (i.e., economic development, conservation, recreational opportunities)?

References

Buckley YM, Han Y (2014) Managing the side effects of invasion control. Science 344(6187):975–976

Colpitts TM, Conway MJ, Montgomery RR, Fikrig E (2012) West Nile virus: biology, transmission, and human infection. Clin Microbiol Rev 25(4):635–648

COSEWIC (2003) COSEWIC assessment and status report on the Rocky Mountain Ridged Mussel *Gonidea angulata* in Canada. Committee on the Status of Endangered Wildlife in Canada, Ottawa, p 29

De Vos JM, Joppa LN, Gittleman JL, Stephens PR, Pimm SL (2015) Estimating the normal background rate of species extinction. Conserv Biol 29(2):452–462

Destination BC News (2017) Thompson-Okanagan regional tourism profile. Research, planning and evaluation, p 4

Eiswerth ME, Donaldson SG, Johnson WS (2000) Potential environmental impacts and economic damages of Eurasian watermilfoil (*Myriophyllum spicatum*) in Western Nevada and northeastern California1. Weed Technol 14(3):511–518

Fisheries and Oceans Canada (2011) Management plan for the Rocky Mountain ridged mussel (Gonidea angulata) in Canada. In: Species at risk act management plan series. Fisheries and Oceans Canada, Vancouver, p 52

Global Invasive Species Database (GISD) (2017) Species profile: *Dreissena polymorpha*. http://www.iucngisd.org/gisd/speciesname/Dreissena+polymorpha. Accessed 27 Apr 2021

Hanley N, Roberts M (2019) The economic benefits of invasive species management. People Nat 1(2):124–137

Howard BR, Francis FT, Côté IM, Therriault TW (2019) Habitat alteration by invasive European green crab (*Carcinus maenas*) causes eelgrass loss in British Columbia. Canada Biol Inv 21(12):3607–3618

Jones CG, Lawton JH, Schachak M (1994) Organisms as ecosystem engineers. Oikos 69:373–386

Kilpatrick AM, Wheeler SS (2019) Impact of West Nile virus on bird populations: limited lasting effects, evidence for recovery, and gaps in our understanding of impacts on ecosystems. J Med Entomol 56(6):1491–1497

McCune JL, Harrower WL, Avery-Gomm S, Brogan JM, Csergő AM, Davidson LN, Garani A, Halpin LR, Lipsen LP, Lee C, Nelson JC (2013) Threats to Canadian species at risk: an analysis of finalized recovery strategies. Biol Conserv 166:254–265

Naito G (2000) Presence and distribution of exotic fish species in the Okanagan watershed. Prepared for the Okanagan Nation Fisheries Commission, pp 1–11

Seebacher F, Post E (2015) Climate change impacts on animal migration. Climate Change Responses 2(1):1–2

Seebens H, Blackburn TM, Dyer EE, Genovesi P, Hulme PE, Jeschke JM, Pagad S, Pyšek P, van Kleunen M, Winter M, Ansong M (2018) Global rise in emerging alien species results from increased accessibility of new source pools. Proc Natl Acad Sci 115(10):2264–2273

Smith CS, Barko JW (1990) Ecology of Eurasian watermilfoil. J Aquat Plant Manage 28(2):55–64

Smith RM, Baker RH, Collins DW, Korycinska A, Malumphy CP, Ostojá-Starzewski JC, Prior T, Pye D, Reid S (2018) Recent trends in non-native, invertebrate, plant pest establishments in Great Britain, accounting for time lags in reporting. Agric For Entomol 20(4):496–504

Stanton L, Lauzier R, MacConnachie S, Nield L, Pollard S, Heron J, Davies S (2012) Exploratory surveys and directed studies of Rocky Mountain ridged mussel (*Gonidea angulata* lea, 1839) in British Columbia. Can Manuscr Rep Fish Aquat Sci 3003:3

Swan KD, Lloyd NA, Moehrenschlager A (2018) Projecting further increases in conservation translocations: a Canadian case study. Biol Conserv 1(228):175–182

Symonds BJ (2000) Background and history of water management of Okanagan lake and river. In: Water management. Ministry of Environment, Lands and Parks, Penticton, Vancouver, p 8p

Wade J, Stephen C, Robertson C (2020) Mapping Rocky Mountain ridged mussel beds with preliminary identification of overlapping Eurasian watermilfoil within the Canadian range. Nature Conservation 42:19–31

Williams JD, Warren ML Jr, Cummings KS, Harris JL, Neves RJ (1993) Conservation status of freshwater mussels of the United States and Canada. Fisheries 18(9):6–22

Part IV

Wildlife Health Promotion Concepts

Human Dimensions of Wildlife Health Management

Craig Stephen

Abstract

Wildlife health cannot be understood or managed without understanding the people who affect or are affected by a state of wildlife health. Key factors that shape decisions about wildlife health management lie outside the domain of science and technical solutions. Human dimensions of wildlife health refer to how and why people value wildlife or perceive wildlife health, what they want to be done to manage health, and how they affect or are affected by health management decisions. Human dimensions information helps us better understand the role of people in the origins and control of wildlife health, the significance of poor wildlife health on people, and how human behaviors and values can impact management options and success. This chapter introduces that understanding how people perceive a wildlife health event as a risk (or not), how they decide to act on a health threat, and how to help people work together toward a wildlife health goal is critical to effective, acceptable, and sustainable management interventions.

C. Stephen (✉)
School of Population and Public Health, University of British Columbia, Vancouver, BC, Canada

Keywords

Human dimensions · Wildlife · Harm reduction · Health

1 Introduction

At its most basic, wildlife health management is influenced by three parts: (1) the prevalence of the problem; (2) public or professional concern about the problem; and (3) management objectives and actions to address people's perceptions of the impacts of the problem (Decker et al. 2012). Two of those three parts are focused on concern, objectives, and actions, which are human constructs. Chapter 2 introduced the concept of wildlife diseases such as social illnesses. The disease is the objective term referring to diagnosable abnormalities in organs or body systems, whereas illness is a subjective term referring to an individual's perception of the origin and significance of a health event. In other words, disease is a biological condition and illness is the social meaning of the condition. In the absence of a social meaning, such as risk to public health, impact on economies, effect on valued ecosystem components, or implications for perceived inherent worth of an animal, there would be no motivation for people to respond to wildlife disease. Chapter 2 also introduced four pillars of wildlife population health management: (1) building wildlife management and health policies that protect

C. Stephen (ed.), *Wildlife Population Health*,
https://doi.org/10.1007/978-3-030-90510-1_19

the determinants of health; (2) focusing on creating healthy environments that support sustainable access to the determinants of health; (3) strengthening collaborative action; (4) and reorienting wildlife health from only disease to a providing a continuum of care. Each of these involves getting people to do something. Wildlife health, therefore, cannot be understood or managed without understanding the people who affect or are affected by a state of wildlife health. However, the bulk of research and response to wildlife disease has historically medicalized the problems, ignoring the social context of wildlife health.

The rapidly changing environmental, disease, and economic pressures facing wildlife are multiplying and compounding each other and threatening the sustainability of an ever-increasing number of species. Biomedical interventions, while critical tools for wildlife health management, are not designed to remedy the situations that lead to the disease, problem, or illness in the first place. Wildlife health has usually been viewed as a technical topic under the domain of technical experts like veterinarians or biologists. Plentiful experience has, however, shown that key factors that shape decisions about environmental risk management lie outside the domain of science and technical solutions (Gregory et al. 2006).

Emerging diseases are not microbiological events. The extinction crisis is not an environmental issue. Climate change is not due to the type of fuels we use. They are the results of the constraints we place on human decisions and choices. Wildlife health managers need to understand the ways decisions involve difficult and sometimes controversial trade-offs between ecological, health, and social objectives if they hope to succeed in identifying feasible and socially acceptable ways that enable collective actions to meet wildlife health goals. The gaps in information and problems finding the 'one right answer' due to the complexity of many wildlife health issues leave room for different interpretations of the best course of action. No longer is it possible, or advisable, to "just let the science" lead the way. Population health specialists must be able to thrive at the intersection where science meets the human dimensions of wildlife health.

2 Human Dimensions of Wildlife Health

Scientific evidence may be the foundation of wildlife health policies and practices but "all wildlife management is motivated by human values" (Decker et al. 2006). Society, not scientists, decides what we want to protect, how much we want to invest, and what we are willing to lose. Human dimensions of wildlife health refer to how and why people value wildlife or perceive wildlife health, what they want to be done to manage health, and how they affect or are affected by health management decisions. Human dimensions researchers study how values, individual and social behaviors, institutional and decision-making frameworks, communication, and knowledge affect wildlife management (Decker et al. 2006). The purpose is to better understand the role of people in the origins and control of wildlife health, the significance of poor wildlife health on people, and how human behaviors and values can impact management options and success (Box 1).

Box 1 Three Most Important Human Dimensions for Wildlife Health Managers (Adapted from Leong and Decker 2020)

Understand how beliefs, attitudes, and norms influence how people perceive risks or benefits arising from wildlife health status or outcomes.

Work with stakeholders and rights holders to better understand their concerns and values to co-design policy and effective management response that will gain public approval or consent.

Communicate risks and options to empower and inspire people to reduce human contribution to wildlife disease or other health harms, increase

(continued)

protective behaviors, and avoid misunderstandings about management intentions.

Human behaviors and choices are sometimes the most viable targets of wildlife health interventions or are at the root of many problems. For example, the frequency and adequacy of cleaning bird feeders can influence the exposure of songbirds to pathogens such as *Salmonella* spp. (Feliciano et al. 2018). The desire for wildlife in the pet trade and food can drive the spread of emerging zoonotic pathogens in the wildlife trade (Smith et al. 2009). Urbanization and agriculture shift wildlife ecology in ways that alter infectious disease dynamics (Becker et al. 2015). Human activities can spread wildlife pathogens such as Ranavirus in amphibians (Price et al. 2016). Looking beyond infectious diseases reveals even more ways human decisions influence wildlife health. For example, road design and traffic speed affect automobile accidents involving wildlife (Glista et al. 2009). Industrial practices determine the extent of wildlife exposure to contaminants or the degree of degradation of habitat quality or availability needed to meet wild animals' daily needs.

Wildlife health management and research need to be informed and prioritized by stakeholder needs, values, and expectations to ensure that evidence is translated to policy and actions that are acceptable and feasible, and therefore, likely to be implemented (see Chap. 21). Human values shape how we perceive wildlife health issues and establish our tolerance of the impacts of the issue and acceptability of management actions. Managers should be equally reluctant to make decisions without biological and ecological information as when they are faced with inadequate information about the public and other interest groups (Vaske et al. 2009). Understanding how stakeholders perceive a wildlife health issue is crucial in crafting effective management responses (Wobeser 1994).

3 Risk Perception

Population health practitioners often want to influence people's risk perceptions, believing that if they had the "right" perceptions, they would opt for behaviors that protect their health or wildlife health. But what technical experts perceive to be the right risk perception may not be the same as others. The complexity, uncertainty, and ambiguity of many wildlife health issues often result in stakeholder and public risk perceptions differing significantly from risk estimations determined through formal risk analysis (Decker et al. 2006). "People judge a risk not only by what they think about it but also by how they feel about it" (Slovic and Peters 2006).

Volumes have been written about risk perception without one single best explanation. Some generalizations can be made (Table 1), but they are filtered and modulated by gender, age, education, experience, and culture. People are not generally optimistic or pessimistic about risks but instead tailor their perceptions to a specific risk (Ferrer and Klein 2015). If the goal of risk communication is to help people make health protection decisions on behalf of wildlife, risk communication efforts must involve not just giving people data and information but also accounting for the audiences who hear this information.

Risk communication (see Chap. 25) has several functions (Rickard 2021), one of which is to alert people to a circumstance. Usually, it also strives to direct information to people who can

Table 1 Questions influencing how people perceive a risk (Slovic 1987)

Are the risks sufficiently balanced by benefits that arise from the risky situation?
Is the risk imposed or voluntary?
Is there the potential for a catastrophic outcome?
Will there be effects on future generations?
Are the risks and benefits equitably distributed?
Will children be affected?
Are the risk or its impacts controllable?
Do associated outcomes produce a feeling of dread?
Is a small or large number of individuals likely to be impacted?
Is there potential for a broad social impact?

take or influence actions to avoid an underside outcome. Risk communication can help put information into an appropriate social context that allows people to better understand how they relate to the issue. Effective risk communication is not only about what you say but also how you say it. Whose voice is heard, who gets to speak at meetings, how decisions on assessing the risk are made, and even seating arrangements at meetings can all affect the real and perceived fairness and trustworthiness of the risk communication messages. People are more likely to make informed decisions about risk and take appropriate actions when they can voice their concerns and are given due consideration in the process. Efforts to solve wicked problems need attention to the messages, messengers, and audiences involved in the risk management process (Rickard 2021).

4 Healthy Decision-Making

Health management strives to turn knowledge into action by targeting behaviors, situations, and circumstances influencing vulnerability and risk. Virtually everything we try to influence to promote and protect wildlife health requires someone to decide or to act in a certain way. Perceptions, motivations, skills, and environments are key influencers of if or how we change our decisions or actions. There are two broad aspects to changing: thinking about it and doing it (Stephen 2020). Helping people find and understand information, become comfortable with the value, feasibility, and acceptability of change, and believe they can change are critical first steps in helping people think about making a change. Creating or finding opportunities to act, showing the value or benefits of change, and developing the social support to motivate maintaining the change all help people turn their intentions into action.

Changing a behavior is not a discrete or single event. Change requires patience and persistence. It is a process. There are many models and theories of change. None are ideal and suited to all situations and contexts. The Health Belief Model (Champion and Skinner 2008) argues that to change, people must first perceive that the problem they are trying to avoid is serious and that they are susceptible to being affected by the problem. They must believe that the proposed change can reduce risk and that the barriers to taking the action are outweighed by the benefits. These barriers can be logistic, financial, social, or others. The Model recognizes that people need a cue to action before the change will happen. These may be internal cues (e.g., their access to a wild food source is limited) or external cues (e.g., media messages about risks from the wildlife trade). The last component of the Model deals with a persons' confidence in their capacity to act and make the change. Perceptions and beliefs influencing the various stages of this Model are impacted by modifying variables such as age, gender, personality, socioeconomic status, and knowledge. Perceived barriers and perceived susceptibility can be the most powerful predictors of the likelihood that a person will adopt a behavior. Concepts from the Health Belief Model can be used to formulate questions that can help understand the human dimensions of a management recommendation (Table 2).

The concepts and theories described earlier for supporting individual change are also relevant to supporting community and organizational change. Although we often need organizational change to enable our desired health management action, it is people who change and not organizations. The literature on leading change in socioecological systems tells us that complex systems cannot be changed by the top-down, command, and control management. Instead, it needs a collaborative, inclusive, and participatory approach that allows individuals to act independently and make their own free choices (Westley et al. 2013). A key to community change is finding out what is important to people in the community and helping them reach their goals rather than organizing people to do something you think should be done. Community change happens when local people work together to transform the conditions and outcomes that matter to them. The Centers for Disease Control and Prevention in the United States of America

Table 2 Some guiding questions based on the Health Belief Models that can help population health practitioners appreciate human dimensions of a proposed management action (adapted from Stephen et al. 2020)

Perceptions of susceptibility, seriousness, benefits, and barriers	• Who/what is at risk? • Does the population at risk (or those in a position to make decisions on behalf of animals or ecosystems at risk) have access to accurate, trustworthy information presented in a manner suited to their circumstances and characteristics? • Have the benefits of the actions been explained in a manner relevant to those being asked to change? • Do the perceived risks of change outweigh perceived benefits and what is the basis for that 'calculation'? • Can barriers to action be feasibly and acceptably overcome?
Cues to action	• What is the preferred and trusted medium and method of the target audience to provide cues to action? • Who is a trusted voice that can provide cues?
Self-efficacy	• Have people been shown how to perform the desired behavior or trained or assisted in implementing the change? • Is there a series of incremental steps that can encourage change?

Table 3 Five stages of community change described by the US Centers for Disease Control and Prevention (USCDC 2018)

Stage	Features
Commitment	A coalition of community members and other agencies is established to give participants ownership of the process and create a pool of resources to support change
Assessment	Information on what the community needs is gathered in a manner that gives the community a voice. This helps to organize the community around the issue and can significantly influence program design
Planning	The coalition works with key partners to collectively develop a plan to implement the change
Implementation	Stakeholders and partners collaborate with the community team to implement and maintain the change by securing commitment and ownership of the actions
Evaluation	Evaluation runs throughout the entire change process to determine if you are implementing the right strategies and if the desired impacts are being reached

described five phases in the process of community change (Table 3) (USCDC 2018).

5 Collaborative Policy and Planning

Wildlife health depends on multiple partners, stakeholders, and the public working together to achieve health goals. Collaborative partnerships are, therefore, needed to improve wildlife health conditions and outcomes. Such collaborations are not easy to establish or continue. Developing a clear vision and mission is widely regarded as an essential aspect of collaborative partnerships (Roussos and Fawcett 2000). Seven other elements for positive partnerships are (1) a broad range of participation from diverse partners and a balance of human and financial resources; (2) leadership that inspires trust, confidence, and inclusiveness; (3) effective multi-way communication; (4) clear formal and informal roles; (5) trust; (6) attention to the political, economic, cultural, social and organizational impacts; and (7) evaluation for continuous improvement (Corbin 2017).

Action planning has also been associated with increases in rates of change (Roussos and Fawcett 2000). This is the process of identifying what changes to facilitate, who will produce them and by when, and how to gain support and minimize opposition in the process of bringing about the change. Effective communication, meeting facilitation, negotiation, and networking are needed to avoid conflicts and increase the chance of an agreed-to plan. The more the planned outcome matters to community members, influential leaders, and stakeholders, the more likely there will be human and financial support for progress toward those outcomes (Roussos and Fawcett 2000).

6 Summary

How we value wildlife, our willingness to act, and how we judge risks play a much bigger role in the outcome of a health management action than does biology or medicine. This chapter, along with Chap. 5 (participatory epidemiology), Chap. 7 (socio-ecological field investigation), Chap. 20 (policy), Chap. 21 (engaging community, government, and researchers in collaborative wildlife health management), Chap. 22 (leading change with diverse stakeholders), and Chap. 25 (describing and communicating risk to diverse audiences), provides some insights and approaches to dealing with the human dimensions of wildlife health. But, together, they only scratch the surface of human dimensions research, methods, and concepts that are critical for wildlife health management. If only one message is remembered from these chapters it is that wildlife health managers are in the business of human belief, behavior, and values. Wildlife population health practitioners are strongly encouraged to form alliances and partnerships with social scientists who are experts in understanding and facilitating collaboration with the people who affect and are affected by wildlife health.

References

Becker DJ, Streicker DG, Altizer S (2015) Linking anthropogenic resources to wildlife–pathogen dynamics: a review and meta-analysis. Ecol Lett 18(5):483–495

Champion VL, Skinner CS (2008) The health belief model. In: Health behavior and health education: theory, research, and practice, vol 4. Wiley, Hoboken, pp 45–65

Corbin JH (2017) Health promotion, partnership and intersectoral action. Health Promot Int 329:23–929

Decker DJ, Riley SJ, Siemer WF (2012) Human dimensions of wildlife management. JHU Press, New York

Decker DJ, Wild MA, Riley SJ, Siemer WF, Miller MM, Leong KM, Powers JG, Rhyan JC (2006) Wildlife disease management: a manager's model. Hum Dimens Wildl 11(3):151–158

Feliciano LM, Underwood TJ, Aruscavage DF (2018 Mar) The effectiveness of bird feeder cleaning methods with and without debris. Wilson J Ornithol 130(1):313–320

Ferrer RA, Klein WM (2015) Risk perceptions and health behavior. Curr Opin Psychol 1(5):85–89

Glista DJ, DeVault TL, DeWoody JA (2009) A review of mitigation measures for reducing wildlife mortality on roadways. Landsc Urban Plan 91(1):1–7

Gregory R, Failing L, Ohlson D, McDaniels TL (2006) Some pitfalls of an overemphasis on science in environmental risk management decisions. J Risk Res 9(7): 717–735

Leong K, Decker DJ (2020) Human dimensions considerations in wildlife disease management. Techniques in disease surveillance, and investigation. In: Field manual of wildlife diseases. US Geological Survey, Washington

Price SJ, Garner TW, Cunningham AA, Langton TE, Nichols RA (2016) Reconstructing the emergence of a lethal infectious disease of wildlife supports a key role for spread through translocations by humans. Proc R Soc B Biol Sci 283(1839):20160952

Rickard LN (2021) Pragmatic and (or) constitutive? On the foundations of contemporary risk communication research. Risk Anal 41(3):466–479

Roussos ST, Fawcett SB (2000) A review of collaborative partnerships as a strategy for improving community health. Annu Rev Public Health 21(1):369–402

Slovic P (1987) Perception of risk. Science 236(4799): 280–285

Slovic P, Peters E (2006) Risk perception and affect. Curr Dir Psychol Sci 15(6):322–325

Smith KF, Behrens M, Schloegel LM, Marano N, Burgiel S, Daszak P (2009) Reducing the risks of the wildlife trade. Science 324(5927):594–595

Stephen C (2020) Animals, health, and society: health promotion, harm reduction, and health equity in a one health world. CRC Press, New York

USCDC (2018) Community change process and the CHANGE tool. Unites States Centers for Disease Control and Prevention, Washington

Vaske JJ, Shelby LB, Needham MD, Manfredo MJ (2009) Preparing for the next disease: The human-wildlife connection. In: Manfredo MJ (ed) Wildlife and society: the science of human dimensions. Island Press, Washington, pp 244–261

Westley FR, Tjornbo O, Schultz L, Olsson P, Folke C, Crona B, Bodin Ö (2013) A theory of transformative agency in linked social-ecological systems. Ecol Soc 18:3

Wobeser GA (1994) Investigation and management of disease in wild animals. Plenum Press, New York, p 265

Healthy Policy for Healthy Wildlife

Bob Petrie

Abstract

Policy is a tool to put knowledge into practice. Policies are courses or principles of action adopted or proposed by a government, business, or individual that balance social and technical knowledge and turn it into action. Wildlife management is often conducted by government or government-appointed agencies, therefore, policies are developed in the public interest by government agencies. There are, however, multiple policies outside of wildlife agencies that influence the determinants of wildlife health. Healthy policies for wildlife improve the conditions under which wildlife live (thus ensuring their access to their determinants of health), protect public interests in wildlife (for cultural, economic, or ethical reasons), and/or manage risks to people from wildlife or economic interests (such as from zoonotic or exotic infectious diseases). The aim is to improve health outcomes through collaboration between wildlife health managers and those nontraditional partners who have influence over vulnerability and adaptive capacity animals through innovative policy. This chapter reviews the context and processes through which policies for healthy wildlife can be created.

Keywords

Policy · Knowledge-to-action · Wildlife · Healthy policy · Public interest · Preparedness · Decision-making

1 Introduction

The pathology, epidemiology, and environmental elements of wildlife health all inform our understanding of what is happening, but do not answer the question of what should we *do* about it? The challenge in managing wildlife is often in managing people, not wildlife. Our interactions with wildlife, their habitats, and the quality of the environment determine their health. Ecosystems, without human intervention, will function within their range of natural variability, but since human environmental disturbance is now a dominant influence, the questions we have before us are if, how and when to intervene, in what way do we intervene, and how do we manage human opportunity to address wildlife health issues? This is where the role of policy becomes relevant, since without the tools to put knowledge into action, the knowledge is literally and unfortunately, only academic.

B. Petrie (✉)
Department of Natural Resources and Renewables, Government of Nova Scotia, Nova Scotia, Canada
e-mail: Bob.Petrie@novascotia.ca

C. Stephen (ed.), *Wildlife Population Health*,
https://doi.org/10.1007/978-3-030-90510-1_20

2 Context and Governance

Wildlife management rests largely, but not exclusively, within the responsibility of government agencies, or bodies appointed by governments. It exists within a much larger, almost overwhelming context of other social, environmental, and economic challenges that compete for the attention and budgets of elected officials and policy makers. Wildlife agencies exist in their own ecosystem of other government agencies with different mandates and priorities, and with limits on what they can do amid high public expectations.

Scientists may often think that there should be a direct line between research outputs and action. When discoveries or findings are made concerning a particular disease or wildlife health issue, or if the status of species-at-risk changes, it is believed that action must automatically be taken. In reality, resources are limited, and public policy challenges are increasing and complex, making it difficult to balance budgets, time, and other resources between various government responsibilities such as education, public health care, infrastructure, and social programs to name a few. Policymaking is a social process that is ideally informed by, but not dictated by science (Sienkiewicz and Mair 2020). This highlights how important it is for wildlife health and other natural science practitioners to understand the various approaches by which knowledge is used to make the arguments for, and to inform policy, and to look for innovative ways to create and implement policy.

2.1 What is "Policy?"

A useful enough definition of policy is "a course or principle of action adopted or proposed by a government, party, business, or individual" (Oxford Languages 2021). We can think of policy as an umbrella term for describing what we *should* do. Policy can be expressed and implemented in various forms, from the general to the specific, including strategies, policies, laws, plans, procedures, guidelines, and technical standards (Table 1). For example, it is the policy

Table 1 The policy continuum ranging from more general at the top of the table to specific policy mechanisms at the bottom

Policy form	Description	
Strategy	a plan of action or policy designed to achieve a major or overall aim	General
Policy	a course or principle of action adopted or proposed by a government	
Legislation	overarching laws governing a particular topic	
Regulation	more specific rules made under legislation	
Plans	detailed proposals for doing or achieving an objective	
Guidelines	general rules, principles, or advice that may not be legally binding	
Procedures	a series of actions conducted in a certain order or manner	
Technical standards	an established norm or requirement for a repeatable technical task	Specific

of most governments that harmful interactions between wildlife and people should be minimized. One approach used to accomplish this could be to establish legislation and regulations that prohibit feeding wild animals. The public can be educated with guidelines on how to interact with wildlife (or not). Government staff can be given specific procedures that they follow when responding to a report of a wild animal attack or a diseased animal. If an animal must be drugged or euthanized in response, technical standards and guidelines would exist that ensure this is done safely and humanely, and that biological samples are taken to look for any disease.

Healthy wildlife policy provides leadership, direction, preparedness, and consistency in the management of wildlife and health issues. Good policy outlines who is in charge, what they are trying to accomplish, and how, such that all parties know what is expected of them. A frequent concern cited by business sectors is the inconsistency in how government rules are sometimes applied, or differences in the implementation of programs, between jurisdictions, or sometimes within a jurisdiction. Many business sectors will accept policies and regulations, so long as they are applied fairly and consistently.

2.2 Policy Roles

Whether the perspective is to conserve wildlife for their inherent existence value, utilitarian reasons or to manage risks to society, ultimately, policy for healthy wildlife is in the public interest. Wildlife health issues can affect numerous and varying public interests. For example, diseases are becoming a more frequent cause of declines in wildlife abundance and extinctions, and wildlife has been the source of numerous emerging infectious diseases of public health, agricultural, or conservation concern. Climate change, resource extraction, and landscape change are increasing, with increasing impact on wildlife populations. Social license and public demand for ecological goods and services require society to have the capacity to measure, monitor, maintain, and respond to wildlife health concerns (Anon 2018).

Healthy policy is needed to address the specific dimensions and determinants of wildlife health. In a general sense, policy can influence:

- Decisions affecting wildlife habitat—quantity, quality, and distribution.
- The introduction and movement of alien species and pathogens and the surveillance and preparedness for and response to these.
- Harvest levels, including the utilization rights of indigenous peoples.
- Government responses to wildlife disease and related public health issues.
- Funding priorities for wildlife management and research.
- The promotion or regulation of business sectors that may influence wildlife health (i.e., exotic pet trade, agriculture, forestry, mining).

Regulatory policy need not always be seen as a burden to economic development. Policy can support economic activity when the market requires that an absence of infection or harm be demonstrated in order to sell a product (Anon 2018). For example, the Agreement on International Humane Trapping Standards sets out welfare thresholds for trap performance requirements as agreed to by Canada, the European Union, and Russia. Meeting these requirements enables trade in fur products between these jurisdictions.

Policies affecting wildlife health may be enacted by agencies unrelated to wildlife management. For example, policy will determine how wildlife habitat is managed (i.e., whether wetlands can be filled in, how much of what type of forest must be retained during timber harvesting), whether specific diseases must be reported by veterinarians and physicians, or how governments are prepared (or not) to respond to diseases. In gathering health intelligence, general government policy may dictate how information may be gathered, by whom, if and how it is shared. For example, information management policies dictate how the public may request access to government data. Newer policy, for example, increasingly calls for government data to be posted openly online where it is available to the public and researchers.

Policy can inform or affect the capacity of governments to manage wildlife health and the manner in which they do so. Ideally, when a policy and a course of action are decided upon, it provides a business case for the necessary budget and human resources to be assigned to meet the objectives.

Policy also informs what specific roles and responsibilities are assigned to which government agencies. This has implications for determining priorities for work, how risks are assessed and addressed, and by whom. In many instances, multiple policies across several regulatory agencies can affect the same wildlife health issue. For example, in some jurisdictions, cervid farming is managed as an agricultural activity due to the economic benefits it produces, while the wild cervid herds and the consequent hunting and conservation are managed by the wildlife agency. This means that deliberate effort is needed to communicate and coordinate between agencies on the joint management of the same species, literally depending on which side of the fence it is on. Measures necessary to protect the wild and farmed herds from diseases such as Chronic Wasting Disease should be developed and implemented jointly. As another example, in jurisdictions that regulate wildlife in captivity, the inclusion of exotic wildlife within those responsibilities may vary between jurisdictions. This means that the health and welfare of exotic wildlife in the pet trade, or within zoos, the probability of zoonotic disease transmission to owners or handlers or the public, or the risks associated with the escape of potentially invasive or dangerous animals would be handled differently depending on which country, province, state, or even town and county the animals happen to be in.

The direct responsibility for wildlife health policy often rests with the wildlife management agency. That is where the biologists, wildlife managers, veterinarians, and ecologists work, so this seems logical. However other agencies make decisions that determine, for example, where highways are built that increase wildlife collisions, how and when wastewater treatment plants are upgraded to reduce pollution, and how agricultural operations manage livestock and waste. These examples illustrate that: (a) there is no one set of policies that determine the health of wildlife and (b) wildlife policy makers must understand and develop relationships, formally and informally, across a breadth of government agencies, industry groups, academic, and non-government organizations to build awareness and influence others in ways that promote healthy wildlife.

3 Making Policy

Policies are usually made when there is uncertainty in decision-making that needs to be resolved, or when people are making different decisions in similar circumstances and a common approach is needed. This often arises because of a new problem that has emerged or changing circumstances for which old policies and approaches do not suffice. In beginning to determine what should be done, it is always necessary to accurately *define the problem*, as this will affect all subsequent steps in policymaking. In the case of wildlife health, this can be a particularly tricky step in the process. There are many determinants of healthy populations (see Chapter 1), and if we define the problem too broadly, we risk ending up with policy that provides no specific or useful direction. However, when we look at problems in isolation, we risk missing the connections and interactions between issues that cumulatively or synergistically lead to the challenge at hand.

Once the problem is defined, a decision is made whether to intervene, and the specific goals and objectives are outlined. This is typically completed by wildlife professionals and approved by senior staff within government departments. At this stage, a risk assessment might also be completed to determine the likelihood and the severity of consequences if no action were taken. Once the decision is made to proceed with the development of a policy tool, consultation must be completed. As mentioned above, managing wildlife is about managing people, their interactions, and responses to wildlife. Government decisions affect people, in fact they must, to take effect. However, if decisions are made without consulting the people that are either affected by the decision, or who may play a key role in determining its success, then the policy tool runs the risk of either not receiving ultimate government approval due to public opposition (i.e., in the passage of a law), or of being designed in a way that doesn't reflect the reality of the targeted sector or process. Consultation with indigenous peoples is particularly important. It will help determine whether indigenous rights are affected by the proposed policy and serve to gather traditional knowledge which may add value to the design of the policy instrument. For example, a consultation with indigenous people concerning the joint management of a moose harvest in a Canadian province helped build a relationship that led to indigenous harvesters submitting biological specimens to the wildlife agency during their harvest. These specimens supplemented the sample collection already underway with non-indigenous harvesters and added to the knowledge base concerning the health of the moose population.

Once consultation with the right stakeholders and rights holders, the public and other government agencies are complete, a full analysis of the options is typically conducted and the instrument (i.e., a new law, a policy, or procedure) is designed. Final approval is often up to government decision makers whether they be at the executive levels of government agencies, or elected leaders, depending on the significance of the policy. In most democratic societies, the ultimate authority to make major policy decisions rests with elected officials who rely on the professionals within their departments to analyze issues, and present evidence, options, and recommendations. In these situations, the process by which scientific evidence is presented is key to the ability to get new policies and initiatives approved. As mentioned at the beginning of the chapter, governments and elected officials routinely deal with an overwhelming variety of issues that compete for time, attention, and money. Scientific evidence that supports a course of action is one of several kinds of information that will influence decisions. Other factors involved in decision-making may include the personal experiences of decision makers; information as presented in the media; and opinions and experiences of constituents and the public as voiced to the decision maker, most often now as conveyed through (and amplified by) social media. It is arguable whether all these factors are helpful in making informed policy decisions, but they must realistically be acknowledged (see Chapter 8 on evidence-based decision-making).

All of this often culminates in the briefing, which is the meeting with executive or elected officials in which the problem, context, evidence, and consultation results are presented to inform the discussion, confirm a course of action, or seek approval. As the name implies, these are usually not lengthy discussions about the issues (such discussions arise from preceding efforts completed by professional and bureaucratic staff). It is important to follow some best practices that will help ensure at the very least, the information has been presented effectively in a policy briefing, and at best, decisions are made that are well informed. Features of effective policy briefings are in Table 2.

Governments typically have high-level strategic priorities that are intended to communicate to the public what the government will focus on during their term in office. Examples of these often include health care, education, economic development, and infrastructure. Environmental

Table 2 Key attributes of effective policy briefings (adapted from Professional Policy Making Workshop—Dalhousie University, School of Public Administration, Centre for Advanced Management Education)

Clearly identifies the purpose, intended outcome, and decision to be made
Summarizes the most important contextual information
Summarizes critical issues (opportunities, risks, consultation results);
Acknowledges risks and trade-offs
Presents conclusions, advice, or recommendations
Concludes with a clear "ask" (a request or recommendation for action)

issues such as climate change are increasingly found in these high-level priorities, but one would not often find wildlife health among them. When possible, it helps to demonstrate how the wildlife health policy initiative being proposed can support the government's higher level goals. For example, a program of monitoring wild birds for avian influenza may also be seen as reducing risks for the poultry farming sector. New laws restricting the movement or possession of exotic animals as pets can contribute to public health by reducing the probability of zoonoses.

Once a course of action has been decided upon, work will continue to implement the chosen policy instrument. This may entail passing a new law by a legislature or parliament, formal approval of a new policy, or a technical standard created by an organization. The policy makers then need to effectively communicate the new policy. Depending on the significance of the policy change, it may be necessary to communicate it broadly through the media, hold specific meetings or workshops with the affected stakeholders, or conduct training for staff and/or stakeholders. New laws or policy instruments should be posted online where the people who need to know can access them. When the policy change is a significant departure from the status quo, it is sometimes necessary to allow a transition period so that affected parties have time to make the changes needed before the new policy or law takes full effect.

4 Collaboration and Coordination

Wildlife health involves a wide variety of government agencies, academic interests, industry groups, and nongovernment organizations. Indigenous peoples at the community, regional and national government levels are particularly important because of the knowledge they can contribute, and the importance of wildlife to both their cultures and livelihoods. Geographically, the scale at which policy can affect wildlife health ranges from the community level to subnational (i.e., provincial, state), national, and even international levels (Table 3). Policy instruments are applied at all these levels. Working arrangements are needed that bring the right people together to collaborate and coordinate at the right scales.

Although wildlife does not respect borders, policies and laws are passed by governments with borders. Consequently, various arrangements have evolved over time to enable governments to work together. Internationally, agreements such as the Convention on Biodiversity provide a forum

Table 3 Examples of policies that are found across geographic scale to influence how wildlife health is managed

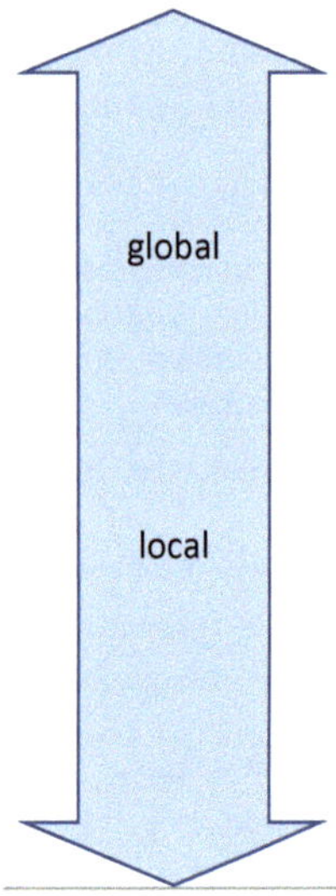

Treaties and International Agreements	Convention on Biodiversity OIE—World Organization for Animal Health
National	Health of Animals Act–Canada
Subnational	Provincial/State Wildlife Acts
Local	Municipal by laws on exotic pets

for nations to collectively agree on common goals and objectives. While international agreements such as this are difficult to enforce, they do have a trickle-down effect on national governments and how they set goals to adopt within their own jurisdictions and for which there is clearer authority to implement and enforce. Realistically however, the success of higher level (national/international) goals and objectives is often dependant on subnational governments to implement.

5 Summary

Policy making sits at the crossroads of information, institutions, and interest. Whether the role of the wildlife health professional is that of a researcher, pathologist, veterinarian, or wildlife biologist; whether they work in an organization focussed on wildlife, environmental management, public health, or agriculture, the value of integrated and cross-disciplinary collaboration and relationships should not be underestimated in developing more robust, relevant policy options that gain acceptance and most importantly, action by a diversity of partners.

References

Anon (2018) Pan-Canadian approach to wildlife health. http://www.cwhc-rcsf.ca/docs/technical_reports/EN_PanCanadian%20Approach%20to%20Wildlife%20Health%20Final.pdf

Oxford Languages (2021) Oxford University Press. https://languages.oup.com/

Sienkiewicz M, Mair D (2020) Against the science–policy binary separation: science for policy 1.0. In: Science for policy handbook. Elsevier, Amsterdam, pp 2–13

Stakeholder Engagement for Collaborative Wildlife Health Management

Matilde Tomaselli and Ryan W. Barry

Abstract

By presenting examples of comanagement regimes for natural resources and lessons learned, this chapter highlights why engaging with multiple stakeholders—from government representatives to resource user groups—is essential for effective and collaborative management around wildlife health. Wildlife professionals will learn to recognize the social and technical skills required to effectively develop, participate in, and maintain comanaged programs for wildlife health. This chapter specifically complements the wildlife surveillance and participatory epidemiology chapters of this book and provides practical considerations for successful stakeholder engagement, enabling wildlife health practitioners around the world to better influence and achieve collaborative wildlife population health management.

Keywords

Collaboration · Comanagement · Community-based management · Joint decision-making · Joint problem solving · Participatory management · Power sharing

M. Tomaselli (✉)
Canadian High Arctic Research Station, Polar Knowledge Canada, Cambridge Bay, Nunavut, Canada
e-mail: matilde.tomaselli@polar-polaire.gc.ca

R. W. Barry
Environmental Consultant, Kensington, Prince Edward Island, Canada

1 Stakeholder Engagement for Wildlife Health Management

Developing an appropriate level of understanding of wildlife health status through the continuous collection of epidemiological data from wildlife populations is the first step for implementing informed wildlife health management (Mörner et al. 2002, see Chaps. 4 and 7). However, developing robust and comprehensive data on wildlife health, including early identification of possible problems, alone is insufficient to support effective and timely evidence-based decision-making. Identifying and working together with relevant stakeholders, including decision makers and local resource users when present, are necessary so that the information generated through either monitoring or surveillance activities is successfully utilized to implement wildlife health actions. Failing to recognize the integral role of stakeholder engagement in programs for biodiversity conservation can hinder their effectiveness and result in suboptimal use of resources (Conallin et al. 2017; Giakoumi et al. 2018), this is also particularly true for wildlife health programs.

Experiences from the comanagement of natural resources have demonstrated that engaging with local resource users, including wildlife

C. Stephen (ed.), *Wildlife Population Health*,
https://doi.org/10.1007/978-3-030-90510-1_21

harvesters, can improve management outcomes (see next section). In the context of wildlife health assessments, programs that adopt participatory approaches, such as participatory epidemiology and surveillance, are specifically designed to engage with local stakeholders and resource users from the outset, offering increased chances of achieving collaborative wildlife management (see Chap. 5). However, to ensure information generated is successfully utilized for targeted interventions, it is essential to also engage decision makers from the wildlife health sector and, as appropriate to the context, also from the public health and domestic animal health sectors. Therefore, identifying who to engage with and how should be tailored to the circumstances on a project-by-project basis with consideration for project objectives and issues that may arise, preferably through the development of a specific written plan.

Formal wildlife comanagement regimes offer well-established frameworks and platforms for engagement with a variety of stakeholders which directly support wildlife management, including governments and resource users. As such they offer insights on the value of and frameworks for working together with a variety of stakeholders to improve decision-making around wildlife health. However, wildlife health practitioners should recognize such regimes as a starting place only. As wildlife health intersects with the health of humans, domestic animals, and the environment, successful interventions around wildlife health often require going beyond the existing engagement structure and being inclusive not only of decision makers from the wildlife or natural resources sectors but also from the public health and domestic animal health sectors as appropriate to the operating context.

2 Wildlife Comanagement: A Framework for Engaging Stakeholders in Collaborative Management

Borrowing from various definitions (i.e., Berkes et al. 1991, IUCN 1997; The World Bank 1998; Carlsson and Berkes 2005), comanagement can be defined as a form of collaborative and decentralized decision-making to manage a specific area or set of resources in which power, responsibilities, and duties are shared between stakeholders, including local resource users, governments, and others (e.g., nongovernmental organizations) as appropriate to each context. While earlier examples of comanagement for natural resources date back to the 1970s, since the 1990s this form of collaborative decision-making has become increasingly adopted and expected (Plummer and Fitzgibbon 2004). Today comanagement for natural resources—alternatively referred to as collaborative management, participatory management, or community-based management—appears in different formats and arrangements along a spectrum, where voices of user groups increasingly shape management interventions as we progressively move from decision-making by centralized governments to more decentralized community decision-making models; thus expectations for stakeholder engagement also progressively evolve from simple consultation to joint decision-making (Berkes et al. 1991; Sen and Nielson 1996).

Formal comanagement regimes for natural resources were developed in response to the recognition that traditional scientific top-down approaches were inadequate to achieve effective and sustainable interventions, which instead required the inclusion of resource users in the management equation. In other words, government managers recognized that local users must be a part of the solution to manage resources sustainably. The recognition of the management role of resource users stemmed from the increased awareness that resource users' knowledge (or ecological knowledge) offered a different but complementary perspective to scientific knowledge (Berkes et al. 2000; Rist and Dahdouh-Guebas 2006) and, when properly documented and utilized, improved the holistic understanding of natural systems, including environmental processes and their impacts (Huntington et al. 2004a, b, see Chap. 5). Natural resource comanagement regimes have also evolved through increased recognition of indigenous rights, devolution of government responsibilities, or negotiation of land claim agreements and similar legislative processes (Plummer and Fitzgibbon 2004).

Examples of comanagement approach applied to different resources can be found throughout the world, most commonly for fisheries (e.g., Sen and Nielson 1996; Pomeroy et al. 2001; Makino and Matsuda 2005), wildlife (e.g., Mayaka 2002; Decker et al. 2005; Popp et al. 2019), and protected areas (e.g., Borrini-Feyerabend 1996; Da Silva 2004; Kepe 2008; Ross et al. 2009; Farrier and Adams 2011). Here we offer two examples from Canada that illustrate different formats in which wildlife comanagement can be operationalized: the Beverly and Qamanirjuaq Caribou Management Board, an interjurisdictional initiative to comanage caribou herds; and the Nunavut wildlife comanagement governance system established through the Nunavut Agreement, a modern-day comprehensive land claims agreement. While each type of framework can offer examples of their specific challenges, successes, and limitations, when considered together they offer similar lessons on the importance of integrating common best practices for stakeholder engagement into project design, with consideration for the operating environment (see next section).

The Beverly and Qamanirjuaq Caribou Management Board (BQCMB, www.arctic-caribou.com) was established in 1982 as the first caribou comanagement board in North America and has been described as one of the most successful and long-standing comanagement institutions in northern Canada (Kendrick 2000). The BQCMB consists of representatives from the government of Canada and both local community members and local government representatives from several provinces and territories which harvest the herds across their vast ranges and have associated management responsibilities. This initiative is not a land claims-based comanagement and has no formal status in Canadian law but was formalized through a specific Management Agreement that clarified the board's overall objective: to safeguard the Beverly and Qaminirjuaq barren-ground caribou herds in the interest of the Indigenous peoples who have traditionally relied upon caribou. Although not without challenges, over the years the BQCMB successfully coordinated monitoring activities and established cooperative management of the herds for caribou conservation, including through habitat protection (Thomas and Schaefer 1991). Bringing a diverse group of stakeholders to the table to comanage caribou has had many benefits but, according to the BQCMB, "of all the strides made throughout the board's history, none is more important than the improved level of trust and respect among different Indigenous and government groups that these meetings have fostered. Before, relations were uneasy as different cultures and knowledge systems collided. But both sides have made tremendous efforts to find common ground, in order to conserve caribou for the use of future generations" (BQCMB 2021a).

Unlike the BQCMB which was created through an interjurisdictional agreement for the management of a single shared wildlife resource, wildlife comanagement in Canada's newest and most northerly territory—Nunavut—was established through a comprehensive land claims agreement and enacted through federal enabling legislation. A multitude of different organizations with distinct mandates and processes share responsibilities for the management of wildlife through a single regime that is applicable to all wildlife across the vast territory of more than 2,000,000 km^2. The objectives of Nunavut's system of wildlife management include fully acknowledging and reflecting the primary role of Inuit—northern Indigenous peoples—in wildlife harvesting, serving and promoting the long-term economic, social, and cultural interests of Inuit harvesters, and inviting public participation and promoting public confidence in wildlife management, particularly among Inuit. The strength of Nunavut's system originates in part from its empowering of community-level Hunters and Trappers Organizations (HTOs) which oversee the exercising of wildlife rights by Inuit in their respective communities, including harvesting practices and techniques and nonquota limitations (Government of Canada and Tungavik Federation of Nunavut 1993). HTOs bring forward the experience and knowledge of individual harvesters into their regular participation in the formal processes of the Nunavut Wildlife Management Board (NWMB), which has primary responsibilities for establishing, modifying, or removing levels of total allowable harvest, conducting harvest studies, supporting wildlife research, approving wildlife management plans, and other discretionary functions associated with wildlife research, management, and conservation (Government of Canada and Tungavik Federation of Nunavut 1993). Nunavut's system is constitutionally protected and reflects a respect for Inuit culture, traditions, and rights; for example, consideration for the Inuit tradition of oral communication and consensus style decision-making, requirements for operation in the Inuit language of Inuktut, consideration for *Inuit Qaujimajatuqangit* (Inuit knowledge), and extensive programs for public participation are all hallmarks of the Nunavut system.

While the BQCMB and Nunavut's wildlife management regime are unique to the jurisdictions and resources they each represent, to be effective, they and other comanagement structures share a common need to ensure the stakeholders they represent are actively engaged throughout their respective functions and processes. The BQMCB grounds the development of its positions and recommendations on how best to contribute to caribou protection and conservation through direct engagement with resource users, traveling to caribou-range communities to host workshops, and talk to residents about caribou issues. The BQCMB places significant emphasis on the development of educational resources and tools including posters, fact sheets, videos, youth programs, and scholarships to "help present and future generations understand the important environmental, cultural, and economic roles that caribou play, and especially to help them thrive in the future" (BQCMB 2021b). Under the Nunavut Agreement, the NWMB can choose how it should carry out a public consultation, including which parties should be able to make submissions and how those submissions are to be made to it (Government of Canada and Tungavik Federation of Nunavut 1993). As a matter of practice, in making decisions about the wildlife that will affect a community, the NWMB will consult with the appropriate HTOs prior to making a decision, typically in a format open to the public with advance notice, simultaneous interpretation into local dialects, and adequate time for discussion and sharing of local perspectives and *Inuit Qaujimajatuqangit* (Wheatley 2003). Like the BQCMB and Nunavut's wildlife management regime, most comanagement structures and initiatives benefit from the incorporation of meaningful stakeholder engagement and offer insight into best practices that can be applied to other jurisdictions, regimes, and issues.

3 Ingredients for Success to Develop and Maintain Comanaged Projects for Wildlife Health

Whether it is the creation of project-specific committees or involvement in an ongoing comanaged project, here are some aspects to consider for effectively developing, participating in, and maintaining comanagement initiatives around wildlife health:

1. *Understand the working context.* As noted above, comanagement may arise through different means (e.g., management agreement, land claim processes, devolution of government responsibilities, etc.) and it is important to develop an understanding of the framework for participation and decision-making, and the roles and responsibilities of the respective stakeholders. Understanding the management framework will allow for the identification of the most effective opportunities for engagement and best use of participants' time. Investing time and effort into understanding the local context is also necessary for demonstrating respect for local cultures and traditions, and for ensuring methods used respect participant's norms and beliefs.
2. *Identify the right stakeholders.* Consult to determine the relevant local knowledge

holders and scientific knowledge holders and establish a framework for respectful acknowledgment of differing viewpoints and consideration and treatment of everyone's knowledge. Be clear about roles, responsibilities, and demands on time to effectively manage expectations. Recognize that some stakeholders may not be regular participants in a management framework. Creating a bridge between parties working in silos may also be necessary (e.g., public health and wildlife management departments in government may not commonly work together).

3. *Demonstrate respect and cultural sensitivity.* Wildlife managers, resource users, and knowledge holders may have very different backgrounds, experiences, cultural outlooks, and viewpoints. Ensure that all participants are encouraged to actively engage, invited to speak and share opinions, ask questions, etc. A nonjudgmental approach should be adopted, understanding our own biases and accepting other viewpoints. This is particularly important when indigenous knowledge holders are at the table. For comanagement frameworks originating in land claims agreements (such as for the Nunavut wildlife management regime referenced above) or other formats with indigenous participants, it is critical to understand and respect indigenous rights, languages, and customs to build trust, establish common ground and develop productive working relationships.
4. *Design a living plan.* Work with participants to develop a written plan that establishes the scope of the program or initiative, with clear identification of objectives, strengths, weaknesses, roles, and responsibilities. Written plans can serve as terms of reference for a project or initiative to keep everyone focused on the goals and their role in helping to realize them. They should be updated on an ongoing basis as needed to fit evolving needs of the initiative and stakeholders, with the input of all those involved. Monitor and evaluate the plan as it is implemented to take stock of what is working well and what could be improved upon and consider whether periodic reporting or formal audits may be helpful to achieve continuous improvement.
5. *Design a communication strategy.* It may not be enough to encourage participants to actively engage. Instead, a more comprehensive communication strategy (included in the 'living plan' discussed above) can be used to guide these efforts, including such items as consideration for interpretation/translation needs; confirming a common understanding of important terminology that will be used and utilize plain language as much as possible; and articulate planned meeting frequency, length of review periods, notice for public meetings, provisions for reporting back and/or reporting publicly. Documenting engagement efforts and communicating on a regular schedule can prevent loss of momentum when established participants exit and new participants enter comanagement initiatives. A strategy should help to ensure information is presented and communicated in formats that resonate with stakeholders and facilitate decision-making—this may require different ways of presenting the same information, depending on the audience. Remember that scientific publications are not necessarily the right format to present information for community members and decision makers as they do not allow for inclusive discussion when people other than scientists are at the table. For example, the BQCMB utilized a multiday workshop to enable a diverse group of stakeholders to effectively share information on caribou and work together for their conservation (see BQCMB 2011).
6. *Practice consensus-building.* Recognize that you may encounter conflicts with how information and results are interpreted or understood and build in the adequate opportunity for reviewing data and results together. Be ready to help organizations or individuals with opposing viewpoints or conflicts to work together, reminding participants with different viewpoints and backgrounds that they are all working together toward common goals

that can help to establish common ground and create a positive working environment. As conflicting views and concerns may arise or develop throughout the process, it is also important to create opportunities to listen to and address stakeholders' concerns, particularly for the concerns of resource users. To be successful, comanagement processes must provide sufficient time for knowledge generation and joint learning amongst participants, with truly collaborative decision-making often only made possible through extensive deliberation and negotiation (Berkes 2009).

7. *Consider opportunities for capacity-building.* Whether initiatives are short term or operate for years or decades, identifying opportunities for education and capacity building (such as the example of the BQCMB noted above) can empower current and future participants, build a positive legacy and improve future outcomes. Some stakeholders in comanagement initiatives may have limited resources and experience capacity challenges which can adversely impact their effective participation. There is growing recognition that research and monitoring programs that rely on and benefit from local knowledge and participation must consider how to give back, leave a positive legacy, and build capacity within the communities of interest (Hacker et al. 2012; Diver and Higgins 2014). Capacity building at the institutional, individual, and community levels also contributes to developing a sense of common purpose and empowering stakeholders to participate more effectively in the collaborative management process (Raik 2002).

4 Summary

Wildlife health practitioners must acknowledge stakeholder engagement as an integral and necessary component of wildlife health programs that should be formalized through a specific plan as early as possible to best influence decision-making around wildlife health. While participatory approaches to wildlife health assessments such as participatory epidemiology and surveillance are designed to meaningfully engage with local stakeholders, wildlife health practitioners must also engage with decision makers from different government sectors, including wildlife, domestic animal, and public health, to make sure the participatory surveillance process and output effectively influence management.

Comanagement regimes of natural resources including wildlife can serve as models for how to bring multiple stakeholders, from government representatives to resource user groups, together at the decision-making table to work toward collaborative wildlife health management. Shared management decisions between government and local resource users can be achieved through meaningful stakeholder engagement even when the jurisdictional framework of operation does not expressly contemplate the joint decision-making model (as seen with the BQCMB). To achieve this goal, wildlife health practitioners must be equipped with social and technical skills to be able to effectively develop, participate in, and maintain comanaged programs for wildlife health. This chapter provides insights on how to do so, highlighting that to be effective stakeholder engagement must also be customized to the operating context. As comanagement approaches are becoming more common throughout the world, in part through the settlement of land claims agreements and growing worldwide recognition of indigenous rights, it is increasingly important that wildlife health practitioners acquire the necessary skills to develop and actively participate in formal multi-stakeholder discussions to be better equipped to effectively influence and achieve collaborative wildlife population health management.

References

Berkes F (2009) Evolution of co-management: role of knowledge generation, bridging organizations and social learning. J Environ Manage 90:1692–1702. https://doi.org/10.1016/j.jenvman.2008.12.001

Berkes F, Colding J, Folke C (2000) Rediscovery of traditional ecological knowledge as adaptive management. Ecol Appl 10:1251–1262. https://doi.org/10.1890/1051-0761(2000)010[1251:ROTEKA]2.0.CO;2

Berkes F, George P, Preston RJ (1991) Co-management: the evolution in theory and practice of the joint administration of living resources. Alternatives 18:12–18

Beverly and Qamanirjuaq Caribou (2011) Management board (BQCMB). BQCMB Caribou workshop February 2010. Overview report: challenges facing the Beverly and Qamanirjuaq Caribou herds and some possible solutions. https://arctic-caribou.com/pdf/BQCMB_Caribou_Workshop_Overview_Report.pdf. Accessed 6 Aug 2021

Beverly and Qamanirjuaq Caribou Management Board (BQCMB) (2021a) BQCMB achievements. Success via cooperation. https://arctic-caribou.com/bqcmb-achievements/. Accessed 9 Aug 09, 2021

Beverly and Qamanirjuaq Caribou Management Board (BQCMB) (2021b) BQCMB education resources and tools. https://arctic-caribou.com/education/. Accessed 9 Aug 2021

Borrini-Feyerabend G (1996) Collaborative management of protected areas: tailoring the approach to the context. IUCN-The World Conservation Union, Gland

Carlsson L, Berkes F (2005) Co-management: concepts and methodological implications. J Environ Manage 75:65–76. https://doi.org/10.1016/j.jenvman.2004.11.008

Conallin JC, Dickens C, Hearne D, Allan C (2017) Stakeholder engagement in environmental water management. In: Horne AC, Webb JA, Stewardson MJ, Richter B, Acreman M (eds) Water for the environment: from policy and science to implementation and management. Academic Press, Amsterdam, pp 129–150

Decker DJ, Raik DA, Carpenter LH, Organ JF, Schulser TM (2005) Collaboration for community-based wildlife management. Urban Ecosyst 8:227–236. https://doi.org/10.1007/s11252-005-3264-6

Diver SW, Higgins MN (2014) Giving back through collaborative research: towards a practice of dynamic reciprocity. J Res Pract 2014(10):M9

Farrier D, Adams M (2011) Indigenous - government co-management of protected areas: Booderee National Park and the National Framework in Australia. In: Lausche B (ed) Guidelines for protected area legislation. Special protected area types. IUCN-The World Conservation Union, Gland, pp 1–40

Giakoumi S, McGowan J, Mills M, Beger M, Bustamante RH, Charles A et al (2018) Revisiting "success" and "failure" of marine protected areas: a conservation scientist perspective. Front Mar Sci 5:223. https://doi.org/10.3389/fmars.2018.00223

Government of Canada and Tungavik Federation of Nunavut (1993) Agreement between the inuit of the Nunavut Settlement area and Her majesty the Queen in right of Canada. Indian and Northern Affairs Canada, Ottawa

Hacker K, Tendulkar SA, Rideout C, Bhuiya N, Trinh-Shevrin C, Savage CP et al (2012) Community capacity building and sustainability: outcomes of community-based participatory research. Prog Community Health Partnersh: Res Educ Action 6:349–360. https://doi.org/10.1353/cpr.2012.0048

Huntington H, Callaghan T, Fox S, Krupnik I (2004a) Matching traditional and scientific observations to detect environmental change: a discussion on Arctic terrestrial ecosystems. Ambio 33:18–23. https://doi.org/10.1007/0044-7447-33.sp13.18

Huntington HP, Suydam RS, Rosenberg DH (2004b) Traditional knowledge and satellite tracking as complementary approaches to ecological understanding. Environ Conserv 31:177–180. https://doi.org/10.1017/S0376892904001559

IUCN (The World Conservation Union) (1997) Resolutions and recommendations, World Conservation Congress, Montreal, 13–23 Oct 1996. https://portals.iucn.org/library/sites/library/files/documents/WCC-1st-002.pdf. Accessed 9 Aug 2021

Kendrick A (2000) Community perceptions of the Beverly-Qamanirjuaq Caribou management board. Can J Native Stud 20:1–33

Kepe T (2008) Land claims and Comanagement of protected areas in South Africa: exploring the challenges. Environ Manag 2008(41):311–321. https://doi.org/10.1007/s00267-007-9034-x

Makino M, Matsuda H (2005) Co-management in Japanese coastal fisheries: institutional features and transaction costs. Mar Policy 29:441–450. https://doi.org/10.1016/j.marpol.2004.07.005

Mayaka TB (2002) Wildlife co-management in the Benoue National Park-Complex, Cameroon: a bumpy road to institutional development. World Dev 30: 2001–2016. https://doi.org/10.1016/S0305-750X(02)00111-0

Mörner T, Obendorf DL, Artois M, Woodford MH (2002) Surveillance and monitoring of wildlife diseases. Rev Sci Tech Off Int Epiz 21:67–76

Plummer R, Fitzgibbon J (2004) Co-management of natural resources: a proposed framework. Environ Manag 33:876–885. https://doi.org/10.1007/s00267-003-3038-y

Pomeroy RS, Katon BM, Harkes I (2001) Conditions affecting the success of fisheries co-management: lessons from Asia. Mar Policy 25:197–208. https://doi.org/10.1016/S0308-597X(01)00010-0

Popp JN, Priadka P, Kozmik C (2019) The rise of moose co-management and integration of indigenous knowledge. Hum Dimens Wildl 24:159–167. https://doi.org/10.1080/10871209.2019.1545953

Raik DB (2002) Capacity building for co-management of wildlife in North America. HDRU Series No. 02-2. https://ecommons.cornell.edu/bitstream/handle/1813/40382/HDRUReport02-2.pdf?sequence=1. Accessed 9 Aug 2021

Rist S, Dahdouh-Guebas F (2006) Ethnosciences–a step towards the integration of scientific and indigenous

forms of knowledge in the management of natural resources for the future. Environ Dev Sustain 8:467–493. https://doi.org/10.1007/s10668-006-9050-7
Ross H, Grant C, Robinson CJ, Izurieta A, Smyth D, Rist P (2009) Co-management and indigenous protected areas in Australia: achievements and ways forward. Australas J Environ Manage 16:242–252. https://doi.org/10.1080/14486563.2009.9725240
Sen S, Neilson R (1996) Fisheries co-management: a comparative analysis. Mar Policy 20:357–438. https://doi.org/10.1016/0308-597X(96)00028-0
Silva D (2004) From common property to co-management: lessons from Brazil's first maritime extractive reserve. Mar Policy 28:419–428. https://doi.org/10.1016/j.marpol.2003.10.017
The World Bank (1998) Report from the international workshop on community-based natural resource management (CBNRM). World Bank, Washington
Thomas D, Schaefer J (1991) Wildlife co-management defined: the Beverly and Kaminuriak caribou management board. Rangifer 11:73–89. https://doi.org/10.7557/2.11.4.997
Wheatley M (2003) Caribou co-management in Nunavut: implementing the Nunavut land claims agreement. Rangifer 14:303–305. https://doi.org/10.7557/2.23.5.1714

Leading Change with Diverse Stakeholders

Catherine Machalaba and Jonathan M. Sleeman

Abstract

The shift to holistic approaches to managing wildlife health, and the complex landscape of partners and stakeholders, has led to a focus on the development of leadership skills in addition to technical expertise. This chapter introduces key elements and core skills for successful cross-sectoral and transdisciplinary leadership that will help wildlife health practitioners effectively lead change toward integrated, mutually beneficial health outcomes for all sectors. Leading change benefits from having good individual leadership skills, including emotional intelligence or the capacity to be aware of, control, and express one's emotions, and to handle interpersonal relationships judiciously and empathetically. Driving multi-sectoral change is facilitated by wildlife health professionals becoming champions for change and being empowered to form and lead teams and participate effectively on governance structures such as interagency committees or working groups. Finally, the four main elements in leading change are envisioning the future state, engaging key stakeholders and coalition building, identifying barriers and breaking down resistance to change, and institutionalizing change. Effective leadership can be enabled by being open to new ideas, constantly questioning the effectiveness of one's approach, and a commitment to lifelong learning. Application of these leadership skills and approaches can help effect change, resulting in improved wildlife population health.

Keywords

Leadership · Change · Authority · Systems · One Health · Partnerships · Multi-sectoral · Collaboration · Wildlife health

1 Introduction

Maintaining the health of wildlife populations and the habitats upon which they depend is growing in importance. There is increasing recognition that global drivers, including landscape and climate change, are degrading the functioning of ecosystems, resulting in poor health outcomes for humans, animals, and the environment (Acevedo-Whitehouse and Duffus 2009; WHO and CBD 2015). Consequently, wildlife health is inherently linked to health and economic

C. Machalaba
EcoHealth Alliance, New York, NY, USA
e-mail: machalaba@ecohealthalliance.org

J. M. Sleeman (✉)
U.S. Geological Survey's National Wildlife Health Center, Madison, WI, USA
e-mail: jsleeman@usgs.gov

C. Stephen (ed.), *Wildlife Population Health*,
https://doi.org/10.1007/978-3-030-90510-1_22

outcomes for a range of sectors (OIE 2021). Traditionally, wildlife health professionals have focused on technical skills and capacities to detect and characterize hazards; however, modern concepts of wildlife health management are shifting from health as the absence of disease to health as the capacity for populations and ecosystems to maintain resiliency in the face of anthropogenically driven changes (Stephen 2014). The shift to holistic approaches to managing wildlife health, and the complex landscape of partners (an entity with an active role) and stakeholders (an entity with an interest or concern), has led to a focus on the development of leadership skills in addition to technical expertise.

Despite this direct and indirect importance, wildlife health professionals find there is often no clear path to advocate for the authorities, resources, and awareness required to lead change and apply effective actions. Realizing positive change in wildlife health requires the ability to lead change while working with a variety of organizations and people who may have shared, complementary or conflicting goals, perspectives, and expectations.

A major challenge for wildlife health, and often environmental health more broadly, is the lack of a dedicated lead organization at global, national, and local levels (Karesh et al. 2020). Mandates are typically fragmented between authorities responsible for veterinary services, environmental protection, agriculture, forestry, fisheries, conservation, parks, and public health, resulting in incomplete and inconsistent coverage and a lack of a shared, overall wildlife health strategy. Wildlife health falls between or across agencies often resulting in a lack of governance and coordination. Consequently, wildlife health considerations often are inadvertently missed in planning, implementation, and evaluation (Berthe et al. 2018). This is evident whenever wildlife health services have limited integration with livestock and public health, resulting in fragmented surveillance systems, variable data sharing, and poor representation in decision-making processes. Effectively leading change would benefit from engaging diverse stakeholders to find entry points for meaningful progress, including advocating for the value of wildlife health (see Case Study 1) (Fathke et al. 2021).

Leadership is a core population health competency. It can serve cross-cutting functions, whether targeted at the adoption of policies, creation of governance structures, refinements in field-based operations, or a shift in viewpoints or processes (Carlin et al. 2019). Leadership to drive and sustain positive changes benefits from both patience and persistence, and adoption of new skills and approaches, but these efforts can break new ground in terms of systems-building and delivering an improved health status for wildlife and other populations. Successful leadership demonstrates the value of wildlife health and makes collaboration more routine as part of wildlife health management and delivery of public health and veterinary services (National Research Council 2013). This chapter introduces key elements and core competencies for successful cross-sectoral and transdisciplinary leadership that can help wildlife health practitioners effectively lead change toward integrated, mutually beneficial health outcomes for all sectors.

Case Study 1: Biodiversity-Sensitive Risk Communication

Concern over perceived or actual disease risk from wild animals has resulted in reports of indiscriminate killing of bats and birds. These well-intentioned but ill-informed responses may be counterproductive and represent an ineffective use of resources, while also harming wild animal populations. To preempt concerns, a locally adapted "Living Safely with Bats" booklet was used by nongovernmental and governmental partners in community engagement for zoonotic disease risk communication efforts in over 13 countries (PREDICT Consortium 2018). Similarly, questions over disease risk to wild animals in the era of COVID-19 resulted in the production of materials on safe research practices for great apes, bats, and free-ranging wildlife generally (OIE and IUCN SSC Wildlife

(continued)

Health Specialist Group 2020). Leadership on resources like these can be valuable to meet the need for publicly available information and guidance for wildlife health managers. Ideally, over the long-term wildlife health considerations will be embedded in processes and regulations to ensure reliable uptake and promote the coordinated, coherent risk communication necessary for maintaining public confidence, effecting positive behavior change, and inter- and intra-agency credibility.

2 Leading Self

To effectively lead others, one must effectively lead oneself. Emotional intelligence is the capacity to be aware of, control, and express one's emotions, and to handle interpersonal relationships judiciously and empathetically (Czabanowska et al. 2014). These are foundational skills for effectively leading change. Key attributes of emotional intelligence include having a better understanding of one's own strengths and limits and gaining a heightened awareness and more objectivity of other people's competence. Gaining self-confidence in one's self-worth and capabilities is also important. Self-management means developing skills and tools to control one's emotions, especially in response to stressful situations. Learning and adapting from experiences and retaining motivation despite setbacks are critical to the personal resiliency needed for long-term success. Sensing others' emotions, understanding others, and acting with empathy help to gain additional perspectives, understand differences, and build personal trust among colleagues and partners (Table 1). Self-assessment tools such as the Myers–Briggs Type Indicator (MBTI) can be helpful in understanding one's own strengths and limits (Firth-Cozens and Mowbray 2001). An awareness of personal strengths and weaknesses is useful in being able to optimize ways of working; for example, by focusing on communication skills, participatory management approaches, and consensus building to help convey ideas, with the goal of building support (Morgan 2009). Changing personality can be successful but requires persistence and commitment (Bleidorn et al. 2019).

Knowing your preferences and the preferences of others can also help understand different viewpoints, reach common understanding, and generate movement toward consensus and shared goals. Seeking feedback, such as through "360-degree feedback" (feedback from direct reports, colleagues, and supervisor, as well as a self-evaluation), can help identify hidden strengths and potential blind spots.

Core values are the essential, authentic, and enduring guiding principles that prescribe and drive the conduct of individuals in an organization and help guide daily work and decision-making (Van Rekom et al. 2006). Establishing one's core values (e.g., integrity, optimism, respect, dependability, honesty) and holding oneself accountable to these values can help ensure one becomes a trusted leader who is able to build trust among others, which is the foundation to the success of any endeavor. We all have leaders we admire and can learn from to shape our own leadership values.

Table 1 Various skills and competencies that are helpful in leading oneself, and leading others

Leading Self	Leading Others
Understanding of one's own strengths and limits	Forming and operating in teams
Heightened awareness and more objectivity of other people's competence	Conflict resolution
Self-confidence in one's self-worth and capabilities	Decision making
Self-management and controlling one's emotions	Negotiation
Learning and adapting from experiences and retaining motivation	Motivating others
Acting with empathy	Setting goals and project management
	Active listening and communication

3 Leading Others

Leading others can occur within an institution or may involve liaising with multiple institutions. For wildlife health and its wide relevance to other sectors, siloed, sector-specific approaches taken by individual institutions often create gaps or result in duplication. Effecting change requires practitioners to expand their work into addressing what needs to be achieved across sectors (Berthe et al. 2018), placing an increased emphasis on building teams, and creating governance structures. Rather than diminish the needs for specialized expertise, this cultural shift requires mutual appreciation of each sectors' contributions and mechanisms that make collaboration routine in both emergencies and peacetime (see Case Study 2).

Driving multi-sectoral change is facilitated by wildlife health professionals becoming champions and being empowered to form and lead teams and participate effectively on governance structures such as inter-agency committees or working groups. This can help build up the sector to fully contribute to initiatives related to population health of all species.

Case Study 2: One Health Coordination to Advance Wildlife Health Leadership

Wildlife health leadership can leverage the "One Health" concept that recognizes the connections between human, animal, and environmental health and signifies a willingness and commitment to work together to solve problems effectively and efficiently for mutual benefit. For example, One Health platforms have been established that are chaired by a neutral representative, with shared or rotating leadership to facilitate genuine multisectoral engagement (Agbo et al. 2019). Such platforms have been mobilized for national action planning processes, surveillance and laboratory strengthening, data sharing, policy development, and disease-specific objectives. Some of these bodies also involve civil society representatives, such as media, providing a way to amplify calls to action and build societal support. These mechanisms may be relevant at national, subnational, regional, or global levels, but reinforce the importance of having a seat at the table and embedding wildlife health firmly into the operationalization of One Health.

In Cameroon, a newly adopted One Health strategy was rapidly tested in practice when an outbreak of monkeypox was reported in chimpanzees. The strategy allowed for mobilization of a multi-sectoral investigation team critical for assessing risks to wild animal and human populations. In this case, wildlife health expertise played a key role in the investigation, but sign-off was required by only one authority, lessening the administrative burden across government. This coordinated approach was estimated to save 10 days in response time and considerable resources compared to if each sector had proceeded independently (PREDICT Consortium 2016).

Increasingly, national and subnational One Health platforms are being established to facilitate genuine multisectoral engagement, and wildlife health managers have an important role to play in decision-making and implementation (Agbo et al. 2019).

Starting partnerships with open discussions about sectoral differences encourages equity, transparency, and mutual benefits three key requirements for maintaining effective partnerships (Sleeman et al. 2019). Transparency builds trust through open and consistent communication that includes clear context and guidelines for interactions, allowing for the sharing of viewpoints, and support for the resulting decisions. Equity ensures that each organization and individual in a partnership, regardless of financial or political capital, has "a seat at the table" and that the contributions are given equal weight to those of other partners. This is particularly important for organizations in the wildlife and environment sector, which are typically smaller and have less funding or political capital than those in the animal or human health sectors. Rubin et al. (2014) found that successful One Health partnerships shared and acknowledged a common threat or purpose, focused on science-based outcomes, had clear mandates, good governance, organization-level support, rotational leadership, defined roles and responsibilities, and a foundation of trust.

Wildlife health professionals may oversee a wide range of species and disciplines, requiring self-reflection (introspection) to be aware of and overcome biases, whether to aquatic or terrestrial species, infectious or toxicological issues, or about certain stakeholder groups (Showry and Manasa 2014). This objectivity is important when it comes to leading others, who may bring a range of expertise, priorities, skills, and competencies. Other key skills and competencies in leading others include forming and operating effectively in teams, conflict resolution, decision-making, negotiation, motivating others, setting goals and project management abilities, and active listening and communication (Table 1). Effective and sustainable partnerships should have a way to measure and celebrate the success, or at least the impacts, of implemented interventions as illustrated in Case Study 2. Population health success relies on the ability of different sectors to build effective partnerships that are focused on mutually beneficial solutions to shared health and environmental challenges as illustrated in Case Study 3.

Mutual edification and understanding are also important components of effective teams. Countries have recently brought together the public health, agriculture, and environmental sectors to jointly plan and prioritize national activities using a One Health approach. For example, the U.S. Centers for Disease Control and Prevention (CDC) has created a One Health Zoonotic Disease Prioritization Tool to help countries identify zoonotic diseases of greatest national concern using equal input from representatives of the human, animal, and environmental sectors (Salyer et al. 2017).

In summary, integrative leadership (leading across sectors) and governance structures are recognized as key components for successful coordination (Nyatanyi et al. 2017) and help create and maintain the cross-sectoral collaboration needed to tackle "wicked" problems, i.e., problems that lack clarity due to incomplete, contradictory, and changing requirements (Crosby and Bryson 2010) (see Chap. 17).

Case Study 3: Motivating Others

Motivating others toward the success of any initiative often requires instilling a sense of collective purpose. For example, the WHISPers (https://whispers.usgs.gov/home) database, managed by the U.S. Geological Survey National Wildlife Health Center, collects and displays verified wildlife disease information from governmental and nongovernmental entities providing summaries on a publicly accessible database that recognizes submitting organizations. The system has been created with stakeholder input from its initiation, through participatory methods such as partner focus groups, ensuring the outputs deliver on partner needs. This feedback loop incentivizes participation, even if benefits are indirect, and has resulted in an incremental increase in voluntary participation, despite the lack of mandates to report wildlife health events.

4 Leading Change

Four main elements can guide the process of leading change (Fig. 1): envisioning the future state, engaging key stakeholders and coalition building, identifying barriers and breaking down resistance to change, and institutionalizing change (Kotter 1996). Consideration of these steps can help to make the process of leading change intentional and have a solid strategy to prepare for and overcome roadblocks.

Given the previously referenced wicked and complex problems faced by wildlife health professionals, application of a systems thinking

Fig. 1 Major steps for leading change. While presented as a generally linear process, some steps may be iterative as expectations may be adjusted from ongoing ground-truthing and stakeholder engagement. Adapted from Kotter, 1996

approach to leading change may be beneficial. Systems approaches focus on the relationships between the parts of the whole, and the patterns of these relationships that are caused by the variability and interdependent nature of the individual components (Mahajan et al. 2019; White et al. 2020). The resulting framework allows one to examine both the deeper structural foundation of complex systems and understand how to influence the future behavior and outcomes of these systems. Systems approaches have rarely been applied in wildlife health challenges; however, case studies in conservation planning illustrate their value in encouraging collaboration and focusing solutions on root causes rather than the immediate concern (Mahajan et al. 2019). The process helps to identify the mindsets or beliefs that generate the current system. The process can then be used to generate new mental models or beliefs by promoting dialogue among all stakeholders, gaining a deeper understanding of the system complexities, and help create new solutions. For example, enhancing the transparent reporting of wildlife mortality events would not be improved by a focus on increasing technical knowledge, if there is, for example, a lack of political will or organizational capacity to conduct such investigations. Mapping this system would help to identify the underlying reasons for a lack of reporting and develop the most effective solution.

4.1 Envisioning the Future State

A primary step in leading change is a clear vision of the future. This desired future state can be generated from systems approaches and strategic planning to provide a road map for how to achieve the desired state. Visioning and strategic thinking can help to extend desired goals into other sectors to "mainstream" wildlife health—whether in conservation, public health, agriculture, or other sectors and processes. For example, building wildlife health roles or technical inputs into environmental protection plans and regulations (e.g., National Biodiversity Strategies and Action Plans), will help to promote new mindsets and approaches, and accountability for reaching shared goals. Participatory approaches to strategic planning are vital, and coordination opportunities may help establish trust essential to relationship building and lead to further opportunities for envisioning new future possibilities. Nonprofit organizations can play a particularly vital role in envisioning and advocating for change, and thereby help bring lasting change.

4.2 Engaging Stakeholders and Coalition Building

Stakeholder analysis can help target the partners needed to effect the change one seeks. Strategies for getting from the current state to the future state can benefit from a theory of change or logic models (see Chap. 3), in which expected inputs, outputs, and outcomes are articulated for reaching the desired state. These can serve as valuable roadmaps and ensure clarity in purpose and expectations across stakeholders. Driving change often involves diplomacy and humility to be open to feedback and adapt approaches as necessary, including from key constituents likely to be impacted from the change and other stakeholders whose support is needed for success. This commitment to continual learning and improvement is captured in concepts such as "Plan-do-check-act," whereby approaches are studied and refined for optimization (Solaimani et al. 2019). For twenty-first-century leaders, lessons learned from leading change—whether from successes or failures—are important to share with others to allow for synergies in progress and modification/upscaling of approaches. Lasting change for complex issues can also benefit from collective impact, i.e., agreement on the agenda and collaboration among multiple entities for solving common problems (Quick 2017).

4.3 Identifying Barriers and Breaking Down Resistance to Change

Given the fragmentation of authority for wildlife health, the "5-whys" technique, an iterative interrogative technique used to explore the cause-and-effect relationships underlying a particular problem, can help identify the root cause(s) of barriers, whether political will, authority, or budgetary (Solaimani et al. 2019). This problem-solving method successively asks the question "Why?" to explore the underlying cause-and-effect relationships to identify the root causes of particular problems.

Mobilizing external pressure can be a constructive way to elevate calls to action and foster champions for a cause. Governmental, intergovernmental, and nongovernmental stakeholder groups or individuals may have the resources to help effect change and lend additional credibility but may require sensibilization about the issue and targeted asks (Table 2).

Documenting the scale of the problem, while sometimes time-intensive, can help demonstrate the need for action while also establishing baselines important for wildlife health management. Quantifying the burden of wildlife disease is a challenge. Putting the challenge into terms stakeholders understand or value (whether negative from a problem, or positive from a solution) is important for gaining political support. Knowing the audience and what they may be receptive to help to ensure clarity of the message and shared understanding. For example, economic impact is one comparable metric across sectors that may help demonstrate return on investment. Wildlife health professionals, therefore, may want to develop fruitful collaborations with economic, epidemiological, and ecological modelers to demonstrate cost-effectiveness of proposed interventions. This can also help avoid potential perceptions about competition for resources, instead prioritizing coordination and shared understanding of how to optimize available resources for best outcomes.

Gaps in knowledge, or in the ability to apply knowledge, can also be a barrier to generating change (see Chap. 23). Knowledge mobilization is the process of creating value for stakeholders or a value stream through the creation, assimilation, leveraging, sharing, and application of focused knowledge to a community (Bennet et al. 2007). Knowledge mobilization brings knowledge, people, and action together to create this value, and provides a process for decision-makers to take an evidence-based approach. Stephen et al. (2021), in a review of evidence on managing the risk of disease emergence in the wildlife trade, used a knowledge mobilization approach to identify actions to overcome impediments to implementation of interventions in the wildlife trade to reduce emerging disease risks. In this example, identified knowledge gaps would result in research efforts to fill these gaps, whereas barriers to sharing and applying knowledge would be managed through knowledge brokering, and barriers to change would be addressed through the study and application of methods to enhance the use of new interventions in routine operations or practice, or implementation science.

4.4 Institutionalizing Change

This step focuses on implementing change. Long-lasting change requires the appropriate enabling environment, the necessary technical capacities, and the functional or organizational capacity to effectively deliver the program. This requires

Table 2 Examples of key stakeholder groups driving policy and practice that may be directly or indirectly relevant for wildlife health, by sector and scale. Each has their own leadership structure and decision-making processes

Scale	Sector			
Global	Animal health and agriculture	Environment	Human health	Other/multi-sectoral
Global and regional	Food and Agriculture Organization of the United Nations (FAO) World Organisation for Animal Health (OIE)	United Nations Environment Programme (UNEP) and multilateral environment agreements, e.g., Convention on Biological Diversity, Convention on International Trade in Endangered Species of Wild Fauna and Flora, Convention on Migratory Species, framework convention on climate change International Union for the Conservation of Nature (IUCN)	World Health Organization (WHO) Regional technical institutions (e.g., Africa CDC, Caribbean Public Health Agency)	Multilateral development banks (e.g., the World Bank, Inter-American Development Bank) United Nations Office for Disaster Risk Reduction (UNDRR) World Trade Organization (WTO) Regional associations and trade agreements (e.g., ASEAN, African Union)
National	Ministries of agriculture (livestock, fisheries, and veterinary services)	Ministries of environment and natural resources (climate, ecosystem, water, forestry, and wildlife services)	Ministries of health (medical and public health services)	National One Health Coordination Platforms Ministries of finance Disaster management agencies Parliamentary bodies
Local	Livestock extension and community animal health officer networks Farmers and farmer associations Laboratories	Rangers Biodiversity and protected area managers Indigenous populations and local communities Ecotourism operators and managers	Clinics Community health worker networks Patients Laboratories	District and town leaders Businesses Academic and research institutions The public (event reporting, passive surveillance, consumer preferences, behavior change)

focus on the practical implementation as well as the ways that the change can be maintained and built into an organization or process long term (vs. an ad hoc initiative). Attention to this step from the beginning can help to set realistic and meaningful targets about what success will look like. This step is important as changes may take time (and refinements) to yield intended results. Institutionalizing change can help to build in the organizational and supportive infrastructure (training, policy, governance, budgets, reporting requirements, etc.) for its success. This makes the desired change less likely to be abandoned if its value is not immediately clear or if a change in administration brings different awareness and priorities.

Successful leadership in wildlife population health will likely look different depending on the objectives, context, and ones' own interest. While championing and developing customized solutions may makc scnsc in some situations, considerations such as reach, uptake, and sustainment may be practical priorities. For the latter, there may be national or international organizations with existing infrastructure where enhancements could be incorporated cost-efficiently and leverage existing training, reporting, and event response or other assistance; this type of leadership may involve extensive technical or advocacy efforts behind the scenes and result in something "owned" by another party. Lastly, the utility of patience is a practical consideration for leading change in wildlife health. Grounding expectations in an understanding of the barriers, targeting the appropriate audience that can enact the necessary change, and expecting that a series of "wins" are likely to be needed for sustained progress will support realistic planning and implementation. The investments in time and energy are likely just as important—and sometimes more impactful—than short-term campaigns for broad changes (see Case Studies 4 and 5).

Case Studies 4 and 5: Examples of Leading Change

Examining system gaps

A major gap in public sector investment planning and prioritization architecture relates to the lack of a capacity assessment tool for wildlife and broader environmental health. Whereas public health and veterinary services have assessment tools (the World Health Organization's Joint External Evaluation and the World Organisation for Animal Health's Performance of Veterinary Services), there is no parallel globally adopted tool available to assess system gaps for wildlife health (Machalaba et al. 2018). As a result, potential needs for national wildlife health programs are not reviewed systematically, or worse, designated mandates, functions, and resources may be completely lacking for some or all wildlife health scopes. To address this gap, two tools have been developed to support practical exercises at the country level to identify deficits and major priority needs. The Country Assessment for Environmental Health Services, drafted by EcoHealth Alliance as an extension of the World Bank One Health Operational Framework, is intended to support countries in baseline capacity assessment and identification of key gaps as well as provide orientation and further reading. The Needs Assessment for National Wildlife Health Programs, developed by the U.S. Geological Survey National Wildlife Health Center, provides a pathway for assessing the current state of the national wildlife health program, helps define the future desired state, and identifies the programmatic gaps and needs in functions and capabilities. The needs assessment has been applied with partners in the Republic of Korea, Republic of Rwanda, and the Kingdom of Thailand.

Emergency movement of diagnostic samples to reference laboratories

Over a 3-year process, leaders of the International Union for the Conservation of Nature (IUCN) Species Survival Commission Wildlife Health Specialist Group worked to address the issue of delays in the movement of emergency diagnostic samples from species of conservation concern to international reference laboratories. The lack of access to comprehensive laboratory services is a major constraint for timely diagnosis of wildlife disease events in many countries, but standard regulations under the Convention on International Trade in Endangered Species of Wild Fauna and Flora (CITES) (intended to reduce endangerment threat from trade) can inadvertently hinder international movement of samples. Major delays have been observed, even during wildlife mass mortality events (Karesh et al. 2016). This issue reinforces the equity challenge for wildlife health given that human and domestic animal health emergency sample movement to reference laboratories is expedited as a central component of event investigation. Through awareness raising and coordination with the CITES Secretariat and delegates, simplified procedures were approved at the 18th Conference of the Parties in 2019. The approved

(continued)

procedures make recognition of the importance of emergency diagnostic samples from fauna and flora more explicit and provide a dedicated pathway for rapid movement. While there are still regulatory, political, and implementation barriers that affect consistency in the process, this example demonstrates how wildlife health leadership can be targeted to strengthen the effectiveness of global processes, including in the context of scope that unintentionally affects wildlife health.

5 Conclusions

Protecting and promoting wildlife health in the twenty-first century can benefit from effective leadership and technical expertise. Leadership skills offer opportunities to drive positive impacts and address challenges in this field in meaningful ways. Thus, wildlife health professionals may choose to embrace leadership opportunities. To date, academic articulation of leadership is poorly translated into day-to-day situations, and prior research found that publications vary widely in how they report on successes, challenges, and outcomes of collaboration, making comparability challenging (Baum et al. 2017; Errecaborde et al. 2019).

Collaborating with diverse stakeholders benefits from developing and championing approaches which in some cases may arrive at solutions that are not perfect but allow for necessary buy-in and progress. These realities also reinforce the importance of diverse stakeholders contributing to success, including addressing the full range of expected and unexpected implementation needs. In general, effective leadership will be supported by being open to new ideas, constantly questioning the effectiveness of one's approach, and a commitment to lifelong learning. Application of leadership skills can effect the change needed to implement these approaches to wildlife population health.

Acknowledgments Any use of trade, firm, or product names is for descriptive purposes only and does not imply endorsement by the U.S. Government.

References

Acevedo-Whitehouse K, Duffus AL (2009) Effects of environmental change on wildlife health. Philos Trans Royal Soc 364(1534):3429–3438

Agbo S, Gbaguidi L, Biliyar C, Sylla S, Fahnbulleh M, Dogba J et al (2019) Establishing national multisectoral coordination and collaboration mechanisms to prevent, detect, and respond to public health threats in Guinea, Liberia, and Sierra Leone 2016-2018. One Health Outlook 1:4

Baum SE, Machalaba C, Daszak P, Salerno RH, Karesh WB (2017) Evaluating one health: are we demonstrating effectiveness? One Health 3:5–10

Bennet A, Bennet D, Fafard K, Fonda M, Vaugeois N (2007) Knowledge mobilization in the social sciences and humanities. MQI Press, Frost

Berthe FCJ, Bouley T, Karesh WB, Le Gall FG, Machalaba C, Plante CA, Seifman RM (2018) Operational framework for strengthening human, animal and environmental public health systems at their interface. World Bank, Washington

Bleidorn W, Hill PL, Back MD, Denissen JJA, Hennecke M, Hopwood CJ, Jokela M, Kandler C, Lucas RE, Luhmann M, Orth U, Wagner J, Wrzus C, Zimmermann J, Roberts B (2019) The policy relevance of personality traits. Am Psychol 74(9):1056–1067

Carlin EP, Machalaba C, Berthe FCJ, Long KC, Karesh WB (2019) Building resilience to biothreats: an assessment of unmet core global health security needs. https://www.ecohealthalliance.org/wp-content/uploads/2019/04/Building-Resilience-to-Biothreats.pdf

Crosby BC, Bryson JM (2010) Integrative leadership and the creation and maintenance of cross-sector collaborations. Leadersh Q 21(2):211–230

Czabanowska K, Malho A, Schroder-Back P, Popa D, Burazeri G (2014) Do we develop public health leaders?- association between public health competencies and emotional intelligence: a cross-sectional study. BMC Med Educ 14:83

Errecaborde KM, Macy KW, Pekol A, Perez S, O'Brien MK, Allen I et al (2019) Factors that enable effective one health collaborations - a scoping review of the literature. PLoS One 14(12):e0224660

Fathke RL, Rao S, Salman M (2021) The COVID-19 pandemic: a time for veterinary leadership in one health. One Health 11:100193

Firth-Cozens J, Mowbray D (2001) Leadership and the quality of care. BMJ Qual Saf 10:3

Karesh WB, Kock R, Machalaba C (2020) It is time for a global wildlife health authority? IUCN Crossroads

Blog. https://www.iucn.org/crossroads-blog/202009/it-time-a-global-wildlife-health-authority
Karesh WB, Kock R, Machalaba CC (2016) CITES: in sickness and in health? Ecohealth 13(3):441–442
Kotter JP (1996) Leading change. Harvard Business School Press, Boston, MA
Machalaba CC, Salerno RH, Barton Behravesh C, Benigno S, Berthe FCJ, Chungong S et al (2018) Institutionalizing one health: from assessment to action. Health Sec 16(S1):S37–S43
Mahajan SL, Glew L, Rieder E, Ahmadia G, Darling E, Fox HE et al (2019) Systems thinking for planning and evaluating conservation interventions. Conserv Sci Pract 1(1):e44
Morgan G (2009) Reflective practice and self-awareness. Perspect Public Health 129(4):161–162
National Research Council (2013) Workforce needs in veterinary medicine. NRC, Washington
Nyatanyi T, Wilkes M, McDermott H, Nzietchueng S, Gafarasi I, Mudakikwa A et al (2017) Implementing one health as an integrated approach to health in Rwanda. BMJ Glob Health 2(1):e000121
OIE (2021) OIE wildlife health framework: 'protecting wildlife health to achieve One Health'. https://www.oie.int/app/uploads/2021/03/a-wildlifehealth-conceptnote.pdf
OIE and IUCN SSC Wildlife Health Specialist Group (2020) Guidelines for working with free-ranging wild mammals in the era of the COVID-19 pandemic. http://www.iucn-whsg.org/COVID-19GuidelinesForWildlifeResearchers
PREDICT Consortium (2016) One Health in action. https://www.ecohealthalliance.org/wp-content/uploads/2016/10/One-Healthin-Action-Case-Study-Booklet_24-October-2016.pdf
PREDICT Consortium (2018) Living safely with bats. PREDICT Consortium, Washington
Quick K (2017) Locating and building collective leadership and impact. Leadership 13(4):445–471
Rubin C, Dunham B, Sleeman J (2014) Making one health a reality—crossing bureaucratic boundaries. Microbiol Spectr 2:1
Salyer SJ, Silver R, Simone K, Barton BC (2017) Prioritizing zoonoses for Global Health Capacity Building-Themes from One Health zoonotic disease workshops in seven countries, 2014–2016. Emerg Infect Dis 23:13
Showry M, Manasa KVL (2014) Self-awareness—key to effective leadership. IUP J Soft Skills 3:1
Sleeman JM, Richgels KLD, White CL, Stephen C (2019) Integration of wildlife and environmental health into a one health approach. Rev Sci Tech 38(1):91–102
Solaimani S, Talab AH, van der Rhee B (2019) An integrative view on lean innovation management. J Bus Res 105:109–120
Stephen C (2014) Toward a modernized definition of wildlife health. J Wildl Dis 50(3):427–430
Stephen C, Berezowki J, Carmo LP, de las Nieves Montano Valle D, Friker B, Maximiano Alves de Sousa FM, Vidondo B (2021) A rapid review of evidence on managing the risk of disease emergence in the wildlife trade. OIE, Paris
Van Rekom J, Van Riel CBM, Wierenga B (2006) A methodology for assessing organizational core values. J Manag Stud 43(2):175–201
White CL, Lankton JS, Walsh DP, Sleeman JM, Stephen C (2020) An ecological and conservation perspective. In: Zinsstag J, Schelling E, Crump L, Whittaker M, Tanner M, Stephen C (eds) One health: the theory and practice of integrated health approaches, 2nd edn. CAB International, Oxfordshire, pp 25–38
World Health Organization and Convention on Biological Diversity (2015) Connecting global priorities: biodiversity and human health, a state of knowledge review. WHO, Geneva and Montreal

Bridging the Knowing-to-Doing Gap in Wildlife Population Health

Craig Stephen

Abstract

It is easier to know that something must be done that it is to effectively get it done. Simply having good evidence is not enough to ensure that decisions are made to effectively act to improve population health. Bridging the knowing-to-doing gap requires a collaborative approach between knowledge producers and knowledge users to identify the right questions to ask that will produce credible and trustworthy evidence that can be feasibly, acceptability, and sustainability implemented under the social and ecological condition of its intended application. Inattention to the individual, organizational, and systemic barriers and enablers to implementing knowledge will affect the impact of new evidence on shaping management decisions, regardless of how good the evidence. Systematic program evaluations are needed to understand both the processes and knowledge that are needed to help people translate evidence into action. Ongoing monitoring and evaluation support adaptive management that allows us to continually learn while doing rather than waiting for certainty before acting. The growing pressures on wildlife and ever-growing competition for management resources demand a learning-based approach to wildlife health management.

C. Stephen (✉)
School of Population and Public Health, University of British Columbia, Vancouver, BC, Canada

Keywords

Evaluation · Implementation · Knowing-to-doing gap · Adaptive management · Wildlife

1 Introduction

The challenges that wildlife face today differ from the past because the scale of human influence has increased, the most significant threats are global in nature, and problems are aggregating and compounding with each other. This is being manifested in emerging diseases, changing patterns of endemic disease, new exposures to a growing array of toxins, and altered access to the needs for daily living. There is growing political and scientific urgency to find new ways to manage wildlife health that account for the complex interactions that affect health outcomes (see Chap. 1). Wildlife health research is increasingly expected to advance knowledge that will eventually be translated into improved population health management decisions. It is often clear that something must be done, but less clear on how to get it done.

It is a quite different task to accumulate evidence on the presence of disease, its spread, and the harms it causes than on effective ways to prevent and mitigate those harms under real-world conditions. The effectiveness, acceptability, and sustainability of management interventions are sensitive to the circumstances

C. Stephen (ed.), *Wildlife Population Health*,
https://doi.org/10.1007/978-3-030-90510-1_23

in which the interventions are conceived, developed, implemented, and evaluated. Population health managers need to be able to discern what works for whom, in what circumstances, in what respects, and how it can be done in an effective and efficient manner to successfully translate their aspirations for an intervention into real change.

Researchers in the past, and to a large extent still, are rewarded for making a scientific breakthrough. But the severity and pace of adverse changes affecting wildlife are driving society to increasingly expect researchers to produce follow-through. The gap between knowing what to do and making the desired changes happen is known as the implementation gap (or the knowing-to-doing gap). This gap can be very wide in wildlife health (Stephen et al. 2018; Stephen and Wade 2020; Stephen 2021). The wildlife health knowledge base is comprised largely of research oriented toward "knowing" rather than "doing." Research aimed at understanding how to apply knowledge is scant. Similar conclusions have been reached in wildlife conservation, invasive species management, and public health (McAteer et al. 2019; Braunisch et al. 2012; Esler et al. 2010).

2 What Helps Turn Knowledge into Action?

It was once assumed that policymaking and decision-making were linear-rational processes, moving directly from problem identification to problem solution, and that we could simply close the implementation gap by increasing data availability and accessibility. It is now well understood that this linear conceptualization does not reflect the real world (Gluckman et al. 2021). The situation is much more complex. A wide variety of individual and organizational factors affect the translation of research into action (Bowen et al. 2009). Workload, type of evidence presented, alignment with organizational agendas, politicized decision-making, leadership support, involvement of knowledge users in the process, agreement on what types of knowledge is needed, resource availability, trust in the knowledge producers and more can all affect the flow of knowledge into action. New scientific discoveries cannot improve population health if we also do not know how to apply these discoveries in effective, acceptable, and feasible ways or understand how changes in the ecological, epidemiological, and social contexts affect our ability to translate a discovery into actions. Translating knowledge into action is, therefore, not a single activity but instead a process dealing with both how knowledge is generated and the decision-making context and processes within which it is applied.

Figure 1 outlines a series of questions to ask when looking for obstacles or opportunities on the path from knowledge production to its implementation for evidence-based actions. To start down this path, it is important to ask, "what types of knowledge do we need to implement?" As discussed in more detail in Chap. 8, simply giving people larger amounts of increasingly precise scientific information will not, on its own, lead to more effective outcomes in the "real world." Successful implementation of knowledge into practice has as much to do with the context where the new knowledge has been introduced and how it was introduced as it has to do with the quality of the evidence (Kitson et al. 1998). Lack of experience being impacted by a problem, lack of trust in authorities, and exposure to false alarms reduce willingness to act on new information (Stephen 2020). Failure to act on new knowledge can also be caused by (1) cross-sectoral communication breakdown, bureaucratic conflicts, and inadequate protocols that make organizations unreceptive to information generated outside of their usual scope of practice; (2) priorities and overcrowded agendas that discourage bringing in new information; and (3) insensitivity to new information, and perceived power dynamics that lead to failure to recognize the value of new knowledge. Evidence-based decisions need various types of evidence that are appropriately used and are relevant to the local context (Bowen et al. 2009). Understanding the relationship between interventions and contexts is critical to understanding implementation success, how

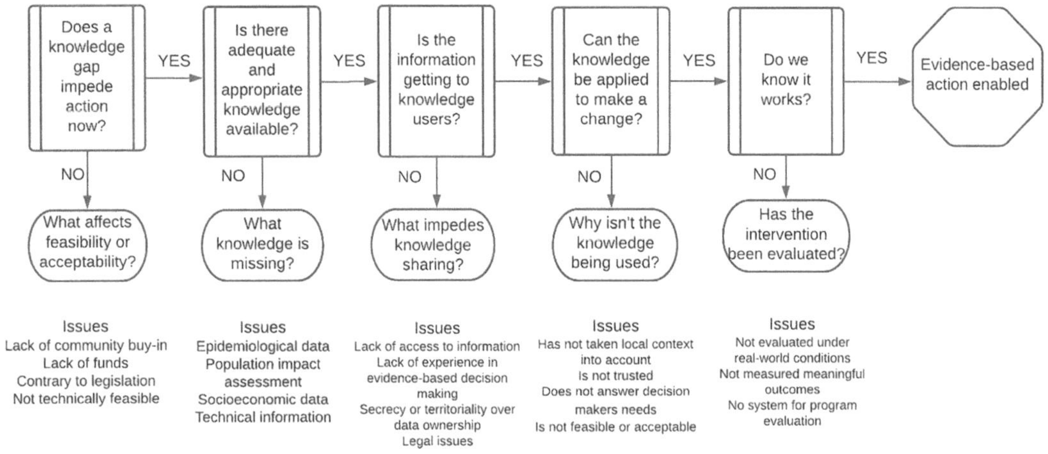

Fig. 1 A set of guiding questions to help bridge the knowing-to-doing gap including illustrative issues causing potential obstacles to translating knowledge into action

interventions achieve impact, why their impacts vary, and whether interventions can be sustained (Craig et al. 2018).

If the appropriate breadth and depth of knowledge are available but knowledge is not being implemented, one needs to look at individual, organizational, and systemic factors that are barriers to action. A program evaluation process (see Sect. 3) can help identify barriers to action in the face of sufficient information. Sometimes, it is not economically or technically feasible to implement an action. In other cases, the social costs or unintended consequences of an action may be unacceptable. The actions might be logical but contravene social conventions, ethics, or laws. In other cases, the knowledge may not be timely, trusted, believable, or developed via a process that considers the values and perspectives of all relevant actors (Cook et al. 2013). Innovations in governance, regulations, technology, community engagement, or collaboration may be needed to overcome barriers at this stage of the knowledge-to-action pathway.

While having new knowledge is not sufficient to make a change, the absence of information to support and inspire decisions will be an impediment to action. Research that effectively and efficiently leads to action is usually borne of collaboration between the knowledge producers and knowledge users. Knowledge users have unique expertise, including knowledge of the context and the potential for implementation, that can help knowledge producers understand what information the users need to build confidence and capacity to act, and in so doing, help shape new research that is more likely to be implemented. Knowledge producers bring methodological and content expertise to the collaboration that can help the knowledge users see how new information can affect their decisions and help users apply their knowledge appropriately and credibly because of their understanding of how it was generated.

Once the evidence is in place and the context is understood, the next step to bridging the knowing-to-doing gap is facilitating knowledge use. Knowledge brokering is the active process of facilitating access and use of new evidence by building knowledge and skills, shifting the culture to evidence-based decision-making and developing infrastructure and mechanisms that support it (Dobbins et al. 2018). The focus of a knowledge broker is on the interactive process between the producers and users of knowledge so that they can coproduce feasible and research-informed intervention options (Van Kammen et al. 2006). Gluckman et al. (2021) provide ten recommendations for effective knowledge brokering (Table 1).

Knowledge producers and knowledge users need to be engaged for the entire process, from identifying the research question to applying the

Table 1 Core principles of effective knowledge brokering (based on Gluckman et al. 2021)

Principle	Details and guiding questions
Consider the demand side	Has anyone asked for help in acquiring new evidence? Are the knowledge users receptive to receiving unsolicited new knowledge? Is the new information too early or too late in the decision-making process? Can a knowledge broker help show the need, place, and timing for presenting new evidence?
Recognize the purpose and evidence needs	Why is someone asking for new information? Are people looking to understand an issue, identify options for action or evaluate impacts? By when is the information needed?
Get the right question	Do researchers and knowledge users agree on what needs to be known to support decision-making? Will the questions being asked help "unfreeze" inaction or help people confidently decide? Is the question being asked answerable given the state of science and the available disciplines, resources, and timelines to answer the question?
Assess the evidence base	What types of social and natural science knowledge are needed? What are the availability, quality, and quantity of existing evidence?
Assess the evidence gaps	Is the available evidence sufficient to make a particular claim despite prevailing unknowns and uncertainties? What is the critical knowledge gap preventing actions or decisions and to whom are they critical?
Communicate the uncertainties, limits, and reliability of evidence	Does the evidence suit the decision-making context? Were the means to produce the evidence reliable and of acceptable quality? What are the implications of the uncertainties?
Identify the gap between available knowledge and conclusions	Is there sufficient knowledge to make a confident conclusion? How do the expectations and timelines for certainty differ for the decision-makers and knowledge producers?
Evaluate the level of consensus	How will the involved parties agree on how to agree if there is no consensus on what the evidence is telling us?
Communicate the trade-offs and nonscientific implications	What are the implications of unintended spillover effects or trade-offs of decisions or actions? How does this affect the types of knowledge and certainties needed?
Present the evidence in a useful and understandable way	Are the format, media, and language being used to present the information tailored to the knowledge users? Can users access the evidence and know how it can be used? Is the knowledge source trusted?

knowledge. The interactions between researchers and knowledge users will vary in intensity and level of engagement depending on the nature of the research results and on the needs of the knowledge user, but these interactions are the key to closing the implementation gap as they build understanding, relevance, credibility, and trust. The bridge over the knowing-to-doing gap is not one-way. Table 2 provides some tips and topics to keep in mind when planning on how to close the knowing-to-doing gap.

3 Program Evaluation

There are many reasons why it is hard to prove wildlife health interventions work. The lack of systematic evaluation of intervention impacts is perhaps the most important. Study design limitations (e.g., researchers use no control, use historical controls, rely on ecological design [in the epidemiological sense]), assumptions that lack of negative events is proof of effectiveness, lack of sufficient time to follow-up to establish medium-to-long-term benefits, using surrogates of risk or benefits rather than directly measuring the impacts on health outcomes, using socially irrelevant end-points, and/or unaccounted confounding variables can also limit the reliability of proclamations that interventions "worked." To make things even more challenging, responses to interventions tackling twenty-first-century health threats are often unpredictable and emergent rather than predictable and planned because

Table 2 Tips and topics to help accelerate the mobilization of knowledge into action. Adapted for wildlife health from the Promoting Action on Research Implementation in Health Service framework (Kitson et al. 1998)

Evidence	Context	Facilitation
Evidence and information provided must be credible, relevant, and trusted	The balance between risks and benefits, losses, and gain from the status quo versus a new action must be understood	The process for moving knowledge into action must be strategically and cooperatively planned by knowledge users and knowledge producers
A wide suite of knowledge is needed including research evidence, local information and experience, prevailing values, and capacity for change	Individual, organizational, and systemic enablers and barriers to implementing knowledge must be managed	Criteria for establishing if an intervention works must be negotiated

Table 3 Standards for program evaluation

Standard	Guiding question
Utility	Do the stakeholders find the evaluation processes and outputs valuable in meeting their needs?
Feasibility	Is the evaluation affordable, do-able, and acceptable to participants and stakeholders?
Propriety	Was the evaluation fair, ethical, legal, and appropriate to the cultural context?
Accuracy	Will the methods and means of evaluation produce adequate and unbiased information about the program or intervention being evaluated?

of the unprecedented rate of social and environmental change and the complexity of interactions between co-occurring threats (Hanlon and Carlisle 2008). Despite these challenges, if we do not systematically evaluate the effectiveness, efficiency, feasibility, acceptability, and sustainability of intervention under the conditions in which they are needed, we will be unable to make evidence-based management recommendations. Instead, we will have to rely on analogy, inference, and limited personal experience to make recommendations.

At its core, a program (or intervention) evaluation simply asks why a program is or isn't meeting its objectives through the originally intended process. Program evaluations are not meant to simply declare if interventions were good or bad. They are learning exercises that systematically investigate the merit, worth, or significance of a program or intervention to improve program effectiveness and inform decisions about the future of the program development. To serve this purpose, program evaluations need to measure more that standard wildlife health outcomes such as morbidity, mortality, or infection rates. They also need to look at the operations of a program, how resources were used, which activities took place and by whom, how variations in context varied adherence to protocols, and how other health and social outcomes changed.

Program evaluations are systemic processes that adhere to four standards (Table 3). Before starting an evaluation, make sure you fully understand the program or intervention being evaluated and who or what will be affected by the interventions/programs. You should then ask (1) what will be evaluated, (2) what criteria will be used to judge success or failure, (3) what types of evidence will be used to measure those criteria, (4) what will be the threshold of evidence that will indicate success or failure, and (5) how will judgments be made about how to integrate or interpret different measured outcomes? Answering these questions may take negotiation and compromises as not all stakeholders will view success the same way. A politician, subsistence hunter, wildlife manager, and an anti-hunting lobbyist may all be concerned about duck conservation, but they might not all judge success the same way. A strong foundational knowledge of the intent of the intervention can help clarify the goals, populations, strategies, activities, outputs, and outcomes that need to be considered in the evaluation. This must be accompanied with an understanding of the interests and expectations of people affected by the interventions and a process to engage them in developing the evaluation questions. Just like in efforts to bridge the knowing-to-doing gap, program evaluations should involve stakeholders in the planning, execution, and interpretation of the evaluation.

Evaluations should assess both the impacts or outcomes that resulted and the processes used to implement the programs. Six questions can help guide an evaluation (Spiegelman 2016; Glascow et al. 2019).

1. Who is willing to participate in an initiative, intervention, or program, and reasons why or why not?
2. To what extent can the intervention be effectively integrated within real-world systems?
3. What are the efficacy (doing things in an optimal way) and effectiveness (producing the desired results) of an intervention in terms of intended and unintended outcomes and their variability across subgroups?
4. Have the processes of the program been evaluated to see if they work and can be improved?
5. How do the impacts on conservation or social outcomes vary with the circumstances under which interventions are applied?
6. Can the program or policy become institutionalized or part of the routine practices and policies?

4 Adaptive Management

Adaptive management is type of learning-based decision making "intended to increase the ability to fashion timely responses in the face of new information and in a setting of varied stakeholder objectives and preferences" (NRC 2004). It is based on the simple premise that there will always be uncertainty and unpredictability in the response of complex social–ecological systems to interventions, but management decisions must still be made, so whenever possible, we should incorporate learning into management (Allen et al. 2011). It is critical for managing biological systems in the presence of uncertainty (Westgate et al. 2013).

Adaptive management combines management actions concurrently with monitoring to characterize the full extent of a problem, anticipate the possible consequences of responses throughout the system being managed, evaluate impacts of interventions, and incorporate lessons learned into future decisions (Ebi 2011). It aims to reduce uncertainty and close the knowing-to-doing gap while managing a problem.

Adaptive management is an approach to learning-by-doing wherein information generated from trying an intervention is used to inform how to modify the intervention to achieve desired outcomes. Learning through management can be achieved in three main ways: (1) interventions can be tried concurrently, evaluating one against another; (2) interventions can be tried in a stepwise fashion wherein one intervention is tried and if it fails a different intervention is launched and monitored; and (3) the same intervention is modified as monitoring and evaluation of its processes and effects are assessed while management is underway (Allen et al. 2011).

Adaptive management can foster resilience and flexibility to cope with an uncertain future by helping managers learn by doing. It does not work in all situations, needs resources and expertise to allow appropriate monitoring and assessment, needs clear stopping rules that must be acted upon if adverse impacts are detected, and needs the engagement of stakeholders to frame assessments that can recognize an agreed-to vision for success. Adaptive management's value is that it prevents us from avoiding action in the face of unknowns and, by learning as we go, continually moves actions to safer and more successful interventions.

5 Summary

The case for implementation science and program evaluation in wildlife health is clear: As management systems work under increasingly dynamic and resource-constrained conditions, and the pressures on wildlife become more widespread, regular, and intense, evidence-based strategies are essential to ensure that research investments are rapidly, reliably, and effectively translated to improve wildlife health. Wildlife population health practitioners need to access and understand the evidence available to them to make changes as well as know how the social and ecological contexts for change create obstacles and opportunities to bridge the knowing-to-doing gap in a timely manner.

References

Allen CR, Fontaine JJ, Pope KL, Garmestani AS (2011) Adaptive management for a turbulent future. J Environ Manage 92(5):1339–1345

Bowen S, Erickson T, Martens PJ, Crockett S (2009) More than "using research": the real challenges in promoting evidence-informed decision-making. Health Policy 4(3):87

Braunisch V, Home R, Pellet J, Arlettaz R (2012 Sep) Conservation science relevant to action: a research agenda identified and prioritized by practitioners. Biol Conserv 1(153):201–210

Cook CN, Mascia MB, Schwartz MW, Possingham HP, Fuller RA (2013) Achieving conservation science that bridges the knowledge–action boundary. Conserv Biol 27(4):669–678

Craig P, Di Ruggiero E, Frolich KL, Mykhalovskiy E, White M, Campbell R, Cummins S, Edwards N, Hunt K, Kee F, Loppie C (2018) Introduction. In: Taking account of context in population health intervention research: guidance for producers, users and funders of research. NIHR, Southampton

Dobbins M, Traynor RL, Workentine S, Yousefi-Nooraie-R, Yost J (2018) Impact of an organization-wide knowledge translation strategy to support evidence-informed public health decision making. BMC Public Health 18(1):1412

Ebi K (2011) Climate change and health risks: assessing and responding to them through 'adaptive management'. Health Aff 30(5):924–930

Esler KJ, Prozesky H, Sharma GP, McGeoch M (2010) How wide is the "knowing-doing" gap in invasion biology? Biol Invasions 12:4065–4075

Glasgow RE, Harden SM, Gaglio B, Rabin B, Smith ML, Porter GC, Ory MG, Estabrooks PA (2019) RE-AIM planning and evaluation framework: adapting to new science and practice with a 20-year review. Front Public Health 29(7):64

Gluckman PD, Bardsley A, Kaiser M (2021) Brokerage at the science–policy interface: from conceptual framework to practical guidance. Human Soc Sci Commun 8(1):1

Hanlon P, Carlisle S (2008) Do we face a third revolution in human history? If so, how will public health respond? J Public Health 30(4):355–361

Kitson A, Harvey G, McCormack B (1998) Enabling the implementation of evidence based practice: a conceptual framework. BMJ Qual Saf 7(3):149–158

McAteer J, Di Ruggiero E, Fraser A, Frank JW (2019) Bridging the academic and practice/policy gap in public health: perspectives from Scotland and Canada. J Public Health 41(3):632–637

NRC (National Research Council) (2004) Adaptive management for water resources project planning. National Academies Press, Washington

Spiegelman D (2016) Evaluating public health interventions: 1. Examples, definitions, and a personal note. Am J Public Health 106(1):70–73

Stephen C (2020) The call to action. In: Animals, health, and society: health promotion, harm reduction, and health equity in a one health world. CRC Press, London

Stephen C, Carmo LP, de las Nieves Montano Valle D, Friker B, Sousa FM, Vidondo B, Berezowski J (2021) The implementation gap in emerging disease risk management in the wildlife trade

Stephen C, Wade J (2020 Sep) Missing in action: sustainable climate change adaptation evidence for animal health. Can Vet J 61(9):966

Stephen C, Wittrock J, Wade J (2018) Using a harm reduction approach in an environmental case study of fish and wildlife health. Ecohealth 15(1):4–7

Van Kammen J, de Savigny D, Sewankambo N (2006) Using knowledge brokering to promote evidence-based policy-making: the need for support structures. Bull World Health Organ 84:608–612

Westgate MJ, Likens GE, Lindenmayer DB (2013) Adaptive management of biological systems: a review. Biol Conserv 1(158):128–139

Preparing for the Unexpected

Craig Stephen

Abstract

The twenty-first century will be characterized by more frequent and more impactful surprises. Adopting a surprise-oriented approach will require concurrent shifts in perspective that support a systems-based approach to protecting assets to be resilient to surprise while building the necessary situational awareness and intelligence through partnerships for early warning. A surprise-oriented wildlife health program will need a multifaceted approach that aims to (1) identify and address preexisting social, ecological, and health circumstances that are conducive to disease or threat emergence and increase vulnerability to their effects; (2) proactively build capacity for individuals and populations to cope with multiple interacting threats and stressors to build resilience against the additional pressure of an unexpected health threat; (3) develop capacity to adaptively respond to surges in unexpected problems; and (4) integrate surprise activities into routine health management. As it will not be possible to be ready for all unexpected events, collaborative, multi-solving solutions that look across a spectrum of potential threats to find root causes of vulnerability and resilience may be an effective strategy for Anthropocene preparedness.

C. Stephen (✉)
School of Population and Public Health, University of British Columbia, Vancouver, BC, Canada

Keywords

Surprise · Preparedness · Complex systems · Emerging · Multi-solving · Situational awareness · Vulnerability · Resilience · Wildlife health

1 Introduction

The challenges that face wildlife health managers today differ from the past. The scale of human influence has increased, the most significant threats are global in nature, and problems are aggregating and compounding with each other. Actions to reduce unexpected wildlife health risks will need to address threats that change over time, locations, species, and pathogens, with high degrees of uncertainty as to the rate and magnitude of changes in response to interventions.

Natural and human-induced hazards have become more prevalent and are expected to cause increasing impacts into the foreseeable future (Anon 2017). The rapid pace of change and its associated prediction of more frequent, impactful, and unpredictable wildlife diseases are requiring governments to reevaluate their strategies to prevent, mitigate, and recover from these diseases that can affect conservation, economies, and public health. The COVID-19 pandemic was a stark reminder that we cannot predict how the future will unfold. But so too

C. Stephen (ed.), *Wildlife Population Health*,
https://doi.org/10.1007/978-3-030-90510-1_24

were global amphibian chytridiomycosis, massive wildlife losses due to Australian wildfires, and the North American epidemic of white-nose syndrome in bats. While eminently understandable with hindsight, we often lack the foresight to anticipate such events in time to curtail or prevent their impacts and spread. Many surprises are rarely a "bolt out of the blue," but rather are events whose warning signals were not tracked, recognized, or appropriately interpreted. Other surprises, such as chronic wasting disease or deformities due to environmental pollutants, were inconceivable before we knew about their presence, persistence, and effects.

Wildlife health workers now operate in a globalized, interconnected, and rapidly changing world replete with "foreseeable unexpected events." They are foreseeable in that our experiences and research reveal the possibility of new emerging diseases, health impacts of climate change, changing pollution dynamics, and more. They are unexpected because we can rarely predict their timing, location, and impacts with accuracy. Therefore, the number of times the conditions conducive to epidemics exist without an epidemic occurring far outnumber the few occasions when one occurs (Stephen et al. 2004).

Despite promises for scientific and technological means to improve our ability to predict impending threats, wildlife health managers need to be prepared for surprise because there is presently not enough predictive power to accurately forecast disease outcomes resulting from the environmental and social changes anticipate in the twenty-first century (Whitmee et al. 2015). When we do not prepare for surprise or wait too long to implement measures to deal with it, the social and ecological costs can be remarkably high. Although predicting future shocks in complex dynamic wildlife health systems may currently be rarely possible, being attentive to what might happen helps manager identify vulnerabilities in advance of harm and consider opportunities to make populations more resilient and thus better able to ward off future surprises.

2 The Origins of Surprise: A Complex Systems Perspective

There is something "comfortable" about planning population health programs by seeing events and the contexts in which they occur as being clearly compartmentalized and seeing cause–effect relationships as stable and linear. But the health and ecological sciences have long recognized that health and resilience are products of complex, dynamic, social, and ecological interactions (see Chap. 1) and that surprise is an expected feature of complex, dynamic systems (Stephen et al. 2020).

Much wildlife health research historically assumed that reality is the sum of components parts that can be separated and studied as isolated entities. This reductionistic approach has been extremely successful in combating diseases caused by single elements that can be remedied by targeting that element alone (e.g., using antibiotics to treat an infection or using rabies mass vaccination for outbreak control). Chronic diseases challenged this reductionist thinking. Genetic, environmental, and social factors were seen to interact in complicated ways over varying time scales to make a chronic disease, like cancer, less likely to be prevented or controlled by attacking only one element. The potential interaction of immunotoxic effects of persistent organic pollutants and marine morbillivirus (Duignan et al. 2014) or the complicated interaction of food web dynamics, habitat change, and Lyme disease in wildlife reservoir hosts (Wood and Lafferty 2013) illustrate how diseases rarely are unidimensional, isolated outcomes. Things got further complicated as our attention turned from disease to health, where health is characterized as the cumulative effect of capacities and resources derived from interacting individual, social, and environmental determinants necessary to adapt to, respond to, or control life's challenges and changes. The potential for additive and multiplicative interactions and synergistic and antagonistic relationships between determinants make it hard to accept that health is the product of simple, linear interactions. To make matters even more complicated, One Health now asks us to consider interactions between different types of health for interacting species and over multiple generations that each change over time, as do the hazards and harms they experience.

Health outcomes are multi-scale phenomena affected by relationships and interactions at small, individual, population, social, and ecosystem scales and by the influences and feedbacks across and between these various scales which change over time (Fig. 1). The scale at which one studies a health or disease system will affect the perspective one has of the system. This means that, when viewed from different perspectives or disciplines at different times and scales, the "same" health phenomenon can be described differently. A comprehensive understanding of a health outcome requires us to look at multiple variables interacting across all levels and across different spatiotemporal scales. This seems an overwhelming task; one to which people are increasingly evoking complexity theory as help.

Complex dynamic systems are built up from large numbers of mutually interacting subunits whose repeated interactions and feedbacks result in collective behavior that in return affect the individual parts of the system. Unlike simple linear systems, a complex system cannot be understood by extrapolating the behavior of the individual parts. In complex systems, one level of organization can determine the level above it, and that level then determines the features of the level above it. The subunits of each level have multiple and changing ways they can interact within and between levels. The long and sometimes convoluted causal chains between upstream and downstream animal, human, and environmental determinants of wildlife health create intricate networks of interactions between the many parts of a system making it hard to foresee new relationships or emerging outcomes. The

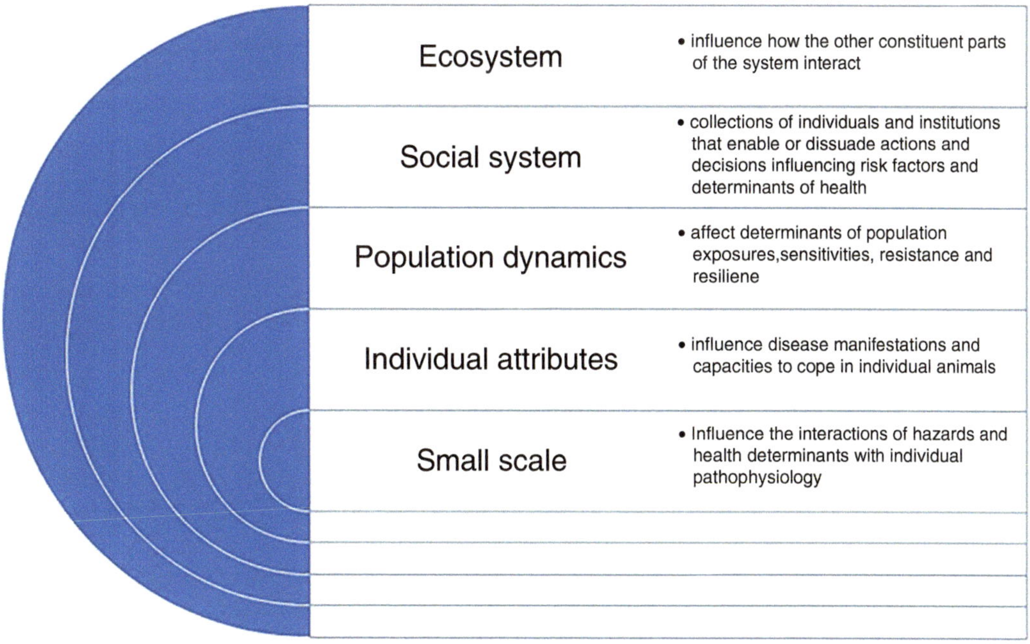

Fig. 1 A simplified wildlife disease model in which outcomes are embedded in an interacting hierarchy of multiple scales of influences, interactions, and relationships

manifestation of a health outcome is not only scale dependent but also is affected by human actions that can impact the likelihood or consequences of an outcome.

3 The Origins of Surprise: An Operational Perspective

If you accept the socio-ecological model of wildlife population health outlined in Chap. 1 and you accept that the drivers of social and ecological change are globalizing and accelerating you need to prepare for surprising wildlife health events and outcomes. Surprise generally comes from two sources: lack of sufficient information or knowledge and the basic dynamics of complex adaptive systems (Gross 2019). Operational surprises arise when there are failures in the links between policy, intelligence, warning, and response (Parker and Sterne 2002). Individual and organizational attributes can predispose us to being surprised (Galaz et al. 2011). Organization can be unreceptive to warning signals outside of their usual scope of practice when cross-sectoral communication breaks down, and when bureaucratic conflicts and inadequate protocols preclude information sharing. Priorities and overcrowded agendas may discourage collaborative actions that extend beyond immediate interests and thus reduce opportunities for novel information to be brought into decision-making processes. This is compounded when there is no investment in developing partnerships to understand or influence socio-ecological causes of wildlife health. Insensitivity to new information and perceived power dynamics can lead to failure to recognize and act on early warning signals. People can become desensitized to warning messages that are too frequent, too many, or too disconnected with their lived experiences or when unaccompanied with solutions.

Warning occurs when a change in risk or vulnerability status is revealed and that change is rapidly communicated to those able to respond (Yamin et al. 2005). Unless warning turns into action, it is not an effective warning. There are five prerequisites for actions to follow from warnings. There must be (1) awareness that a problem exists; (2) a sense that the problem matters; (3) understanding of the causes; (4) the capability to intervene or influence outcomes; and (5) the political will to deal with the problem. Awareness of wildlife emerging disease often arises postemergence (e.g., wildlife surveillance has most often been used to track pathogens and pollutants after they are seen to cause harm). Early on, economic barriers often result in insufficient numbers of samples being submitted to diagnostic laboratories to estimate prevalence, do trend analysis, or conduct risk assessments needed to issue warnings. If the emerged threat is known, actions may follow, but less so for newly emerged hazards. The sense that the problem matters usually only arises when either non-wildlife social values are at risk, large-scale wildlife mortality events occur, or highly endangered species are impacted. Rarely do wildlife surveillance programs have explicit thresholds that signal the need for action nor are they explicitly linked to management interventions. Wildlife often lacks a political voice or private sector advocate to make its health a priority comparable to the health of people or livestock. Thus, these five prerequisites for action are challenging to fulfill in many wildlife health management scenarios.

4 A Taxonomy of Surprise for Wildlife Health

There are four broad types of surprise (Betts 1980; Kates and Clark 1996; Cunha et al. 2006). First, we may be unaware of an event or consequence until it becomes severe or affects a population of special interest. This is frequently the case for emerging wildlife diseases. These surprises are knowable in retrospect but elude detection because of lack of surveillance or interest in a place or population. An example might be a fish die-off in a noncommercial species in a remote body of water that is not subject to surveillance. Second are surprises that come from failure to recognize an actionable signal or respond to that signal despite ample warning.

An example of this type of surprise could be when a public health agency fails to see the detection of a novel zoonotic pathogen in wildlife in the absence of human illness as an early warning sign. Responses to these first two categories of surprise have focused on connecting specialized pools of knowledge and improving access to the larger network of information and expertise.

Unanticipated consequences of socio-ecological interactions are the third category of surprise. These surprises are conceivable in retrospect once additional investigation reveals connections that drove an emergence that was not previously anticipated. For example, shipment of garbage between jurisdictions for disposal unintentionally spread rabid racoons, allowing an epidemic to jump between locations rather than slowly spread (Chipman et al. 2008). The fourth type of surprise is new, previously inconceivable events. Prion diseases like bovine spongiform encephalopathy and chronic wasting diseases are examples. These latter two surprise types arise due to uncertain, ambiguous, or unanticipated circumstances. Strategies for these surprises focus on building population health and resilience against the unforeseen.

5 Managing the Unexpected

There will always be inherent uncertainty and unpredictability in the dynamics and behavior of wildlife health, but management decisions must still be made. There are many methods to try to predict new disease patterns, but all can only act as "the art of the possible" rather than as predictive tools to direct specific risk management actions in specific locations and at specific times. Significant knowledge gaps for even the most studied wildlife diseases along with a simplistic view of diseases disconnected from their social and ecological context reduce the likelihood that we will be able to predict new disease or health threats with accuracy or regularity in the foreseeable future.

The taxonomy of surprise outlined above provides a convenient way to think about how to manage for surprise (Stephen et al. 2015) (Fig. 2). For surprises that elude detection, are not recognized, or are not acted upon, we need to improve our situational awareness so that managers can understand if the circumstance is conducive to a new problem. Situational awareness can be enhanced by better information sharing to detect early warning signals or by including vulnerability surveillance in population monitoring programs (see Chap. 9). The latter two types of surprises are not conducive to enhanced surveillance or situational awareness because managers do not know what to target for in unanticipated or inconceivable events. For these surprises, the best strategy is to invest in making populations healthy so that they can resist the initial event and be resilient enough to recover from the event.

The traditional approach to preparing for surprising and emerging threats has presumed that health is protected by solving problems and

Fig. 2 Categories of surprise and strategies for their management

Category of surprise	Question	Strategy
Knowable in retrospect but **elude detection** **Fail to recognize** actionable signal or not able to **respond** despite warning	**Is a harm possible under these conditions?**	**Broader situational awareness for earlier response or proactive vulnerability reduction**
Unanticipated consequences of socio-ecological interactions Previously **inconceivable** events	**Can the community deal with something unexpected?**	**Building resistance and resilience to cope the next inevitable surprise**

by removing deficits and obstacles. A deficits approach focuses on threats and shortcomings (such as disease, pollution, or habitat degradation). Disaster management, climate change preparation, and pandemic planning are increasingly recognizing that, to deal with a surprising and unpredictable future, communities, populations, and ecosystems must have the assets needed to cope with what may come (Stephen 2020). While there is an urgent need to deal with deficits and depletion driving threat emergence, there is also a need to promote resilience, adaptability, and well-being. An assets-based approach is concerned with identifying and sustaining the shared protective factors that support health. This is a critical approach to dealing with unanticipated and inconceivable surprises.

Adopting a surprise-oriented approach will require concurrent shifts in perspective that support a systems-based approach to protecting assets to be resilient to surprise while also building intelligence through partnerships for early warning. Building a systems-based approach for the unexpected can seem overwhelming due to the breadth of partnerships, capacities, and information required. However, population health program designers can turn to literature on complex systems safety management to identify core principles that act as entry points from which to build their programs (Table 1).

A surprise-oriented health management system will need a multifaceted approach that aims to (1) identify and address preexisting socioeconomic, ecological, and health circumstances that are conducive to disease or threat emergence and increase vulnerability to their effects; (2) proactively build capacity for individuals and populations to cope with multiple interacting threats and stressors to build resilience against the additional pressure of an unexpected health threat; (3) develop a capacity to adaptively respond to surges in unexpected problems; and (4) integrate surprise activities into routine health management (Stephen and Soos 2021).

6 Being Interprobleminary in Preparedness

Detection, monitoring, observation, and early warning systems and technologies that focus on one hazard at a time will not only miss surprising events and changing vulnerabilities but also are not widely available to the most disadvantaged and vulnerable countries (Dominey-Howes 2018). More efficient ways to be ready for surprise are needed. Because the drivers of vulnerability and resilience against multiple global threats overlap, directing actions at shared drivers not only better prepares populations for emerging diseases but also for other threats. It seems reasonable to ask whether a wildlife health risk management program needs to evolve from interdisciplinary teams tackling single issues (e.g., emerging diseases) to "interprobleminary" teams that examine the interactions and implications of multiple problems occurring simultaneously in a place or population. Such thinking is akin to the all-hazards approach to disaster preparedness which focuses on capacities and capabilities that are critical to preparedness for a full spectrum of emergencies or disasters. All-hazard approaches recognize that while hazards vary in source (e.g., pathogens, pollutants, extreme weather), they often challenge health programs in similar ways and resilience against them are rooted in shared determinants.

Multi-solving occurs when where people pool expertise, funding, and will to solve multiple problems with a shared investment of time and resources. Multi-solving is based on the premise that problems might be easier to solve together rather than one by one. A multi-solving approach has the potential to produce win–win situations that improve preparedness against multiple threats by aligning constituencies that might not otherwise see their common interests. The goal with multi-solving is to discover the co-benefits of actions to protect your interests (in this case wildlife health) with the interests of others who

Table 1 Principles for managing for safety in complex adaptive systems (after Reiman et al. 2015)

Principle	Guidance for emerging threat management
Set boundaries, objectives, and priorities	Set explicit boundaries for program activities since there are no natural all-inclusive boundaries between the various overlapping components of the wildlife health and emerging risk. This will help programs focus their unique skills and knowledge, select metrics and means to monitor progress, and develop needed partnerships
Promote prevention as a guiding principle	Work with partners (local to international) to develop a shared value of prevention that will encourage actions in advance of harm
Create standards and processes	Adapt existing programs as well as develop new programs to assess risks holistically and develop standards to assess and apply evidence for action
Monitor the system	Monitoring and surveillance systems must be adapted to support surprise preparedness and should include tracking adverse outcome factors (such as diseases and outbreaks), risk factor (such as pathogen traffic or changes in human uses of wildlife), vulnerability determinants (such as changes in human–wildlife interaction in critical areas), and upstream drivers of threats (such as changes in land use)
Create capacity for situational self-organization	Adjust and interpret priorities and practices as new knowledge arises and situations change. This will require risk managers have sufficient understanding of the possible risk impacts of their actions, building from a good understanding of the core tasks, populations, threats, and hazards that need to be managed within a wildlife health context
Optimize local efficiency	Surprising events must be managed considering other threats and account for differences in local capacities and risk circumstances. For example, emerging diseases cannot be managed in isolation from other unexpected impacts of climate change, biodiversity loss, habitat degradation, and other drivers of harm. Actions to control emerging risks must avoid creating new or additional threats and harms
Facilitate novelty and diversity	Our understanding of the genesis, prevention, and response to surprise health events is changing and can be expected to keep changing under current conditions of social and environmental change. Health managers and researchers must assess, clarify, reinforce, and amplify locally and internationally developed innovations and facilitate their diffusion and adaption to other settings
Facilitate connections, interactions, and collaborations	Interactions in which people trust each other, know and respect each other's competences, and are willing to share information, and learn from each other are needed to foster collaborative interventions and remain innovative. Programs must pay particular attention to human dimensions of the wildlife health management that impede or enable effective risk management

can affect the determinants of vulnerability and resilience. This requires the ability to define problems in ways that include more people in the solutions and optimize many outcomes rather than maximizing one.

7 Summary

Climate change science is concerned with surprise (Schneider 1995), so too is sustainable development (Gladwin et al. 1995), and business management (Taleb 2007); but the focus on prediction, measurement, and detection has comparatively made wildlife health research surprise free. Wildlife health risk management must understand how social and environmental characteristics and circumstances interact, influence, modify, facilitate, or constrain interventions and the effectiveness of their implementation (May et al. 2016).

A complex systems approach to health involves questions different than asking "does pathogen A cause disease B?" or "what risk factors are associated with the transmission of infection?" (Pearce and Merletti 2006). Instead, it asks, "are there circumstances where certain sub-populations are more vulnerable to a disease" or "are there situations where surveillance resources would be more likely to detect an emerging issue," or "which upstream intervention should we target knowing that there are many intervening variables between the intervention and the health outcome that could modify its impact? (Stephen et al. 2020). Instead of asking, can we predict which pathogen will emerge on which day in which locations, complex systems perspectives may help us ask, what are the circumstances that tip a system from being unconducive to conducive to a surprise?

References

Anon (2017) An emergency management framework for Canada. Emergency Management Policy and Outreach Directorate Public Safety, Canada. https://www.publicsafety.gc.ca/cnt/rsrcs/pblctns/2017-mrgnc-mngmnt-frmwrk/2017-mrgnc-mngmnt-frmwrk-en.pdf

Betts RK (1980) Surprise despite warning: why sudden attacks succeed. Political Sci Q 95(4):551–572

Chipman R, Slate D, Rupprecht C, Mendoza M (2008) Downside risk of wildlife translocation. Dev Biol 131: 223–232

Cunha MP, Clegg SR, Kamoche K (2006) Surprises in management and organization: concept, sources and a typology. Br J Manage 17(4):317–329

Dominey-Howes D (2018) Hazards and disasters in the Anthropocene: some critical reflections for the future. Geosci Lett 5(1):1–5

Duignan PJ, Van Bressem MF, Baker JD, Barbieri M, Colegrove KM, De Guise S, De Swart RL, Di Guardo G, Dobson A, Duprex WP, Early G (2014) Phocine distemper virus: current knowledge and future directions. Viruses 6(12):5093–5134

Galaz V, Moberg F, Olsson EK, Paglia E, Parker C (2011) Institutional and political leadership dimensions of cascading ecological crises. Public Adm 89(2):361–380

Gladwin TN, Kennelly JJ, Krause TS (1995) Shifting paradigms for sustainable development: implications for management theory and research. Acad Manage Rev 20(4):874–907

Gross M (2019) The paradox of the unexpected: Normal surprises and living with nonknowledge. Environ Sci Policy Sustain Dev 61(3):20–25

Kates RW, Clark WC (1996) Environmental surprise: expecting the unexpected? Environ Sci Policy Sustain Dev 38(2):6–34

May CR, Johnson M, Finch T (2016) Implementation, context and complexity. Implement Sci 11(1):141

Parker CF, Stern EK (2002) Blindsided? September 11 and the origins of strategic surprise. Polit Psychol 23(3):601–630

Pearce N, Merletti F (2006) Complexity, simplicity, and epidemiology. Int J Epidemiol 35(3):515–519

Reiman T, Rollenhagen C, Pietikäinen E, Heikkilä J (2015) Principles of adaptive management in complex safety–critical organizations. Saf Sci 1(71):80–92

Schneider S, Turner BL (1995) Anticipating global change surprise. In: Hassol SJ, Kasperson J, Katzenberger J (eds) Anticipating global change surprises. AGCI, Basalt

Stephen C (2020) Rethinking pandemic preparedness in the Anthropocene. In: Healthcare management forum. SAGE, Los Angeles, pp 153–157

Stephen C, Artsob H, Bowie WR, Drebot M, Fraser E, Leighton T et al (2004) Perspectives on emerging zoonotic disease research and capacity building in Canada. Can J Infect Dis and Med Micro 15(6): 339–344

Stephen C, Berezowski J, Misra V (2015 Jun) Surprise is a neglected aspect of emerging infectious disease. Ecohealth 12(2):208–211

Stephen C, Soos C (2021) The implications of climate change for veterinary services. Sci Tech Rev 40:421–430

Stephen D, Stephen C, Carmo LP, Berezowski J (2020) Complexity, chaos and systems thinking in health. In: Stephen C (ed) Animals, health, and society: health promotion, harm reduction, and health equity in a one health world. CRC Press, London

Taleb NN (2007) Black swans and the domains of statistics. Am Stat 61(3):198–200

Whitmee S, Haines A, Beyrer C, Boltz F, Capon AG, de Souza Dias BF, Ezeh A, Frumkin H, Gong P, Head P, Horton R (2015) The Rockefeller Foundation–Lancet Commission on planetary health. Safeguarding human health in the Anthropocene epoch: report of The Rockefeller Foundation–Lancet Commission on planetary health. Lancet 386(10007):1973–2028

Wood CL, Lafferty KD (2013) Biodiversity and disease: a synthesis of ecological perspectives on Lyme disease transmission. Trends Ecol Evol 28(4):239–247

Yamin F, Rahman A, Huq S (2005) Vulnerability, adaptation and climate disasters: a conceptual overview. IDS Bull 36:1–14

Wildlife Health Solutions Depend on Effective Risk Communication

Andrew Peters

Abstract

Risk communication is a key tool for wildlife health professionals to engage with society on its role as the ultimate driver of most emerging wildlife diseases, to reduce societal harm from wildlife disease, and to empower society to be an agent of change for the betterment of wildlife health. Risk communication can, if effective, enhance understanding and promote beneficial action in society. If ineffective, including through intentional or unintentional avoidance of communication, it can foster fear, confusion, and disengagement. To be effective, risk communication must recognize that risk is in part a sociocultural construct and that the perception of risk is strongly influenced by sociocultural factors. 'Shadow spaces' and social amplification of risk can lead to deleterious, and sometimes counterintuitive, outcomes. Trust, which underpins effective risk communication, is enhanced through the development of partnerships between experts and societal stakeholders. Understanding the sociocultural context of a wildlife health problem, which can emerge from these partnerships, is needed to inform the framing and delivery of clear, consistent, and concise messages. Effective risk communication is well developed in public health and is founded on the principles and methods of social science.

Keywords

Communication theory · Risk analysis · One Health · Public health · Society · Human-wildlife conflict · Fear factors · Message mapping

1 Wildlife Health and Risk Communication

Humans are the ultimate driver of most emerging wildlife diseases, are subject to harm from wildlife disease, and are an agent of change for the betterment of wildlife health. Wildlife health professionals depend on communication to engage with society on these issues. The effectiveness of our communication shapes the outcome of that engagement, either enhancing understanding and promoting beneficial action or fostering fear, confusion, and disengagement.

Risks (events that have an alterable likelihood and that carry a potentially negative consequence) to society from wildlife disease include intersecting environmental, sociocultural, health, and economic harm. Risk communication is a powerful tool to promote the reduction of such harm through societal action. To be effective, risk communication must recognize that sociocultural

A. Peters (✉)
Institute for Land, Water and Society, Charles Sturt University, Bathurst, NSW, Australia
e-mail: apeters@csu.edu.au

C. Stephen (ed.), *Wildlife Population Health*,
https://doi.org/10.1007/978-3-030-90510-1_25

context strongly influences the perception of risk—that risk is in part a sociocultural construct.

1.1 People Are Central to Wildlife Health Problems

The interaction between wildlife health and society is important and multifaceted. Wildlife health problems very commonly have anthropogenic origins, from the immediate or proximate causation of wildlife disease events to more remote or distal social drivers of disease risk. Wildlife health problems can cause real or potential harm to society. Even where a problem does not have an apparent anthropogenic cause (such as Tasmanian Devil Facial Tumour Disease), its identification as a problem relates to potential environmental, sociocultural, economic, or human health impacts relevant to society. Social dimensions typically shape the response to wildlife disease.

Responding and finding solutions to wildlife health problems is, therefore, most effective when cognizant of the relationship between society and wildlife health, not only to identify how society contributes to a problem or limits the available response or solution options but also to expand the available solution space by looking for opportunities in the social dimensions. This includes identifying the value proposition for society to develop socially and environmentally sustainable solutions to wildlife health problems, potentially leveraging and facilitating community participation, funding support, political support, and behavioral change.

1.2 Risk and Wildlife Health

We can assess the likelihood and potential consequences associated with a risk using risk assessment. The likelihood associated with risks may not be predictable with precision. Similarly, the diverse and interacting consequences associated with risks make a comprehensive prediction of the consequence of a risk challenging, especially in light of the interacting and cascading potential of each consequence to give rise to new risks.

It is important to differentiate risk assessment from risk acceptability (Fischbacher-Smith et al. 2010). While the former can be a technical process, the latter requires a more qualitative understanding of why people respond to a risk, based on the interaction between characteristics of that risk and the societal and personal dimensions associated with it. Risk acceptability is shaped by the *perception* of risk. Notwithstanding this, a robust evidence base facilitates the communication of risk, because of the complexity and potential negative outcomes associated with communicating uncertainty. Establishing the empirical facts should, therefore, be a primary objective of analyzing risk (Fischbacher-Smith et al. 2010; Bennett et al. 2010a, b). This may not always be possible, however, for poorly understood but pressing wildlife health risks.

The way in which individuals understand and perceive risk is very complex. Typically, accumulated information and knowledge from diverse sources are weighted and synthesized heuristically into a point of view that is relatively narrow and more definitive, which is then used to inform the response of the individual to risk or related risks (Fischbacher-Smith et al. 2010) (see Chap. 8). Technical or expert risk assessment is only one contributing source in the development of that point of view.

A number of factors including availability bias, confirmation bias, and overconfidence affect the assessment and perception of risk in both experts and the public (Bennett et al. 2010a, b). Availability bias occurs when familiar risks are ranked more or less significant than unfamiliar risks relative to an independent assessment, while confirmation bias is when the interpretation of the evidence is influenced by prior belief and perception. These factors are pervasive but not ubiquitous in risk assessment and risk perception.

Expert analysis of complex risks, such as those associated with wildlife health, is seldom able to achieve great precision. This creates "shadow spaces" in which other influences on the perception of risk are able to dominate, including political messaging, misinformation, criminal activity, and the advancement of unrelated ideological agendas. Manipulation of risk perception can be used to maintain or control social groups (Slovic 1987). There is a need for caution in undertaking risk communication in these shadow spaces, to avoid supporting hidden agendas and eroding trust (Fischbacher-Smith et al. 2010). Shadow spaces are common in wildlife health risk assessment because of the difficulty in estimating both likelihood and consequence associated with wildlife disease events. This can lead to perverse outcomes, including the demonization of particular wildlife species.

1.3 Effective Risk Communication as an Agent of Change for Solving Wildlife Health Problems

Risk communication is the exchange of information between two or more groups of people for the purpose of promoting, influencing, or reinforcing decision-making relating to events that have an alterable likelihood and that carry a potentially negative consequence. It is the instrument by which societal change in response to wildlife health problems can be facilitated. Often risk communication is thought of as the transfer of information from experts to the public, but it is important to recognize that any two parties can be involved in risk communication, and the direction of information flow in effective risk communication is in both directions.

Risk communication has emerged as a key tool in public health, with valuable lessons for risk communication in wildlife health and conservation (Kidd et al. 2019). If effective, it helps to clarify areas of disagreement, promote inclusivity, and inform better decision-making (Bennett et al. 2010a, b). As wildlife health professionals, we need to consider what we are trying to achieve through risk communication. We also need to recognize that, as experts, we are already participating in risk communication, even if through silence. To inform better decision-making, to benefit wildlife and society, we need to base our approach to risk communication on a sound understanding of how wildlife health challenges are a social problem (see Chap. 19). The perception of risk can create support for wildlife disease management (Hanisch-Kirkbride et al. 2013) but we have a responsibility to consider how our risk communication regarding wildlife health shapes societal values relating to wildlife (Decker et al. 2012) and human–wildlife conflict (Gore and Knuth 2009). Risk communication might be used to address the anthropogenic drivers of wildlife disease or to engage the community in solutions for wildlife health. Either way, our risk communication strategy needs to deeply understand the sociocultural landscape in which it operates so as not to exacerbate risk or promote the emergence of new risks.

2 Principles of Effective Risk Communication

2.1 Risk Communication Is Interpreted Through Sociocultural and Personal Context

Effective communication is an exchange, not a unidirectional flow of information. This is because communication includes the encoding, transmission, and decoding of information (Bennett et al. 2010a, b). Both the communicator and the recipient are participants in these processes, and the outcome of communication is a product of both parties (Smith 2020).

Perceptions of risk can differ significantly between experts, the public, and other stakeholders. Risk is mostly thought of as being both objective, having real and quantifiable properties which may be confirmed post hoc, and constructive, in which the subjective experience of risk defines its properties (Rae and Alexander 2017). The latter can be overlooked by experts during risk communication, even when they themselves interpret risks through their own cultural and professional lens.

The way people perceive and respond to risks is profoundly shaped by their sociocultural context (Slovic 1987), including their values, beliefs, and attitudes (Bennett et al. 2010a, b). To be effective, risk communication needs to consider these factors. One of the most important considerations in risk communication is so-called fear factors (Box 1). These factors are typically associated with feelings of dread (Slovic 1987) and the perception of higher risk by the public.

Box 1 Fear Factors Influential in the Perception of Risk

Fear factors have been well recognized in risk communication for more than three decades, though there is no consensus on all of the factors that increase the perception of risk. Bennett et al. (2010a, b) and Covello (2001) describe the perception of risk as being higher when risk is:

- involuntary rather than voluntary
- controlled by others rather than the individual
- inequitably distributed
- not associated with clear benefits
- inescapable through personal action
- unfamiliar or novel
- anthropogenic rather than natural in its cause
- ethically objectionable or morally wrong
- able to cause hidden or irreversible damage
- able to pose danger to future generations, children, and pregnant women
- able to cause potentially 'dreadful' injury or death
- more likely to affect known rather than anonymous people
- able to have a direct personal impact
- poorly understood or has high uncertainty
- subject to contradictory statements from responsible sources
- associated with untrustworthy individuals or institutions
- able to cause spatiotemporally determined and focused impacts

It is important to recognize that while fear factors are a significant influence on individual risk perception, other factors also exist. These include an individual's preexisting attitude toward the source of risk, the number of people exposed to risk, cultural attitudes, and personality traits such as an appetite for risk-taking (Bennett et al. 2010a, b; Slovic 1987). With regards to natural hazards, prior personal experience with the hazard and level of trust in authorities are the most influential factors shaping risk perception (Wachinger et al. 2013). Strong prior views on risk can be highly influential as confirmation bias shapes the way an individual interprets new information or knowledge relating to that risk (Slovic 1987).

A further, consistent influence on the public perception of personal risk, termed optimism bias, is the individual belief that risk is greater for others than for oneself (Weinstein 1989). Optimism bias can arise when individuals compare themselves to an inappropriate norm when faced with ambiguous risk factors, or through cognitive reliance on groundless risk reduction practices (Weinstein 1989). It can hinder behavioral change for risk reduction. Risk communication needs to, therefore, consider and respond to the potential for optimism bias, especially in high-risk social groups.

One of the more perplexing aspects of risk communication is the social amplification of risk (Bennett et al. 2010a, b). Social amplification describes how sociocultural dimensions either amplify or attenuate perception and response to risks beyond that assessed by technical experts as relatively proportionate (Kasperson et al. 1988). This is an important phenomenon as it can lead to significant impacts on society through ineffective or inappropriate public response to risk. Distinct from, but at times associated with, social amplification is the 'signal' associated with risk. When a risk is considered a portend of other risks, whether objectively true or not, this can highly influence the perception of that risk (Slovic 1987). An example of both phenomena would include the social amplification of risk associated with restrictive public health measures for COVID-19 as a signal for totalitarian governmental control.

How do these influences on the perception of risk inform the wildlife health professional engaged in risk communication? Two broad principles emerge: it is critically important to develop partnerships with those participating in risk communication (Smith 2020), and; a strategic approach to risk communication needs to be taken that focuses on effectively reducing risk through consideration of the sociocultural dimensions of risk perception.

2.2 Partnership: The Foundation of Effective Risk Communication

Fischhoff (1995) described a maturation of process in the field of risk communication. Ultimately, an emphasis on the role of partnerships has emerged, notwithstanding some of the challenges associated with establishing and maintaining these partnerships. The basis for this focus is recognition that expert risk assessment typically only captures some of the parameters influencing risk perception and that the public, especially those who the risk might impact, have valuable insights and knowledge (Fischbacher-Smith et al. 2010; Slovic 1987). Partnerships are now considered by many as the foundation for effective risk communication (Gamhewage 2014).

A key determinant of risk communication is trust (Bennett et al. 2010a, b). In the context of risk communication relating to wildlife health, trust in government agencies, nongovernment organizations and experts typically plays a significant role in shaping the perception of risk (Coombs and Holladay 2006). The social influence of family and friends, mediated through trust, can also strongly influence an individual's perception of risk (Slovic 1987). The need to establish or re-establish public trust in science and scientific governance in order to enable more effective risk communication partly underlies the focus on building partnerships (Fischbacher-Smith et al. 2010). This is clearly important, as low trust reduces the effectiveness of communication between parties, however, the converse, that high trust leads to effective communication, is not always true. This is because strongly held preexisting views can occasionally inform and revise an individual's trust in a communication partner more than preestablished trust influences that individual's views (Coombs and Holladay 2006; Frewer et al. 2003). This is not to be overstated, however. For example, preexisting trust of scientists and doctors is thought to have shaped the emergence of opinion groups around public health compliance early in the COVID-19 pandemic (Maher et al. 2020). Especially on issues where strongly held views don't already exist, the role of establishing trust through partnerships is most often likely to be highly beneficial in risk communication. This may be especially true for new emerging wildlife health risks. An important element of trust is that it is built upon open, two-way communication between experts and the public beyond the specific messaging associated with a risk (Fischbacher-Smith et al. 2010). Specifically, higher trust is associated with communication behaviors including transparency, acknowledgment of uncertainty, meaningful engagement including seeking input from the public, speedy dissemination of understandable scientific communication through multiple channels, and consistent, uniform messaging (World Health Organisation 2017).

One of the key elements of an effective partnership is effective listening. Effective communication is fundamentally dependent on careful listening to resolve the sociocultural dimensions associated with risks and communication (Menzie 1998, Heffner et al. 2003, Macnamara 2016). Authorities and experts have developed sophisticated messaging approaches, especially in public health, but less well known are the architectures of listening that support risk communication. These employ qualitative and quantitative methods that are well known to social scientists (see for example Bennett et al. (2017)), including surveys, focus groups, and interviews. The advancement of information and communication technologies has created new ways of listening to inform risk communication (Table 1) (Arana-Catania et al. 2021). It is important to develop a listening strategy, recognizing the bias and limitations of each method and how these apply to the various partners in communication. For example, traditional or deliberative surveys can

Table 1 Emerging listening tools applicable to developing effective risk communication

Listening tool	Description
Digital participation platforms	These vary in functionality and include an increasingly vast diversity of platforms for meeting, deliberation, and discussion, including open-ended forums with or without "expert" moderation. With high levels of participation, they can create information overload (Arana-Catania et al. 2021).
Deliberative surveys	Deliberative surveys (e.g.Pol.is) are an emerging digital technology that has a more constrained environment while still permitting moderated input from all community members (The Computational Democracy Project 2021). They allow evolving, deliberative participatory discussion and leverages data science to create real-time assessments of patterns of opinion.
Social media data mining	Social media data can provide insights into the perception and effects of risk communication and can inform the real-time development of highly targeted messaging (Merchant et al. 2021).

identify opinion groups and correlating demographic characteristics but are not as effective at exploring the sociocultural basis of risk perception within those groups. Focus groups and interviews can more deeply investigate the qualitative association between sociocultural context and risk perception but might not be able to capture an overview of the relative significance of opinion groups at an all-of-community level. Ultimately, each method can be strongly biased by recruitment methods, heterogeneous patterns of engagement and the questions being asked. Collaboration with social scientists who are familiar with these methods is highly advisable to minimize these biases.

In the process of establishing partnerships, it may also be important to employ active listening (Weger et al. 2014) rather than passive tools such as surveys or data mining. A listening strategy can also be used to test the effectiveness of messaging, enabling adaptive changes to framing and delivery (Bennett et al. 2010a, b; Covello 2006).

2.3 Effective, Solutions-Focused Risk Communication for Wildlife Health Challenges

Risk communication is not an end unto itself. The purpose of risk communication is to reduce risk. The worst-case scenario from risk communication is not that risk reduction doesn't occur, it is that risk is substantially increased, or new risks emerge, often facilitated by social amplification. The latter is particularly relevant to wildlife health, where ineffective risk communication can, for instance, shift risk from human health or economic harm, to environmental or sociocultural harm. Risk communication in wildlife health must be focused on the achievement of real-world beneficial outcomes for society, including through the protection of the earth's biological life support systems. As much as careful listening is critical to build the partnerships and knowledge of the sociocultural dimensions of risk communication, the careful design, and delivery of messaging is critical to the effective exchange of information leading to those beneficial outcomes.

Framing and delivering are key to the effectiveness of messages (Bennett et al. 2010a, b; Covello 2006). Framing is broadly about considering the initial assumptions and sociocultural context of an audience in the development of a message (Fischbacher-Smith et al. 2010). Messages need to be tailored to the language, modes, and channels of communication used by the audience. Framing around wildlife health issues needs to be validated with audiences so as not to unexpectedly create perverse outcomes (Roh et al. 2018). Recognition of sociocultural context and specific stakeholder issues is additionally important in identifying and managing the social amplification or attenuation of risk perception (Pidgeon and Henwood 2010). Care needs to be taken in the communication of uncertainty (Fischbacher-Smith et al. 2010), which can engender trust but also reinforce existing attitudes or open the space for confusion and competing agendas (Frewer et al. 2010). Risk comparisons should also be used in messaging with caution (Bennett et al. 2010a, b), because they rely on a deep understanding of how an audience perceives, sometimes counterintuitively, the risks being compared.

These considerations should be approached systematically. As an example, the tool of message mapping can be used for the effective development and delivery of messaging for risk communication (Box 2). The process of message development and delivery is often most effective by engaging multidisciplinary expertise, for instance, a subject matter expert, a communication specialist, and a policy expert (Covello 2006). Trained facilitators may enhance the outcome of multidisciplinary collaboration in risk communication.

It is important that messaging is consistent, clear, and concise. Messages that are consistent, especially when coming from different sources, are more effective (Covello 2006).

Box 2 Message Mapping: a Tool for Effective Messaging in Risk Communication

Message mapping can be used to provide a unifying framework for risk communication, especially in the context of emergencies or disasters. Covello (2006) describes the following steps in message mapping.

1. Identify and characterize the relevant stakeholders for the risk.
2. Create a comprehensive list of questions and concerns relating to the risk and risk communication for each stakeholder group.
3. Identify common concerns, or categories of concerns, within and across the stakeholder groups.
4. Develop messages that consider accuracy, communication phenomena such as mental noise (the difficulty with which individuals hear, process, and remember a message when they are upset), and audience sociocultural diversity to respond to each stakeholder concern.
5. Prepare so that the proofs (evidence, precedents, etc.) that support the message can be readily pointed to and highlighted.
6. Systemically test the messaging using tools such as focus groups.
7. Plan the role out and delivery of messages.

The potential complexity and pitfalls of communicating risk need to be considered in perspective: society makes countless, mostly effective decisions regarding risk using diverse input and forms of knowledge all of the time (Fischbacher-Smith et al. 2010). Risk communication typically becomes more complex when the risk itself is associated with greater complexity, unfortunately, a common scenario in wildlife health. It needs to be recognized that even where risk communication is effective, and individuals have the knowledge to make informed decisions, behavioral change associated with the reduction of risk can remain stubbornly elusive because of other social and personal dimensions (e.g., social barriers, personal habit) (Frewer et al. 2010).

Social structural barriers and inequality strongly shape access to, and the processing and interpretation of, information in risk communication (Merchant et al. 2021). These need to be addressed when undertaking risk communication in order to effectively reduce risk, especially because the effects of both health and environmental harm from risks are often inequitably distributed and those most at risk are typically excluded from decision-making (Cole et al. 2020, Mohai et al. 2009). It is important therefore to identify those most at risk, because of both their perception of the risk and their vulnerability (Frewer et al. 2010). It is also worth considering that there may be societal differences in the expectation of who is responsible for decision-making (Bennett et al. 2010a, b).

Decisions about risk acceptability weigh both the hazard and perceived benefit or opportunity cost associated with risk (Bennett et al. 2010a, b). This is where wildlife health professionals may be able to reframe society's perception of the risks associated with wildlife health through the One Health and planetary health narrative to achieve better long-term socioecological outcomes.

If we want to create effective solutions to wildlife health challenges, we need to understand that risk communication profoundly shapes outcomes and that risk communication itself is inseparable from the sociocultural landscape that forms the backdrop to those challenges.

3 The Contemporary Future of Wildlife Health Risk Communication: Lessons from the COVID-19 Pandemic

The COVID-19 pandemic has provided an abundance of examples of risk communication, has reinforced current risk communication theory as presented in this chapter, and has provided some additional salient lessons applicable to wildlife health. Of note, the role of public perception of risk in informing public policy, predominantly through the influence of constituents on politics (Abrams and Greenhawt 2020), is worth noting as an example of how the constructionist approach to risk analysis (i.e. that risk is a sociocultural construct) is relevant to real-world outcomes.

The effect of shadow spaces created by imprecise knowledge on the origin of SARS-CoV-2 provided an opening for the proliferation of potentially harmful conspiracy theories (Van Bavel et al. 2020). COVID-19 has been perceived as a signal of various types of environmental and societal change, both with and without empirical support. Social and traditional media have contributed to the social amplification of the highly diverse perceptions of risk associated with the pandemic.

COVID-19 has brought new dimensions to risk communication, especially relating to the rapidly changing and dynamic communication environment created by social media. In addition to creating challenges relating to the dissemination of misinformation (Gabarron et al. 2021), risk communication through social media can have benefits in terms of reaching diverse audiences, engaging and empowering the public, and urgent messaging with speed and reach (Heldman et al. 2013), notwithstanding the need to manage data security, privacy, and perception bias (Merchant et al. 2021). Effective approaches to managing misinformation on social media are rapidly evolving (Merchant et al. 2021; Vraga and Bode 2020). Traditional media is still important in the social amplification of risk (Gore and Knuth 2009), but there can be little doubt that social media will play an increasingly major role in risk communication.

4 Conclusion

Risk communication is a critical tool for positively engaging society in wildlife health, but those employing this tool need to understand that risk has objective attributes (which are typically characterized through a technical risk analysis) but is also a sociocultural construct. Social and personal influences on the perception of risk shape the interpretation of expert risk communication, which, to be effective, must recognize and address those influences.

Trust is an important factor in risk communication, and the development of partnerships between experts and societal stakeholders is a key approach used to establish trust. A deep understanding of the sociocultural context of a wildlife health problem is needed to inform the framing and delivery of clear, consistent, and concise messages, the development of which should be systematic. Message mapping (e.g. Covello (2006)) provides a useful framework for wildlife health professionals to approach the development of messaging in risk communication. The complicating influence on risk communication of shadow spaces and social amplification of risk needs to be understood and dealt with to prevent potentially deleterious outcomes. Effective risk communication can therefore require considerable insight, experience, and expertise. It is well developed in public health (see, e.g., Bennett et al. 2010a, b), and many of the tools used to characterize sociocultural context for risk perception are used widely in the social sciences. Meaningful collaboration with public health risk communicators and social scientists is perhaps the most valuable step in the development of effective risk communication strategies for wildlife health.

References

Abrams EM, Greenhawt M (2020) Risk communication during COVID-19. The journal of allergy and clinical immunology. In Pract 8(6):1791–1794

Arana-Catania M, Lier F-AV, Procter R, Tkachenko N, He Y, Zubiaga A et al (2021) Citizen participation and machine learning for a better democracy. Dig Govern Res Pract 2(3):1–22

Bennett NJ, Roth R, Klain SC, Chan K, Christie P, Clark DA et al (2017) Conservation social science: understanding and integrating human dimensions to improve conservation. Biol Conserv 205:93–108

Bennett P, Calman K, Curtis S, Fischbacher-Smith D (2010a) Understanding public responses to risk: policy and practice. In: Bennett P, Calman K, Curtis S, Fischbacher-Smith D (eds) Risk communication and public health. Oxford University Press, Oxford

Bennett P, Calman K, Curtis S, Fischbacher-Smith D (2010b) Risk communication and public health. Oxford University Press, Oxford

Cole HVS, Anguelovski I, Baró F, García-Lamarca M, Kotsila P, Pérez Del Pulgar C et al (2020) The COVID-19 pandemic: power and privilege, gentrification, and urban environmental justice in the global north. Cities Health:1–5

Coombs WT, Holladay SJ (2006) Unpacking the halo effect: reputation and crisis management. J Commun Manage 10(2):123–137

Covello VT (2001) Risk communication, the West Nile virus epidemic, and bioterrorism: responding to the communication challenges posed by the intentional or unintentional release of a pathogen in an urban setting. J Urban Health 78(2):382–391

Covello VT (2006) Risk communication and message mapping: a new tool for communicating effectively in public health emergencies and disasters. J Emerg Manage 4(3):25–40

Decker DJ, Siemer WF, Evenson DTN, Stedman RC, McComas KA, Wild MA et al (2012) Public perceptions of wildlife-associated disease: risk communication matters. Hum-Wildl Interact 6(1):13

Fischbacher-Smith D, Irwin A, Fischbacher-Smith M (2010) Bringing light to the shadows and shadows to the light: risk, risk management, and risk communication. In: Bennett P, Calman K, Curtis S, Fischbacher-Smith D (eds) Risk communication and public health. Oxford University Press, Oxford

Fischhoff B (1995) Risk perception and communication unplugged: twenty years of process. Risk Anal 15(2): 137–145

Frewer L, van Dijk H, Fischer A (2010) Consumer perceptions of the risks and benefits associated with food hazards. In: Bennett P, Calman K, Curtis S, Fischbacher-Smith D (eds) Risk communication and public health. Oxford University Press, Oxford

Frewer LJ, Scholderer J, Bredahl L (2003) Communicating about the risks and benefits of genetically modified foods: the mediating role of trust. Risk Anal 23(6): 1117–1133

Gabarron E, Oyeyemi SO, Wynn R (2021) COVID-19-related misinformation on social media: a systematic review. Bull World Health Organ 99(6):455–63A

Gamhewage GM (2014) Complex, confused, and challenging: communicating risk in the modern world. J Commun Healthc 7(4):252–254

Gore ML, Knuth BA (2009) Mass media effect on the operating environment of a wildlife-related risk-communication campaign. J Wildl Manage 73(8):1407–1413

Hanisch-Kirkbride SL, Riley SJ, Gore ML (2013) Wildlife disease and risk perception. J Wildl Dis 49(4):841–849

Heffner GG, Zandee GL, Schwander L (2003) Listening to community voices: community-based research, a first step in partnership and outreach. J Higher Educ Outreach Engage 8(1):127–139

Heldman AB, Schindelar J, Weaver JB (2013) Social media engagement and public health communication: implications for public health organizations being truly "social". Public Health Rev 35(1):13

Kasperson RE, Renn O, Slovic P, Brown HS, Emel J, Goble R et al (1988) The social amplification of risk: a conceptual framework. Risk Anal 8(2):177–187

Kidd LR, Garrard GE, Bekessy SA, Mills M, Camilleri AR, Fidler F et al (2019) Messaging matters: a systematic review of the conservation messaging literature. Biol Conserv 236:92–99

Macnamara J (2016) Organizational listening: the missing essential in public communication. Peter Lang Publishing, Bern

Maher PJ, MacCarron P, Quayle M (2020) Mapping public health responses with attitude networks: the emergence of opinion-based groups in the UK's early COVID-19 response phase. Br J Soc Psychol 59(3): 641–652

Menzie CA (1998) Risk communication and careful listening — resolving alternative world views. Hum Ecol Risk Assess Int J 4(3):619–622

Merchant RM, South EC, Lurie N (2021) Public health messaging in an era of social media. JAMA 325(3): 223–224

Mohai P, Pellow D, Roberts JT (2009) Environmental justice. Annu Rev Env Resour 34(1):405–430

Pidgeon N, Henwood K (2010) The social amplification of risk framework (SARF): theory, critiques, and policy implications. In: Bennett P, Calman K, Curtis S, Fischbacher-Smith D (eds) Risk communication and public health. Oxford University Press, Oxford

Rae A, Alexander R (2017) Forecasts or fortune-telling: when are expert judgements of safety risk valid? Saf Sci 99:156–165

Roh S, Rickard LN, McComas KA, Decker DJ (2018) Public understanding of one health messages: the role of temporal framing. Public Underst Sci 27(2):185–196

Slovic P (1987) Perception of risk. Science 236(4799): 280–285

Smith RD (2020) Strategic planning for public relations, 6th edn. Routledge, New York

The Computational Democracy Project (2021) Polis. https://compdemocracy.org/Polis/

Van Bavel JJ, Baicker K, Boggio PS, Capraro V, Cichocka A, Cikara M et al (2020) Using social and behavioural science to support COVID-19 pandemic response. Nat Hum Behav 4(5):460–471

Vraga EK, Bode L (2020) Correction as a solution for health misinformation on social media. Am J Public Health 110(S3):S278–SS80

Wachinger G, Renn O, Begg C, Kuhlicke C (2013) The risk perception paradox - implications for governance and communication of natural hazards. Risk Anal 33(6):1049–1065

Weger H, Castle Bell G, Minei EM, Robinson MC (2014) The relative effectiveness of active listening in initial interactions. Int J Listen 28(1):13–31

Weinstein ND (1989) Optimistic biases about personal risks. Science 246(4935):1232–1234

World Health Organisation (2017) Communicating risk in public health emergencies: a WHO guideline for emergency risk communication (ERC) policy and practice. World Health Organisation, Geneva

Index

C. Stephen (ed.), *Wildlife Population Health*,
https://doi.org/10.1007/978-3-030-90510-1